GOD, SCIENCE, AND SOCIETY

The Origin of the Universe,
Intelligent Life,
and Free Societies

by
Anthony Walsh
Boise State University

Series in Philosophy of Religion

VERNON PRESS

Copyright © 2020 Vernon Press, an imprint of Vernon Art and Science Inc, on behalf of the author.

All rights reserved. No part of this publication may be reproduced, stored in a retrieval system, or transmitted in any form or by any means, electronic, mechanical, photocopying, recording, or otherwise, without the prior permission of Vernon Art and Science Inc.

www.vernonpress.com

In the Americas:
Vernon Press
1000 N West Street,
Suite 1200, Wilmington,
Delaware 19801
United States

In the rest of the world:
Vernon Press
C/Sancti Espiritu 17,
Malaga, 29006
Spain

Series in Philosophy of Religion

Library of Congress Control Number: 2020931389

ISBN: 978-1-62273-941-7

Also available: 978-1-62273-907-3 [Hardback]

Product and company names mentioned in this work are the trademarks of their respective owners. While every care has been taken in preparing this work, neither the authors nor Vernon Art and Science Inc. may be held responsible for any loss or damage caused or alleged to be caused directly or indirectly by the information contained in it.

Every effort has been made to trace all copyright holders, but if any have been inadvertently overlooked the publisher will be pleased to include any necessary credits in any subsequent reprint or edition.

Cover design by Vernon Press. Image by beate bachmann from Pixabay.

Table of contents

List of Figures — v

Acknowledgments — vii

Preface — ix

CHAPTER ONE
Christianity, Rationality, and the New Atheists — 1

CHAPTER TWO
Science Declares the Glory of God — 13

CHAPTER THREE
The Anthropic Principle and Scientific Explanation — 23

CHAPTER FOUR
The Big Bang: The Event that Traumatized Physics — 33

CHAPTER FIVE
Fine-Tuning and Stellar Alchemy — 45

CHAPTER SIX
The Universe Engineered "Just Right" for Us — 55

CHAPTER SEVEN
The "Just Right" Galaxy and Solar System — 65

CHAPTER EIGHT
Earth: Our Privileged and Improbable Home I — 77

CHAPTER NINE
Earth: Our Privileged and Improbable Home II — 87

CHAPTER TEN
Losing God in the Multiverse — 97

CHAPTER ELEVEN
Finding God in the Multiverse — 107

CHAPTER TWELVE
DNA: God's Book of Life — 117

CHAPTER THIRTEEN
Abiogenesis: The Mother of all Scientific Puzzles — 129

CHAPTER FOURTEEN
Molecules, Membranes, and Information   141

CHAPTER FIFTEEN
Cracks in Neo-Darwinism: Micro is not Macro   153

CHAPTER SIXTEEN
The Problems of Information, Devolution, and Time   165

CHAPTER SEVENTEEN
God of the Gaps, Intelligent Design, and Theistic Evolution   177

CHAPTER EIGHTEEN
The Human Body: The Temple of God I   189

CHAPTER NINETEEN
The Human Body: The Temple of God II   199

CHAPTER TWENTY
The Brain: The Little Universe Within   209

CHAPTER TWENTY-ONE
Mind/Soul, Consciousness, and Love   219

CHAPTER TWENTY-TWO
Free Will and Determinism   231

CHAPTER TWENTY-THREE
The Problem of Evil: Atheism's Best Argument   241

CHAPTER TWENTY-FOUR
Christianity, Atheism, and Morality   251

CHAPTER TWENTY-FIVE
Christianity and the Healthy, Happy, Loving Life   261

CHAPTER TWENTY-SIX
Christianity, Freedom, Democracy, and Human Rights   273

CHAPTER TWENTY-SEVEN
The Modern American Assault on Christianity   283

References   293

Index   325

List of Figures

Figure 1.1: Percentage of Nobel Laureates who were/are "Atheists, Agnostics, and Freethinkers" Between 1901 and 2000	8
Figure 4.1: The Timeline for the Expansion of Space from the Big Bang	40
Figure 5.1: Matter and the Carriers of the Fundamental Forces	46
Figure 5.2: The Triple-Alpha Process	53
Figure 6.1: Curved, Hyperbolic, and Flat Geographies of the Universe	60
Figure 8.1: Illustrating the Process of Plate tectonics and Volcanic Activity	82
Figure 9.1: The Nitrogen Cycle	92
Figure 12.1: The Making of a Protein	120
Figure 12.2: Protein Folding	122
Figure 12.3: The Human Cell and its Major Parts	125
Figure 13.1: Illustrating Chirality	136
Figure 14.1: Part of a Human Cell Membrane	146
Figure 15.1: Methylation and Acetylation (Histone Modification)	161
Figure 18.1: Mitosis and Meiosis	197
Figure 19.1: Organs of the Immune System	203
Figure 20.1: Major Areas of the Brain	211
Figure 20.2: Neurons, Axons, Dendrites and Synapses	213
Figure 20.3: Major Pathways of the "Stop-Go" Neurotransmitters	215

Acknowledgments

I would first of all like to thank commissioning editor Victoria Echegaray for her faith in this project. Thanks also to the very able, efficient, and cheerful editorial manager Argiris Legatos, and to Rosario Batana, director. They are all very helpful and pleasant people with whom to work.

I want to acknowledge those who have read all or part of the manuscript. First is prominent physicist/astrobiologist Dr. Guillermo Gonzales, co-author (with Dr. Jay Richards) of the acclaimed book, *The Privileged Planet*. Dr. Giuseppe Pellegrini, president of Observa Science in Society and member of the Public Communication on Science and Technology network, made several useful suggestions, as did Dr. Scott Yenor, professor of political science. My brother, Robert J. Walsh, a retired English professor and active lay theologian, kept me on-track theologically and grammatically. I also want to thank my expert indexer, Hailey Johnson. Hailey completed a brilliant master's thesis under my supervision on epigenetics. I have harassed a number of people in many different areas on which I needed clarification. I thank them one and all. Of course, whatever errors remain are entirely mine.

Last but not least, I acknowledge and dedicate this book to my beautiful wife, Grace. Grace is the nicest of women and the center of my universe.

Preface

Two quotes from eminent physicists provide a good summary of this book. The first is from Lord Willian Kelvin, who devised the absolute temperature scale (the Kelvin scale) and formulated the second law of thermodynamics. Kelvin wrote: "Do not be afraid of being free thinkers! If you think strongly enough, you will be forced by science to the belief in God, which is the foundation of all religion. You will find science not antagonistic but helpful to religion."[1] The second quote is from astrophysicist Paul Davies. Davies has won many awards, including the Kelvin Medal, Faraday Prize, and the Templeton Prize. He writes: "It may seem bizarre, but in my opinion science offers a surer path to God than religion... science has actually advanced to the point where what were formerly religious questions can be seriously tackled."[2]

These quotes are the springboard from which I dive in to answer the atheist claim that God does not exist, and Christopher Hitchens' assertion that "religion poisons everything." I counter these claims with evidence from the physical, natural, and social sciences, and from philosophy, history, and theology, highlighting signposts to God and the many benefits of Christianity to society. Such breadth comes with the problem of how to limit discussion of the details to allow for the appreciation of the whole. My goal is to provide enough detail for the reader to stand in awe at the many thousands of seemingly impossible conditions that must exist for life on this planet, while at the same time realizing that too much detail detracts from this goal. I want to provide Christians with scientific "ammunition" to battle claims against God made by a militant breed of atheists dubbed the "New Atheists." Arguments from theology will not cut it with these folks, who insist that science is on their side. The only theology that a committed atheist may grant a hearing is natural theology. Natural theology offers indirect proof of God's existence and divine purpose through the scientific observation of nature and the use of human reason.

Good natural theology requires knowledge of many different areas of science, which also poses problems. My academic disciplines are biosocial science and statistics, and my research involves exploring the genetic and neurobiological bases of behavior, particularly criminal behavior. Outside these areas, I claim no expertise. Because scientific knowledge increases exponentially, one can only claim to be an expert in specific areas of one's discipline. As science expands, the pressure for specialization in one tiny area of a discipline grows ever stronger. Every science spawns subfields that then may become separate disciplines, which in turn spawns further specialized areas of research. This increases the efficiency and speed with which new knowledge is produced, but it carries with it the danger of intellectual isolation as fewer and fewer scientists are able to critically evaluate work done outside of their own narrow area of specialization.

The kind of extreme specialization demanded today can lead to a kind of general scientific illiteracy among scientists whose work does not overlap with adjacent sciences. For instance, to assess the degree of literacy outside their own disciplines, Hazen and Trefil asked a sample of physicists and geologists if they could explain the difference between DNA and RNA and found that only 12.5 percent could do so.[3] Physicists are arguably the brightest folks on the planet, yet only 1 in 8 were able to explain this basic piece of biological information. This is obviously not due to any lack of intelligence or inquisitiveness, but rather to their strict concentration on one or two problems in their subfield. There is just not enough available time to think about non-related scientific areas, let alone the arts, humanities, and the social sciences.

Universities encourage expertise in limited areas of a scientific discipline because that's where the grant funding, promotions, and salary increases are, but it comes at a cost. Former editor of the journal *Nature*, Philip Ball, mourns the fact that there is no space for the emergence of young Einsteins today, noting that: "The gravitational waves theorist saw physics as no one else did, but if he was around today his time would be spent chasing grants or tenures."[4] I shudder at the thought of being called an expert in anything. That would mean that I had spent my career boring holes into the minutia; sub-problems of sub-problems of sub-problems. It is great that most scientists do this because it is how science advances; I just don't care for it myself. I would rather be a jack of all trades than master of one. If I am puzzled about something I can easily go to experts at my university who have done the heavy lifting for me to get answers.

Having said all that about specialization being the norm, to write this book I have poked around in every science from physics to sociology, as well as philosophy, history, and theology. It has been a grand journey, and to undertake it I had to set aside my own research agenda (the privilege of a tenured full professor) and had two years free of teaching. I have reviewed high stacks of peer-reviewed articles and books written by top-notch scientists and philosophers, and have pestered scientists from many fields to keep myself on track. Scientific writings are highly technical; full of strange words and mathematical equations. I make every effort to explain their content in terms understandable to readers without formal scientific training, making no assumptions about prior knowledge. The nuances present in many topics were new to me when I began writing, and I have had to seek expert correction and verification of my own understanding of them.

I am not a theologian with an intimate acquaintance with the Bible. I do know enough about science and religion, however, to know that science points the way to understanding God's creation and to how Christianity had led to a free, moral, and prosperous society, and even to science itself. This runs against the grain of radical atheist claims that science and religion are in conflict, but the great Albert Einstein disagrees, stating that "Science without religion is lame, religion without science is blind."[5] As we shall see, science itself (particularly physics) has forced many brilliant scientists—including Nobel Laureates—who have taken the time to think deeply about the ultimate meaning of their work, to accept God as the Creator.

It is my belief, formed from years of respectful debates with knowledgeable atheists, that Christians should acquaint themselves with a basic understanding of science, since atheists claim that it is science that has buried God. Christianity is retreating in the Western world in the face of secular attacks, and if we are to halt this retreat we must make robust arguments for our faith from science. The new Atheists view science as "God's undertaker," to steal a phrase from the brilliant, witty, and inspirational Oxford mathematician, philosopher, and theologian Dr. John Lennox, but they are egregiously wrong.

I have written with the strong conviction that anyone espousing a position on any matter be it scientific, political, or religious, should be able to stoutly defend it. I do not say this in the spirit of one-upmanship, but rather in the hope that an informed defense may convince your atheist friends to examine their position with an open mind and come to know God. Most atheists never give much thought to their position, simply believing it to be a reasonable position that puts them on the side of science. If a believer can show how reasoning to the best explanation *from science* leads to theism and not atheism, perhaps some atheists will be led to abandon their empty, hopeless, and nihilistic worldview for the love of God.

Endnotes

1. In Smith, W. 1981, pp. 307-308.
2. Davies, P. 1983, p. ix.
3. Ball, P. 2016, np.
4. Hazen, R. and Trefil, J. 2009.
5. Einstein, A. 1994, p. 49.

CHAPTER ONE

Christianity, Rationality, and the New Atheists

> "The first gulp from the glass of natural sciences will turn you into an atheist, but at the bottom of the glass God is waiting for you."
> Werner Heisenberg, Nobel Laureate physicist

The New Atheism and Turning a Blind Eye

British atheist, mathematician, and philosopher, Bertrand Russell, was asked at his 90th birthday party what he would say to God if God did exist and Russell met Him at the judgment. Russell replied: "Why I should say, 'God, you gave us insufficient evidence!'" God's reply could have been: "Why, Bertie, you didn't look hard enough." Matthew 7:7 assures us, the evidence is everywhere if one bothers to seek it: "Ask, and it shall be given you; seek, and ye shall find; knock, and it shall be opened unto you." Russell assumed that evidence should be direct and immediately obvious, such as God floating down to Buckingham Palace on the first Sunday of each month to dine with the queen. God is purposely in the shadows, but there is enough light for those who wish to see, and enough obscurity for those who do not. There was abundant evidence for God in Russell's day (he died in 1970), and there is even more today: "Like it or not, a significant and growing number of scientists, historians of science and philosophers of science see more scientific evidence now for a personal creator and designer than was available fifty years ago."[1]

Despite the ever-increasing evidence for the Creator—or perhaps because of it—the Christian foundations of American faith, morality, and freedom have become increasingly under attack by a loose confederation of public intellectuals dubbed the "New Atheists." The New Atheists are militantly pushing their Godless agenda on America and other Western societies, and their "in your face" propaganda is having an effect. According to the Pew Research Center's Religious Landscape Study, the number of self-identified atheists in the United States almost doubled from 1.6 percent in 2007 to 3.1% in 2016.[2] This number may be greater, since some people who nominally identify with a religion may not believe in God. Whatever the case may be, atheism is creeping upwards in America, much of it driven by aggressive New

Atheist preaching in best-selling books, and is having a negative effect on America's cultural landscape and on its foundational values.

Citing events that occurred from 2012 to 2015, the 2016 International Christian Concern (ICC) *Hall of Shame Report* included the United States for the first time among the nations persecuting Christians. The report stated that the "ICC sees these worrying trends as an alarming indication of the decline in religious liberty in the United States."[3] A 2017 report from the First Liberty Institute also documents people battling for their religious freedom in all walks of life. Examples include a high school football coach fired for kneeling in prayer after games; wedding vendors subjected to financial ruin for refusing to lend their talents to celebrate same-sex weddings; an air force veteran manhandled off the stage for daring to use the word "God" at a friend's retirement ceremony; the Department of veteran's Affairs banning prayers at funerals containing the words "Jesus" or "God;" a navy chaplain disciplined for counseling sailors using biblical wisdom, and school children threatened with jail if they mention Jesus at graduation ceremonies. The First Liberty report lists an astounding 1,400 such incidents from 2005 to 2016.[4]

To counter the alarming increase in government anti-Christian activity during the Obama administration, President Trump signed the *First Amendment Defense Act* that bars the federal government from discriminating against individuals and organizations based on their religious convictions. Such an Act, however, may only be good as long as the president who signs it remains in office. Not surprisingly, Democratic politicians and various anti-Christian groups have mounted a campaign against the Act. What is surprising (and shocking) is that an Act to protect a First Amendment right would be necessary in 21st-century America. When a Constitutional right held sacred by previous generations of Americans needs a Congressional act to back it up, you know religious freedom is in trouble.

The atheist agenda is given support by ultra-liberal organizations such as the Southern Poverty Law Center (SPLC). The SPLC's reason for being seems to consist of finding new groups to call hate groups. As of 2018, the SPLC website lists 1,020 such groups in the United States. Some groups, such as the KKK and the American Nazi Party, justly deserve to be called hate groups, but the SPLC includes organizations such as the Alliance Defending Religious Freedom, First Liberty Institute, and the American Family Association in their ever-growing list. The SPLC even includes The American College of Pediatricians because it published an article warning young men of the possible medical consequences of a gay lifestyle. It seems that every group in the United States except those who share the SPLC's anti-Christian and anti-family agenda is a hate group. Putting religious and family values groups in the same box as Klansmen and

Nazis serves the purpose of demonizing them, and thus making attacks on them more acceptable to more people.[5]

New Atheists not only reject God, but they want to make the rest of us reject Him too. They are even attacking the Constitutional basis of religious liberty. In his book *Why Tolerate Religion?*, law professor Brian Leiter argues that the religious liberty clause of the First Amendment of the Constitution is no longer tenable. He claims that religion is insulated from "reason and evidence" which he defines as "believing in something notwithstanding the evidence and reasons that fail to support it or even contradict it."[6] Leiter believes that God has been given his walking papers by reason, but the Founders of this nation were hardly irrational. They believed that religion was so important to the moral foundation of society that the very first words in the Bill of Rights are: "Congress shall make no law respecting an establishment of religion, or prohibiting the free exercise thereof."

The Four Horsemen of New Atheism

The New Atheists are led by the intellectually gifted "Four Horsemen" of atheism: Americans Daniel Dennett and Sam Harris, and Britons Richard Dawkins and Christopher Hitchens. In the hands of these clever wordsmiths, atheism has repackaged atheism's old philosophical arguments as doctrinal commitment; delivering them in an aggressively dogmatic way so people sit up and take notice (and buy their books). The basic tenet of the movement is that religion is irrational, should not be tolerated, and must be attacked by rational folk who take science as their guide.

Not all atheists are anti-Christian. Perhaps most just say that they don't believe in God and leave it at that, and may even wish well those who do believe. Some may believe that Christianity is good for society and would never attack it, but they just do not buy into it themselves. Many of them do not bother to examine either theistic or atheistic arguments, and simply dismiss God in the same way they dismiss the tooth fairy. They are indifferent to all arguments for or against God's existence, and thus I define this weak form of atheism as simply the unexamined absence of belief in God. These people harm only themselves by depriving themselves of God's love and spiritual sustenance, but do not seek to harm Christianity. People who attack Christianity are better described as anti-theists rather than atheists. It seems that their message boils down to a hateful: "So many Christians, so few lions."

There is very little new in New Atheism except its political activism and confrontational tone. Political scientist Marcus Schulzke maintains that New Atheists view traditional atheists as too willing to let their worldview coexist as a separate discourse and to keep their non-belief a private thing. New Atheists don't want to accommodate such a notion and urge atheists to "come out of the closet" and let the world know what they think.[7] Schulzke is a supporter of New Atheism, and views it as a political movement designed to spread a left-wing worldview.

The major figure pushing the overtly politicized side of atheism was the recently deceased journalist and fiercely combative public intellectual, Christopher Hitchens. This is evidenced by the disingenuous subtitle of his 2007 book, *God Is Not Great: Religion Poisons Everything.* Hitchens forecloses on any possibility of a constructive dialog with such abusive and incendiary words, but they do sell books! Hitchens called himself a "conservative Marxist" (whatever that means) and was an admirer of blood-thirsty communists, Lenin and Che Guevara—no wonder he hated God. He debated theists from a position that: "Extraordinary claims [regarding the existence of God] require extraordinary evidence and that what can be asserted without evidence can also be dismissed without evidence."[8] In other words, he didn't need to present any contrary evidence since theism offers no credible evidence for him to refute. This is the kind of hubris Ian Markham describes as "fundamentalist atheism" where "there is no room for ambiguity or humility or nuance."[9]

Understanding and Confronting Militant Atheism

Marcus Schulzke is right; militant atheism is an attempt to foist a Godless left-wing agenda on Western civilization. The New Atheists have declared Christians to be their enemy, and in the spirit of the ancient Chinese sage Sun Tzu, to effectively engage the enemy one must know him. There are many atheist websites that Christians should not be afraid to visit in the spirit of "know thy enemy." If you do, you will not be stepping into Satan's lair, but you will encounter arguments you won't like. This is good, because criticism of our faith makes us more intellectually muscular as we wrestle with it. The New Atheists have certainly awakened the theologians from a long slumber, and theism is the better for it because many more of them are now motivated to engage with atheistic arguments. If there is no enemy at the gates, we grow complacent in our faith. To find these websites you may visit the *Born Atheist* website, which lists 11 well-funded atheist organizations. The largest of these organizations is the American Atheist, with 70 national and international affiliates. Some of these groups identify as "Brights" or "Freethinkers," which makes their members feel they have the intellectual high ground looking down on irrational theists, who are evidently seen as "Dulls" or "Enslaved Thinkers." The irrationality of Christianity and the rationality of atheism is taken for granted, as is in evidence by the American Atheist website definition of atheism as "the mental attitude which unreservedly accepts the supremacy of reason."

Marxist geneticist Richard Lewontin emphasizes the supposed irrationality of theists in his *New York Times* review of famous astronomer Carl Sagan's book *The Demon-Haunted World.* In this 1995 book, Sagan equated belief in God with the belief in little green men from Mars, UFOs, Santa Claus, and other such fictions. Sagan decries such beliefs and offers the lack of science education as the root of it all. Equating belief in God with belief in fictions such as Santa Clause is a massive non sequitur. No great scientist has ever been led to accept "Santa Clauseism" through their science the way those mentioned throughout this book have been led to accept the Creator. Nevertheless, Lewontin writes, "The only explanation that he [Sagan] offers for the dogged resistance of the masses to the obvious virtues of the scientific way of knowing is that 'through indifference, inattention, incompetence, or fear of skepticism, we discourage children from science.'"[10] I hope this is not the case with committed Christians, since the wonders that science has revealed point more convincingly to the Creator today than ever before.

It is doubtless true that few of us come to our faith through long philosophical conversations with ourselves about the truths of theism. Most people inherit their faith from their parents, and either they grow in it or it withers. Many among the lapsed may return to their religious roots after their youthful years of rebellion and doubt are over. Others become committed Christians through deeply emotional experience that takes root. Once it has taken root, we need to be able to defend it on objective rational grounds. We need to learn the best Christian apologetics, which is perhaps an unfortunate word that has nothing to do with apologizing. It comes from a Greek meaning "to make a defense," and if we cannot make a robust defense for our faith, atheists may be forgiven for not taking us seriously.

Atheists need not necessarily defend their position since it is an entirely negative one. Atheism is like the prisoner in the dock; he need not say anything. It is we who are prosecuting the prisoner and making affirmative claims about the existence of God who must speak. We must offer sufficient probative evidence for our case for the jury to reject the assumption that atheism is innocent "beyond a reasonable doubt." To do this we must have a solid offense as well as an equally solid defense. If God exists, it is logical to assume that He has given us ways by which He can be known. For the typical Christian, God is known subjectively through the work of the Holy Spirit, and needs no evidence apart from that. The agnostic and the atheist naturally demand more, and we should be ready and able to supply it.

The tools God has provided for those who affirm Him is the intellect, and its greatest achievement—science. The atheist who says that God does not exist sets himself a conundrum, for his belief can never be proven. I am not calling on the old adage that we cannot prove a negative (that is, proving something does not exist) because we do that all the time. The fossil record proves beyond a reasonable doubt that unicorns and griffins never existed; physicists have proved that cold fusion does not exist; chemists proved the same about phlogiston, and you can pinch yourself to disprove the negative "I don't exist." These examples, and a multitude of others, relate to the natural world in which the negative is accepted with regularity. However, God is outside of the natural realm, so His non-existence is a negative that cannot be proven. Atheists therefore affirm a belief that cannot, even in principle, be verified. In effect, atheists demand that theists must rigorously employ science to uphold their position while claiming an exemption from the process for themselves. It is true that theists believe in something that cannot be *directly* tested but can be verified indirectly from the scientific exploration of His creation and reasoning from this to the best explanation. This places the burden of proof on we who affirm His existence to disprove the negative claim that He does not.

It may be commendable to accept God on faith alone and to have a purely emotional attachment to one's faith, but appeals to emotion is hardly useful when answering an aggressive atheist. A person holding any position, religious or otherwise, should be able to defend it intellectually. As Paul put it in Romans 10:2: "For I bear them record [I hear them] that they have a zeal of God, but not according to knowledge." Although intellectual knowledge of God is necessary for the Christian to answer atheists on their own terms, justifying one's faith through science is not sufficient for one's self. A saving faith requires more than intellectual assent; it requires commitment to God at the intellectual, emotional, and spiritual levels. As the philosopher and theologian Martin Buber observed, viewing God only intellectually as the Creator is to place Him in an I-It (subject-object) relationship with you in which He is wholly "other," cold, distant, and inaccessible. A Christian accepts God as both the Creator and the Father with whom we can enter an I-Thou subject-subject relationship.[11]

Emotional attachment to a position is not confined to theists; it is true of many strong atheists as well. Although philosopher Thomas Nagel is quite able to defend his atheism with reasoned arguments, he also admits his emotional attachment to it: "I want atheism to be true and am made uneasy by the fact that some of the most intelligent and well-informed people I know are religious believers. It is just that I don't believe in God and, naturally, hope that I am right in that belief. It's that I hope there is no God! I don't want there to be a God; I don't want the universe to be like that."[12] It is sad to see someone locate his hope in a negative. It reminds me of a statement made by the genial genius John Lennox in one of his YouTube debates. He relates that an atheist at a scientific conference once told him that: "You Christians believe in God because you are afraid of the dark;" to which Lennox replied: "You atheists don't believe in God because you are afraid of the light."

If the atheist takes the stand to protest his innocence with an affirmative stance proclaiming that there is no God, his position requires justification, and you, the Christian prosecutor, must be ready and able to vigorously interrogate him. Just as we cannot know with absolute certainty if the man in the dock is guilty as charged, even though all the available evidence points in that direction, when arguing with a committed atheist we need only reason to the best explanation given the evidence science presents to us to convict him. Science can tell us *how* the universe came into existence which, incidentally, coincides with Genesis, but it cannot tell us *why*. If we are here simply by the blind random chance, as atheists claim, there is no purpose and no meaning in life, so asking "why" makes no sense. Science may not be the best reason for the committed Christian's faith, but it is the only way of knowing a committed atheist will listen to, which is why we need to embrace it. Philosopher Beatrice Bruteau shares this opinion, noting that contemplative Christians should be excited and knowledgeable about science because it "is part of our religious life, our practice, the way we live divine life."[13]

Engaging science to provide evidence for God's existence is known as natural theology. Natural theology investigates the existence of God by momentarily setting aside scripture or appeals to divine revelation and engaging only with evidence supplied by science, history, and philosophical reasoning. This might sound like a good idea, but there are Christians who find apologetics based on natural theology to be unbiblical, and that preaching the Gospel is enough. I feel that these are theologians who are uncomfortable with science, and believe with atheists that natural theology is a "God-of-the-gaps" theology, and that science is opposed to theism. There are eminent scientists, including many Nobel Laureates, who beg to differ. We will meet a cascade of such scientists in this book, but let us begin with William Bragg, a Nobel Laureate physicist, who says: "From religion comes a man's purpose; from science, his power to achieve it. Sometimes people ask if religion and science are not opposed to one another. They are: in the sense that the thumb and fingers of my hand are opposed to one another. It is an opposition by means of which anything can be grasped."[14]

To those theists who distrust natural theology, let me point out that Ephesians 4:15 tells us to "speak the truth in love" with Christians and unbelievers alike. And 1 Peter 3:15 commands "Always be prepared to give an answer to everyone who asks you to give the reason for the hope that you have. But do this with gentleness and respect." The New Atheists' case, they say, rests on science, so if we are to effectively engage them, or those whom they have convinced to follow their path, it is necessary to enter their territory and speak the truths of science, and then perhaps we can begin to speak the truths of the Gospel. Evidence for this is provided by a study of the reasons why 111 former atheists became Christians. In just over 50 percent of the cases, the subjects' primary reason for conversion was intellectual. They mentioned that by studying subjects such as cosmology and intelligent design, as well as philosophical arguments, they became convinced of the inherent rationality of Christianity.[15] Surely this is a good thing.

Figure 1.1: Percentage of Nobel Laureates who were/are "Atheists, Agnostics, and Freethinkers" Between 1901 and 2000

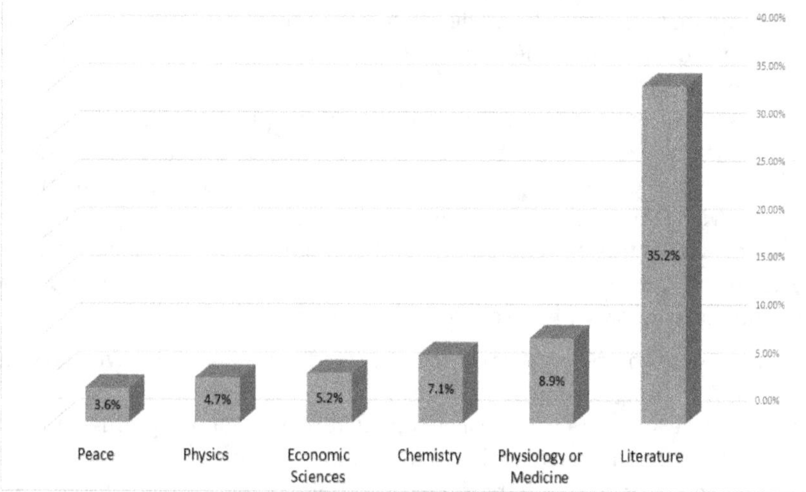

Category	Percentage
Peace	3.6%
Physics	4.7%
Economic Sciences	5.2%
Chemistry	7.1%
Physiology or Medicine	8.9%
Literature	35.2%

Rationality, Theism, Atheism, and Science

If atheism is so rational, and if scientists are the most rational of beings, we should see that most scientists are atheists. Geneticist Baruch Shalev's book documenting the religious views of all 719 Nobel Prize winners from 1901 to 2000 found that only 10.5 percent of these brilliant men and women fell into an atheist, agnostic, or freethinker category. It was winners in literature, not science, who make up by far the biggest category of non-believers (see Figure 1.1). Christians and Jews won 92 percent of Nobel Prizes in science.[16] From 2001 to 2017, Christians or Jews won 95 (85%) of the 112 Nobel Prizes for science, with heavily Western-influenced Japan taking 15 of the remaining prizes.[17] We cannot know how devout those who were not atheists, agnostics, or freethinkers were, but it helps us to understand prize-winning mathematical physicist Robert Griffiths' words: "If we need an atheist for a debate, we go to the philosophy department. The physics department isn't much use."[18]

Christian Anfinsen, a Nobel Laureate in chemistry, has said: "I think only an idiot can be an atheist. We must admit that there exists an incomprehensible power or force with limitless foresight and knowledge that started the whole universe going in the first place."[19] Defending your faith does not mean that you should go around calling your atheist friends and colleagues idiots. We can defend our faith without becoming emotional and cantankerous. When you resort to name-calling the battle is lost, and you have lost it. You will find the more you have a grasp of the scientific arguments that support your faith, the less defensive and quarrelsome you will become, and the more effective you will be.

But let us return to the atheist claim that believers in God are irrational clods who accept their faith blindly (just like all those Nobel Laureates). Commenting on a Gallop study titled *What Americans Really Believe,* Mollie Ziegler Hemingway wrote a telling piece about rationality among Christians and atheists in a *Wall Street Journal* piece aptly titled "Look who's irrational now." She writes that the report decisively shows that: "traditional Christian religion greatly decreases belief in fictions such as the efficacy of palm readers to the usefulness of astrology." The report also shows that the irreligious "tend to be much more likely to believe in the paranormal and in pseudoscience than evangelical Christians."[20] The Gallop study asked a number of questions about people's belief in such things as Bigfoot, the Loch Ness Monster, Atlantis, haunted houses, dreams foretelling the future, and so forth. These beliefs were placed in a cumulative index of "belief in the paranormal." Much to the chagrin of atheists, almost four times (31%) as many people who never attend church expressed strong belief in the paranormal than did weekly church attenders (8%). In fact, the more traditional and evangelical the respondent, the less likely he or she was to believe in the paranormal or in pseudoscience.

To explain these findings, Ziegler Hemingway referenced another study showing that the decline of Christian belief among college students is a major cause for the increase in cults and weird superstitions and that atheist college students are by far the most likely to embrace these things, and practicing Christian college students the least likely.[21] Francis Bacon's observation is instructive here: "A little philosophy inclineth man's mind to atheism, but depth in philosophy bringeth men's minds about to religion. For while the mind of man looketh upon second causes scattered, it may sometimes rest in them, and go no further; but when it beholdeth the chain of them, confederate and linked together, it must needs fly to Providence and Deity."[22] Because atheistic college students have had a "little philosophy," or have taken Heisenberg's "first gulp of natural science," as quoted in this chapter's epigraph, they perhaps felt that it was more intellectually respectable to drop religion and embrace some other form of belief.

A different poll taken by the Pew Research Center rubbed more salt into atheist arrogance. It showed that Christians who attend church weekly, conservatives, and Republicans, express much lower levels of belief in ghosts, fortune tellers, astrology, the "evil eye" (some people can cast spells on others), yoga (as a spiritual practice), and spiritual energy located in physical things (mountains, trees, and so on) than liberals, Democrats, and non-church attenders. Overall, practicing Christians, conservatives and Republicans, are about half as likely as non-attenders, liberals, and Democrats, to harbor "New Age" beliefs. [23]

Despite all this contrary evidence, atheists still like to think that Christian beliefs are irrational, yet it is indisputable that bizarre New Age beliefs have grown stronger among atheists the more science has progressed, and that they do not take hold nearly as strongly among practicing Christians, conservatives, and Republicans. Having forsaken God, atheists, and hard leftists in general, attach themselves not only to relatively harmless New Age beliefs but to more harmful moonshine such as socialism, multiculturalism, radical feminism, and moral relativism. They do so because they need to believe in something greater than themselves to which they can pledge allegiance. Practicing Christians are almost always conservatives because they already have a religion that has stood the test of time and have no need for destructive alternatives. G. K. Chesterton said it well in one of his *Father Brown* books. He had his famous fictional detective say: "It's the first effect of not believing in God that you lose your common sense. It's drowning all your old rationalism and scepticism, it's coming in like a sea; and the name of it is superstition. The first effect of not believing in God is to believe in anything. And a dog is an omen and a cat is a mystery" [24]

Endnotes

1. Wilkins, M. and Moreland, J. 2010, p. 10.
2. Pew Research Center, Religious Landscape Study, 2016.
3. International Christian Concern, 2017, p.11.
4. Shackelford, K., 2017.
5. Walsh, A., 2018.
6. Leiter, B., 2014, p.39.
7. Schulzke, M., 2013.
8. Hitchens, C., 2003.
9. Markham, I., 2010, p.141.
10. Lewontin, R., 1997, np.
11. Buber. M. (2002). I am indebted to Robert J. Walsh for this insight.
12. Nagel, T., 1997, p. 130.
13. Bruteau, B. 1997, p. 7.
14. In Gonzalo, J. 2008, p 121.
15. Langston, J., Powers, H., and Facciani, M. 2019.
16. Shalev, B., 2003, pp. 57–59.

17. The Nobel Prize, nd.
18. Kainz, H., 2010, p.21.
19. In Margenau and Varghese, 1997, p. 139.
20. Ziegler Hemingway, M. 2008, np.
21. Ibid., np.
22. Bacon, F. 1889, p. 57.
23. Pew Forum on Religion & Public Life, 2009.
24. In Cammaer, E., 1937, p, 211.

Chapter Two
Science Declares the Glory of God

> "Everyone who is seriously involved in the pursuit of science becomes convinced that a spirit is manifest in the laws of the Universe–a spirit vastly superior to that of man, and one in the face of which we with our modest powers must feel humble."
> Albert Einstein, Nobel Laureate physicist

The God Hypothesis

In 1798, the French mathematician and physicist Pierre-Simon Laplace published the first of his five-volume work, *Treatise on Celestial Mechanics*. He proudly presented a copy to his friend, Emperor Napoleon Bonaparte, who asked Laplace why, in a book explaining the mechanics of the universe, he had not mentioned its Creator. Laplace is said to have replied, "Sir, I had no need of that hypothesis."[1] When Napoleon told mathematician and astronomer Joseph-Louis Lagrange of Laplace's response, Lagrange exclaimed: "Ah, it is a fine hypothesis; it explains many things." Upon hearing of this, Laplace remarked: "This hypothesis, Sir, explains in fact everything, but does not permit me to predict anything. As a scholar, I must provide you with works permitting predictions."[2]

Laplace's remarks were not made with atheistic intent since he was a practicing Catholic, and nowhere in his public or private writings did he deny God. But technically speaking, he was absolutely correct; physicists do not insert a "God term" in an equation to figure out, for instance, the frequency of an average yellow light wave. But what if Napoleon had asked Laplace the origin of that light, why we are here on this Earth, what is the purpose of life, or one of the biggest philosophical question of all: "Why is there something rather than nothing?" These are questions science cannot answer, so God is hardly irrelevant if we wish answers to these profound questions.

As science has progressed, we have found that Lagrange's God hypothesis is right on the money. But God doesn't just "explain many things," he explains everything in an ultimate sense. Science has discovered immensely more marvelous things since LaGrange's time, and the more it discovers the more it finds mystery, and the more the mind is forced to contemplate God. Even if in the distant future some genius came up with the final equation of everything explaining the *how* the universe works on both the macrocosmic and micro quantum scales, he or she would still not answered the *why*; why the abstract mathematics has such an uncanny relationship with the physical universe, or why he or she has an immaterial mind that brought the equation to life.

God's existence is not contingent on science's current inability to explain something or other in the natural world. Inserting God into a gap in scientific knowledge does theism a disservice, because if and/or when science does explain the phenomenon in question, those guilty of "God-of-the-gaps" arguments get egg on their faces, and play into the hands of atheists claiming that religion and science are in conflict. Scientists readily acknowledge that the big questions of meaning are outside of their purview, so that rather than inserting God into scientific gaps to argue from what we *don't* know about the workings of the universe, theistic scientists argue from what we *do* know, and how it logically leads to the transcendent Creator of the universe who is the ground of all explanation.

To interpret Laplace's answer as implying that God is as unnecessary to explain the most meaningful "why" questions as he is to explain the more mundane "how" questions of the workings of creation is to make the category mistake of confusing impersonal principles with personal agency. Science searches for *how* God created the universe; theology searches for *why* He did. Oxford mathematician and philosopher John Lennox uses the simple but effective example of a Ford automobile to make this point. An engineer, he says, can fully explain *how* a car works using the principles of internal combustion, but if he wanted to know *why* a car exists, he would have to invoke agency. That is, why Henry Ford chose to manufacture automobiles. Ford's agency would have no place whatsoever in the description of how the car works, but it is necessary to explain why it exists for engineers to explain. We need both explanations to have a necessary and sufficient explanation of the car. Likewise, to have a necessary and sufficient explanation of all that exists in the cosmos, we need both science and the Creator.

Lennox's evaluation of Laplace's remark is also revealing: "Considered as a serious observation, his remark could scarcely have been more misleading. Laplace and his colleagues had not learned to do without theology; they had merely learned to mind their own business."[3]

"Their own business" is the business of science, which can be pursued by even the most devout scientists without God entering their work. Affirming science does not imply disclaiming God, as countless first-rate scientists attest. However, there are scientists who refuse to let God into their world at all and place all their faith in materialism/naturalism. Atheist geneticist Richard Lewontin honestly reported his commitment to materialism:

It is not that the methods and institutions of science somehow compel us to accept a material explanation of the phenomenal world, but, on the contrary, that we are forced by our *a priori* adherence to material causes to create an apparatus of investigation and a set of concepts that produce material explanations, no matter how counter-intuitive, no matter how mystifying to the uninitiated. Moreover, that materialism is absolute, for we cannot allow a Divine Foot in the door. [4]

If scientists believe they cannot allow a Divine Foot in the door, they must struggle to explain the most meaningful questions in science, such as the origin of the universe and the origin of life. To avoid a creation event, scientists have posited everything from a static and eternal universe to countless trillions of other universes beyond our ability to ever perceive. It is interesting to note the words of Nobel Laureate physicist George P. Thomson, who remarks that, based on modern evidence: "Probably every physicist would believe in a creation if the Bible had not unfortunately said something about it many years ago and made it seem old-fashioned."[5] As for the origin of life, some of the best chemists and biologists working on the problem have thrown up their hands. It has been noted that there were 150 naturalistic theories of the origin of life in the literature between 1950 and 2000, and the list is still growing.[6] These theories invoke everything from space aliens to superhot thermal undersea vents as possible naturalistic mechanisms for kick-starting life. The creation of the universe and the origin of life are the big—very BIG—questions of existence that theology has answered for centuries, and with which science is slowly recognizing. But first, what of the atheist claim of conflict between Christianity and science?

Christianity and Science: Conflict or Concord?

Lewontin takes for granted that science is in a struggle with the supernatural. He makes it plain that even in areas where the big questions are, scientists must stick to their materialistic guns, no matter what. Lewontin admits that it is not the demands of the scientific method that compel scientists to accept only materialist explanations, but rather it is their faith in materialism that forces them. It is obvious that he sees science and religion as oil and water. Atheists push the notion that they believe only in science and reason, and that Christians reject reason in favor of faith. Yet Lewontin says that scientists *must* accept materialism—on faith. It is true that faith is subjective and emotional, but subjectivism and emotion are not necessarily incompatible with reason. Albert Einstein, the greatest reasoner of all, who revolutionized science with pure thought, wrote of the sensation of the emotional and mystical basis of faith: "The most beautiful and most profound emotion we can experience is the sensation of the mystical. It is the sower of all true science. He to whom this emotion is a stranger, who can no longer stand rapt in awe, is as good as dead.

That deeply emotional conviction of the presence of a superior Reasoning Power, which is revealed in the incomprehensible Universe, forms my idea of God."[7]

There are two classical incidents atheists invoke to argue that science and Christianity are in conflict, the first being Galileo and the Catholic Church. It is little known that Galileo considered scientists, not the Church, to be the chief opponents of the heliocentric (Sun-centered) theory advanced by Nicolaus Copernicus 90 years earlier. Galileo wrote to his friend, German mathematician and astronomer, Johann Kepler, asking him: "What do you have to say about the principal philosophers of this academy who are filled with the stubbornness of an asp and do not want to look at either the planets, the moon or the telescope, even though I have freely and deliberately offered them the opportunity a thousand times?"[8]

In Galileo's time, most natural philosophers (as scientists were called before the word "scientist" existed) subscribed to the geocentric model of the solar system; that is, the Earth was at its center and all else revolved around it. It was the Jesuits, a religious order with a solid scientific reputation, who questioned the geocentric view.[9] A colleague of Galileo's, Cesare Cremonini, took up Galileo's challenge and looked through the telescope. He complained it gave him a headache and would hear no more about it. For Cremonini to have endorsed the evidence before his eyes would have called into question his life's work. But in Cremonini's defense, the heliocentric model not only defied common sense, it was ridiculed by fellow scientists. We can understand that ridicule because we would all subscribe to the geocentric model if all we had to go by were our unaided sensory observations. The geocentric theory comports with our immediate sense experiences in a way that the heliocentric theory does not. After all, we don't feel the Earth moving as it spins on its axis at just over 1,000 miles per hour while hurtling through space in orbit around the Sun at about 67,000 miles per hour, and we see the Sun rise in the east, move across the sky, and then set in the west.

The famous 1860 debate between Thomas Huxley ("Darwin's bulldog") and Bishop Samuel Wilberforce on evolution is the second instance. Atheistic accounts of the debate paint Wilberforce as an ignoramus, but he had degrees in mathematics and the classics, and Darwin himself regarded Wilberforce's 50-page review of his work "uncommonly clever... It quizzes me most splendidly."[10] The debate was not science versus religion, as atheists claim, but, at Wilberforce's insistence, it was science against science. As was the case with the Galileo incident, a number of scientists, including the leading anatomist of the day, Richard Owen, and the eminent physicist Lord Kelvin, were among those who opposed Darwin. Thus, the two events that atheists use as props for their view that science and religion are at loggerheads have been kicked out from under them. Historian of science, Colin Russell, comments on the notion of conflict and hostility between science and religion: "The common belief that the actual relations between religion and science over the last few centuries have been marked by deep and enduring hostility...is not only historically inaccurate, but actually a caricature so grotesque that what needs to be explained is how it could possibly have achieved any degree of respectability."[11]

The Christian Origin of the Science

Far from being in conflict, Christianity and science are intimately linked. While it is true that some *scientists* are at war with God, this is not the same as saying that *science* is at war with God. Atheists may be surprised to learn that so many of the advances in early science were made by men of God. Friar Roger Bacon is the father of the scientific method; Jesuit priest and mathematician, Roger Boscovich, produced the precursor of atomic theory; Gregor Mendel, a monk, founded the science of genetics; Nicolas Steno, the father of geology was a priest; as was Jean-Baptiste Carnoy, the father of cell biology, and Georges Lemaître, the father of the Big Bang theory. As Albert Einstein asserted, the deep study of science offers a clearer path to God for the skeptic than arguments from theology: "The more I study science the more I believe in God."[12] He also said: "Everyone who is seriously involved in the pursuit of science becomes convinced that a spirit is manifest in the laws of the Universe–a spirit vastly superior to that of man, and one in the face of which we with our modest powers must feel humble."[13]

It is thus hard to take seriously that science and Christian theism are in conflict. Many historians, philosophers, and scientists claim that the spirit of science grew out of the Christian belief in a rational and orderly God who created us in His image. Einstein once wrote that: "The eternal mystery of the world is its comprehensibility...The fact that it is comprehensible is a miracle."[14] The fact that it is comprehensible and elegantly described by mathematics is proof positive that God wants us to understand His creation, and He gives us the intelligence to do so. Nobel Laureate biochemist Melvin Calvin comments on his understanding of the origin of the scientific conviction that the universe is orderly and knowable: "I seem to find it in a basic notion discovered 2,000 or 3,000 years ago, and enunciated first in the Western world by the ancient Hebrews: namely that the universe is governed by a single God and is not the product of the whims of many gods, each governing his own province according to his own laws. This monotheistic view seems to be the historical foundation for modern science."[15]

This God of order and reason designed a predictable universe intelligible to the human mind so that we can to know Him by His creation. We see the spirit reason articulated in Solomon's prayer in Wisdom 7:17-22 in which he praises

God for giving him the ability to do what he could to discover what God has hidden for us to find:

> For He gave me sound knowledge of what exists, that I might know the structure of the universe and the force of its elements, The beginning and the end and the midpoint of times, the changes in the sun's course and the variations of the seasons, Cycles of years, positions of stars, natures of living things, tempers of beasts, Powers of the winds and thoughts of human beings, uses of plants and virtues of roots—Whatever is hidden or plain I learned, for Wisdom, the artisan of all, taught me.

From the very earliest days of Christianity, the Church taught that reason is a unique gift of God, and that we must use this gift to come to know Him. As early as the second century AD, theologian Quintus Tertullian of Carthage, gifted with the title of the founder of Western theology, informed us that: "Reason is a thing of God, inasmuch as there is nothing which God the Maker of all has not provided, disposed, ordained by reason—nothing which He has not willed should be handled and understood by reason."[16] The Catholic Church founded the first Western university—the University of Bologna—in 1088 and made mathematics and natural philosophy (science) compulsory parts of the education of anyone wanting to study theology.[17] Christian theology thus created the fertile intellectual soil for science to grow.

Ultimately, the great contribution of Christian theology to science lies in its conviction that there are laws of nature front-loaded by God at the beginning of time awaiting discovery. The 13th-century patron saint of science, Albertus Magnus, wrote: "It is the task of natural science not simply to accept what we are told but to inquire into the causes of things."[18] In another work, Magnus stated that: "In studying nature we have not to inquire how God the Creator may, as He freely wills, use His creatures to work miracles and thereby show forth His power; we have rather to inquire what Nature with its immanent causes can naturally bring to pass."[19] This hardly sounds like unreasoned faith. Physicist Paul Davies notes the influence of this belief in order and reason on the great Sir Isaac Newton: "Isaac Newton first got the idea of absolute, universal, perfect, immutable laws from the Christian doctrine that God created the world and ordered it in a rational way."[20] Other scientists have found their work to be a form of revelation and worship. Geneticist, physician, and former atheist, Francis Collins, is one who finds God in his science: "I have found there is a wonderful harmony in the complementary truths of science and faith. The God of the Bible is also the God of the genome. God can be found in the cathedral or in the laboratory. By investigating God's majestic and awesome creation, science can actually be a means of worship."[21]

The Judeo-Christian exhortation to explore the fingerprints of God in the natural world is absent in the theology of other religions. Although Islam accepts God as the all-powerful, all-knowing, Creator of the universe, it asserts that humans exist solely to surrender (this is what "Islam" means) to God and worship him, and that we cannot come to know Him through His creation. According to Rodney Stark: "Allah is not presented as a lawful creator but has been conceived of as an extremely active God who intrudes on the world as he deems it appropriate. Consequently, there soon arose a major theological bloc within Islam that condemned all efforts to formulate natural laws as blasphemy insofar as they denied Allah's freedom to act."[22] Thus, Islamic theology does not provide the fundamental assumptions necessary for the emergence of empirical science devoted to discovering God's natural laws. This may be the reason why, despite all the excellent scholarship of medieval Muslims in philosophy, mathematics, and medicine, the Islamic world never produced anything like the science of the Western world. If Christianity has dogmatically opposed science across the centuries, how is it that science only developed in Christian countries?

Materialism and Naturalism

Recall that Lewontin noted that the supposed struggle between science and theism involves science's obligatory commitment to naturalism/materialism. There is no conflict between science and theism, but there is conflict between theism and materialism/naturalism. Atheistic scientists are committed to the metaphysics of ultimate meaninglessness; viewing everything that exists as either due to random processes or necessity; that is, the combination of chance and the laws of physics. However, scientists can be thoroughgoing naturalists in their daily work, but still reject the notion that the matter of nature they work with is not all that there is. These scientists agree with Nobel Laureate physicist Max Plank's words delivered in a lecture titled *Religion and Natural Science*: "Both Religion and science require a belief in God. For believers, God is in the beginning, and for physicists He is at the end of all considerations... To the former, He is the foundation, to the latter, the crown of the edifice of every generalized world view."[23]

Materialism and naturalism are closely linked, but there are differences. Materialism is the most dogmatic of the two in its ontology (the philosophy of becoming, existence, or reality). Materialist ontology avers that all existence is matter, only matter, the physical "stuff" you can see, touch, measure, and manipulate, is real, and that there is no metaphysical reality. Naturalism agrees with materialism in its denial of causal mechanisms outside the natural, which means that it denies the supernatural, but naturalists also deny that all effects have material causes. The philosophical position that naturalists and materialists agree on is that there is nothing separate, above, or prior to the natural world. Theism, of course, affirms that there is a supernatural reality outside of nature, and calls that reality God.

Some naturalists affirm the existence of a non-material mind separate from the brain, but for the ontological materialist, mental phenomena are illusionary, and merely reflect electrical energy moving stuff around in the brain. Francis Crick, one of the discoverers of the genetic code, has infamously written that, "'You,' your joys and your sorrows, your memories and your ambitions, your sense of personal identity and free will, are in fact no more than the behavior of a vast assembly of nerve cells and their associated molecules. Who you are is nothing but a pack of neurons."[24] If Crick is going to travel that route, he might have gone further because neurons are made of atoms, and atoms are made of even smaller particles, so he might have said that we are nothing but a pack of quarks. Mental phenomena require a physical platform called the brain, but the mind cannot be reduced to it without remainder. Crick's reductionist words come from his book, *The Astonishing Hypothesis*. It is indeed astonishing, and one wonders why he didn't subtitle it *The Zombie Within*.

Take the phenomenon of love as an example of Crick's folly. Neuroscientists have looked at the neurochemical correlates of romantic love in a soup of neurotransmitters lighting up the brain's pleasure centers.[25] However, we cannot reduce the intoxication of romantic love to the soup and sparks of brain activity. These physical things do not come remotely close to explaining why Romeo fell in love with Juliet; they merely tell us what happened in his brain when he did. This is an example of "top-down" causation because love came first, and only then came the soup and sparks. To be sure, the soup and sparks are necessary for love to exist, but the physical product is a consequence of the mental process and not the other way around.

I have no problem with materialistic/naturalistic science as a working assumption; we call this *methodological* materialism/naturalism. As a regulative principle for science, it has been enormously successful in our understanding, prediction, and control of natural phenomena. The problem is when we jump from a working assumption to a comprehensive philosophy; that is, the assumption that there is nothing beyond the materialist/naturalistic realm of being. The inherent atheism of materialism/naturalism as an ontological philosophy is made transparent in a speech made by the infamous Madalyn Murray O'Hair, founder of the American Atheists: "Atheism is based upon a materialist philosophy which holds that nothing exists but natural phenomena. There are no supernatural forces or entities, nor can there be any. Nature simply exists."[26] A chain of rational reasoning about human existence that ends abruptly when the materialist arrives at the beginning of creation with "Well, that's just the way it is," is expecting us to accept the notion that the whole of existence is ultimately reasonless. This is not a very satisfactory explanation at all; it is simply a discussion ender.

The materialist world view does not address how the concept of matter was affected by scientific discoveries such as quantum mechanics and Einstein's famous E = MC² in which matter and immaterial energy were shown to be different manifestations of the same thing. As Nobel Laureate physicist Max Planck, the father of quantum mechanics, has noted:

> As a man who has devoted his whole life to the most clear-headed science, to the study of matter, I can tell you as a result of my research about atoms this much: There is no matter as such. All matter originates and exists only by virtue of a force which brings the particle of an atom to vibration and holds this most minute solar system of the atom together. We must assume behind this force the existence of a conscious and intelligent Mind. This Mind is the matrix of all matter.[27]

As previously noted, some naturalists argue that there are emergent properties not reducible to matter. Planck's mentalism—the belief that at the most fundamental level that the universe is made of "mind stuff"—is shared by a number of other physicists. The great astrophysicist Sir James Jeans famously wrote that: "The stream of knowledge is heading towards a non-mechanical reality; the Universe begins to look more like a great thought than like a great machine [can you read God here?]. Mind no longer appears to be an accidental intruder into the realm of matter...we ought rather hail it as the creator and governor of the realm of matter."[28] Another great British astrophysicist, Sir Arthur Eddington, wrote: "The universe is of the nature of a thought or sensation in a universal Mind."[29] When we think about the almost infinite divisibility of matter we have to wonder what it ultimately is; is it ever smaller pieces of "stuff," or does it point to John 1:1 "In the beginning was the Word, and the Word was with God, and the Word was God"?

Arguing against ontological materialism and naturalism is not arguing against science. Christianity-inspired science is humanity's greatest intellectual achievement, enabling us to perceive, understand, and manipulate the natural world. It has lifted us to a level of health, prosperity, freedom, and comfort beyond the wildest imagination of people living in pre-scientific days. It has so richly transformed our material lives that it is reasonable to argue that the average Westerner enjoys far better health, comfort, and diversity of experience than any ancient monarch. Science can do this because its way of knowing yields justified beliefs verifiable across all cultures. If it gets things wrong, and it often does, scientists know that their work is tentative and self-correcting. Science is a process by which the answers lead to more questions; it feeds on ignorance for what is already known is boring, but it does not claim to answer the big questions of existence. Christian theism has always done that without science, but we are in an age when science's evidential force is too strong to ignore. It is for this reason that Christians must know how science points to God so that they can bear witness to His glory in the only way atheists will accept.

Endnotes

1. In Keyser, C., 1915, p. 28.
2. In Jennings, B, 2015, p. 59.
3. Lennox, J., 2009, p. 46.
4. Lewontin, R., 1977, np.
5. In Singh, S., 2004, pp. 361-362.
6. Świeżyński, A., 2016.
7. In Christian, J., 2011, p. 608.
8. In Duck, M., and Duck, E., 2014, p. 32.
9. Lennox, J., 2009, p. 25.
10. Ibid., p. 27.
11. Ibid., p. 28.
12. In Holt, J., 1997, np.
13. In Marsh, J., 2012, p. 72.
14. In Galison, P., Holton, G. & Schweber, S., 2008, np.
15. In Lennox, J., 2009, p 20.
16. In Coyne and Heller, 2008, p. 42.
17. Walsh, A., 2018.
18. Kennedy, D. 1907, p. 265.
19. Ibid., p. 265.
20. Davies, P., 2007, np.
21. Collins, F. 2007, np.
22. Stark, R., 2003, p. 154.
23. Planck, M., 1949, p. 184.
24. Crick, F., 1994, p. 3.
25. Esch, T. and G. Stefano, 2005.
26. In Weitnauer, C., 2013, p. 28.
27. Planck quoted in Olsen, 2013, p.382.
28. Jeans, J., 1930, p. 137.
29. In Schafer, L., 2006, p. 509.

CHAPTER THREE

The Anthropic Principle and Scientific Explanation

> "The clearer we see that in the very admission of our ignorance and finiteness, we recognize the existence of a Something, a Power, a Being in whom and because of whom we live and move and have our being—a Creator by whatever name we may call Him."
> Robert Millikan, Nobel Laureate physicist

The Anthropic Principle

The literature is awash with Nobel Laureates like Millikan who marvel at the beauty of our universe and how extraordinarily improbable it is that we are here to appreciate it. Many come to the conclusion that there is a powerful and incredibly intelligent Mind with a purpose behind it all. When discussing meaning and purpose in the universe and the razor-edge fine-tuning of its many parameters for the emergence of intelligent life, it is not long before someone brings up the Anthropic ("human-centered") Principle. Astronomer Brandon Carter coined the concept to counter the so-called "Copernican Principle," or "Principle of Mediocrity," that asserts that there is nothing special or privileged about us or our planet; we are just accidental creatures in an accidental universe. Atheists love the Copernican Principle because it aids their efforts to relegate God to history's dustbin, and despise the Anthropic Principle because it points to a universe with purpose, which implies we humans are very privileged indeed.

Beginning in the late 1960s, physicists started to ponder the many exquisitely fine-tuned parameters of the universe. Fine-tuning means that the parameters or physical constants of the universe must be adjusted with mind-boggling precision in order for intelligent life to exist. Many physicists who gave serious thought to this began to believe that the cosmological "coincidences" that make out existence so astronomically improbable are not the result of blind chance, but are rather part of a purposeful universe. There are different of versions of the Anthropic Principle, starting with the Weak Anthropic Principle (WAP). The WAP is defined by Carter as, "we must be prepared to take account of the fact that our location in the universe is necessarily privileged to the extent of being compatible with our existence as observers."[1]

It has been pointed out that this compatibility is not at all surprising since if the universe were not so we wouldn't be here to discuss it. While this is obvious, it does not inform us of *why* we are here to discuss it. Philosopher John Leslie rebutted the atheist argument with his famous firing squad analogy. He asks us to imagine a condemned man facing a firing squad composed of 50 expert marksmen. The order to fire is given, the shots ring out, but they all miss and the condemned man walks away. It is entirely possible that one marksman missed, but it is hardly possible that all 50 did. It would not make sense to say that it is not at all surprising that they all missed since if they had not, the condemned man would not be alive to walk away. It is more sensible to conclude that something intentional was afoot; that is, the firing squad designed it such that the condemned man should go on living.[2] We can apply the same reasoning to all our lives—there is something intentional afoot.

Why would any physicist find such an apparent truism as the WAP useful as it applies to his or her daily work? Physicist Frank Tipler, one of the pioneers of the Anthropic Principle, observes: "But the Weak Anthropic Principle is not trivial, for it leads to unexpected relationships between observed quantities that appear to be unrelated!"[3] The remarkably gifted Stephen Hawking says that the "Anthropic Principle is essential, if one is to pick out a solution to represent the universe," and another great physicist, Andrei Linde, opined that: "Those who dislike anthropic principles are simply in denial...One may hate the Anthropic Principle or love it, but I bet that eventually everyone is going to use it."[4]

Carter then added the Strong Anthropic Principle (SAP): "The universe (and thus the fundamental parameters on which it depends) must be such as to admit the creation of observers within it at some stage"[5] This statement strongly implies some purpose and deliberate design behind the universe and human existence. The reason the universe seems tailor-made for our existence is that it *is* tailor-made for our existence. Philosopher of science Michael Corey agrees, and provides an Anthropic Principle of his own he calls the Design-Centered Anthropic Principle (DCAP). The DCAP is stated as: "The universe possesses life-supporting configuration because it was deliberately infused with these properties by a higher power."[6] There is no other reasonable explanation of why the universe had to admit the creation of observers; an endless trail of wildly improbable coincidences just doesn't fit the bill.

Astrophysicist Luke Barnes came more or less to the same conclusion: "The anthropic coincidences are so arresting because we are accustomed to thinking of physical laws and initial conditions as being unconcerned with how things turn out. Physical laws are material and efficient causes, not final causes."[7] A material cause is the physical properties of a thing (matter, energy, and the laws of physics); an efficient cause is the agent which brings it about (Almighty God), and the final cause is the ultimate purpose for the thing's being. That purpose is revealed in Isaiah 43:7: "Even every one that is called by my name: for I have created him for my glory, I have formed him; yea, I have made him." The final cause, or purpose, for the existence of the universe and for our being is thus to honor and love God.

Barrow and Tipler then proposed the Final Anthropic Principle (FAP): "Intelligent information-processing must come into existence in the universe, and, once it comes into existence, it will never die out."[8] SAP and FAP are the polar opposites of the Copernican principle because they invoke meaningfulness and purpose in the universe. Indeed, the FAP is reminiscent of a basic tenet of Christian faith as set forth in John 3:16: "For God so loved the world that He gave His only begotten Son, that whoever believes in Him shall not perish, but have eternal life." Theoretical physicist Heinz Pagels has written that the idea that a Supreme Being created the universe as a home for intelligent life is most unattractive to atheists, and notes: "Faced with questions that do not neatly fit into the framework of science, they are loath to resort to religious explanation; yet their curiosity will not let them leave matters unaddressed. Hence, the anthropic principle. It is the closest that some atheists can get to God."[9]

The Anthropic Principle is not a predictive theory; it is rather a form of reasoning about nature that accounts for the fine-tuning we observe. It is not predictive because it looks backward to explain what already exists in a logically coherent way, although some confirmed anthropic predictions have been made. Anthropic reasoning is similar to abductive reasoning; that is, *post hoc* explanations of what we observe, such as if we observe that the street is wet and conclude that it has been raining. It could also be wet if the street cleaners had just gone past your house, or a water pipe had burst close by. All three hypotheses (rain, street cleaners, a broken water pipe) have explanatory power; if any were true, it would explain why the street is wet. Intuitively, however, the rain hypothesis is better than the others, especially if we seek further evidence. If we find that the grass is wet in your back yard and there is fresh water in the rain gutters, we can reject the other possibilities and conclude that the street is wet because it rained. This is a backward-looking conclusion only because the wet street will not allow you to predict that it will rain on the same date next month, but it is the best explanation of the current state of the street.

The Anthropic Principle is a powerful argument for design and purpose in the universe and for the notion that we humans are very privileged indeed. Physicist Josip Planinić views the SAP precisely in this manner: "The anthropic principle, or the fine-tuned universe argument, can also be put forward as a design argument...It seems that the universe is arranged (tuned) exclusively to be agreeable to man. This thought on the notion of purposefulness implies the existence of a Creator of the universe."[10] Albert Einstein also believed in a purposeful universe, as his words attest: "The religious inclination lies in the dim consciousness that dwells in humans that all nature, including the humans in it, is in no way an accidental game, but a work of lawfulness that there is a fundamental cause of all existence."[11]

Explanation in Science

An explanation in science is not something divorced from explanations in other domains of inquiry. The difference is that scientific explanations are more precise and rigorous, and rest on many lines of evidence obtained by following a formal set of procedures, although they still follow the rules of logic we use when trying to explain anything. Science relies on three methods of reasoning to arrive at explanations of phenomena it explores: deduction, induction, and abduction. Deduction, the most reliable of the three, is a "top down" method that reasons from a premise that is self-evidently true ("All men are mortal.") to a minor premise ("Plato is a man."), and on to a conclusion ("Therefore, Plato is mortal."). There are people belonging to a school of thought called rationalism who contend that the world can only be understood as *it is* through the intellect because the senses allow us only to see it as *it appears*. They say that the phenomena of the world come to us through the buzzing confusion of sense perceptions and must be filtered and organized by the intellect.

While it is true that our perceptions are organized by the mind and that our senses may deceive us, rationalists appear to be saying that our intellect cannot deceive us, which is a serious error because it has deceived even the greatest of minds. Nevertheless, rationalists idealize mathematics as the only true paradigm of truth because mathematical thinking rests on a priori knowledge that is true by definition: if $x = 2$ and $y = 3$, then $(x)(y) = 6$ is absolute in all possible instances. Deductive "top-down" reasoning from truths considered self-evident had been taken as the ideal path to knowledge ever since Plato. It is considered ideal because it guarantees the truth of the conclusion given that it is already present in the premise ("All crimes are against the law."), and any denial of it is self-contradictory.

Once we leave behind the certainty of mathematics we run into trouble with deductive reasoning because except in the most trivial sense ("All mothers are females") we have precious few major premises that are self-evidently true. We cannot simply "rationalize" ourselves into knowing; knowledge must be gained by observation and experiment. This is "bottom-up" reasoning from the specific to the general and is called induction. A conclusion in a philosopher's deductive mode is a hypothesis in a scientist's inductive mode; an assertion to be tested experimentally. To conduct experiments and make observations, scientists are guided by theories from which hypotheses are logically deduced. Unlike mathematical axioms, however, theories are not true by definition. Deductions from theory must presuppose broad inductions to validate their major premises. Knowledge of the world can only be achieved with some degree of confidence when we test our concepts in the world outside our own minds. Empirical science cannot produce the absolute certainty demanded by those who identify all true knowledge with mathematics, but the experimental-observational inductive method is the bedrock of science.

The third method of reasoning is abduction. Abductive reasoning starts with all available observations relevant to a particular phenomenon and offers the most reasonable explanation for it, but leaves space for other possible explanations. Philosopher Peter Lipton offers a simple example of abductive reasoning in the form of Sherlock Holmes zeroing in on his arch-enemy, Professor Moriarty, as the one guilty of a murder. Holmes infers that Moriarty is guilty because his inference best explains all the evidence gathered, such as fingerprints, bloodstains, and other such evidence. Lipton says that Holmes' belief is not arrived at deductively because: "The evidence will not entail that Moriarty is to blame, since it always remains possible that someone else was the perpetrator. Nevertheless, Holmes is right to make his inference, since Moriarty's guilt would provide a better explanation of the evidence than would anyone else's."[12]

Given the sum of the evidence presented, any neutral jury would have to conclude unanimously that Moriarty is guilty "beyond reasonable doubt," but not beyond all doubt. Unlike deductive reasoning, whereby the conclusion is guaranteed by the axiom, the jury will have to reach the simplest and most logical conclusion it could draw from multiple lines of evidence, just as you inferred that the street was wet because it rained. Science sets up such a decision-making process the same way the Anglo-American legal system tests the guilt or innocence of the accused—it assumes that the person in the dock is innocent. Science sets up what is called the null hypothesis, which is a position the scientist no more believes in than the prosecutor believes that the accused is innocent. The assumption of innocence and the null hypothesis are precautionary measures that require stringent evidence to reject. Some of the evidence may carry little weight, but piling evidence upon evidence, upon evidence, will eventually build such a weighty case that few reasonable people will reject on rational grounds. When we have collected all relevant evidence, we test it against competing hypotheses (ruling out other possible suspects) and make an "inference to the best explanation." As we weigh the evidence supplied by the multiple instances of fine-tuning required for conscious life in the universe, the God hypothesis wins hands-down.

We are all members of a very important jury in which we must make the most important decision of our lives—the existence of God and whether or not we will accept Him. Luckily, we do not have to rule out multiple hypotheses; the only alternative to our hypothesis that God exists is that He does not (the null hypothesis). Reasoning from multiple observations to the best explanation is a method graced with simplicity and elegance and in accordance with Occam's razor ("Entities are not to be multiplied without necessity"). This simply means that among competing explanations, the one with the fewest assumptions that fits all the evidence should be chosen. We have only two mutually exclusive options to choose from for the most plausible explanation for the origin of our life-bearing universe: either it was created by an Intelligent Designer or it created itself from nothing. Abductive reasoning—reasoning from multiple observations from diverse fields of inquiry—points unerringly to the first option.

It is inherently impossible to prove in any absolute sense that an unseen metaphysical being exists. God is a spiritual Being forever undetectable by the methods of science. If we demand absolute proof for our beliefs, then many phenomena are beyond our belief. We often take things to be true on the basis of strong circumstantial evidence; if we did not, we would be in a perpetual state of indecision. As one who has taught statistics for 34 years and written three books on the subject, I know that most hypotheses in science are accepted or rejected on the basis of probability, and that scientists never say that they have proven their hypotheses beyond all possible doubt. After all, even extremely improbable events do occur. We will see that many things atheists call "coincidences" have a probability of occurring much lower than one time in a trillion. We can accept one extremely unlikely phenomenon as a lucky coincidence, but when we see so many parameters in the universe falling into that finely tuned "just right" category, we have to be open the assertion that such stupendous precision could only come from an omniscient and omnipotent God who crafted the universe for our benefit.

Geophysicist and philosopher of science Stephen Meyer illustrates how abduction is used in science and how science is pointing to the existence of the Creator "beyond a reasonable doubt." He notes that multiple lines of evidence from the natural sciences cosmology, physics, and the various branches of the life sciences, show that theism has scope and power to explain a very wide range of scientific evidence "more simply, adequately, and comprehensively than other major competing worldviews or metaphysical systems." Meyer realizes that this does not prove God's existence because it "does not constitute deductive certainty. It does suggest, however, that the natural sciences now provide strong epistemological support for the existence of God as affirmed by both a theistic and Judeo-Christian worldview."[13]

Do we need an Explanation of the Explanation?

We have abductively chosen the power of Almighty God—the First Cause—as the best explanation for the origin of the universe. Atheistic scientists do not like to think in terms of a first cause because it leads them into areas they would rather not go. Prior to the early 1930s, the standard position of science was that the universe is past eternal, static, and uncaused. This was a convenient position because it relieved scientists of having to ponder questions of its origin. Science now knows the universe had a beginning, but is in a quandary regarding how or why it began, and the theistic implications of even thinking about it. Christianity has always asserted a beginning as revealed in Genesis.

It is now accepted science that the universe commenced its existence with the Big Bang, at which time, physicists tell us, their laws break down. If the laws of physics cannot supply an explanation due to this impasse, then the explanation must be beyond physics; that is, it must be *meta*physical. Yes, this is philosophy and not data-driven empirical science, but science depends upon philosophy to justify its presuppositions and methods, and to keep our thinking on an even keel. Some of the greatest physicists in history—Planck, Gödel, Schrodinger, Heisenberg, Eddington, Bohr —were deeply immersed in philosophy and the greatest of them all, Albert Einstein, wrote that philosophical insight is "the mark of distinction between a mere artisan or specialist and a real seeker of truth."[14]

If we assert that everything has a cause and, by inference to the best explanation, that God is the cause of the universe, the standard atheist "checkmate" response is: "If everything requires a cause, then what caused God?" This is asking for an explanation of the explanation, and demonstrates a faulty understanding of God as Christians know Him. Christians do not believe in a *created* God such as Zeus; we believe in a *creating* God. In his last book before his death, the brilliant physicist Stephen Hawking was guilty of a quite unbrilliant, non sequitur when he gave us his thoughts on why God could not possibly exist. He wrote: "When people ask me if a God created the universe, I tell them that the question itself makes no sense. Time didn't exist before the Big Bang so there is no time for God to make the universe in."[15] Hawking assumes that causal priority presupposes temporal priority, which of course it does in the everyday reality of a universe that already exists. There was indeed no time for the Creator to exist in before the creation; there were no physical events of any kind before time existed. But God exists outside of time, which began, as Hawking says, at the first event—the Big Bang. God is causally prior to creation, but not temporally prior to it because God created time.

As for the question of who or what created God, Christians maintain that only things that *begin to exist* have a cause. After all, the universe is finite, so we can't keep pushing contingent causes back forever; we have to stop at something that is a sufficient explanation for its own existence. "Who made God" is a meaningless question because God is not bound by the naturalistic parameters of time, space, beginning, and causality—He is the beginningless Uncaused Cause. Arguing that it is contradictory to claim both that everything has a cause and that God has no cause misses the point entirely. When we assert that everything is caused; we mean that everything *contingent*, everything *material*; everything in *time*, and everything *imperfect* requires a cause. God, who is unconditional, immaterial, timeless, and perfect, does not. God is God precisely because He does not have a creator. He exists in and of Himself, independent of anything else. This is the *aseity* ("of oneself") of God.

Another reason why asking Christians to explain the explanation is simply not logical, is that all explanations must arrive at a stopping point beyond which it is impossible to go, a point Christians call God. As philosopher of science and theology William Craig, puts it: "one needn't have an explanation of the explanation. This is an elementary point concerning inference to the best explanation as practiced in the philosophy of science."[16] Craig tells us that an attempt to explain the explanation leads to an infinite regress: If X created God; who or what created X? Y created X. Who or what then created Y? And so on *ad infinitum*. We could go on repeating this regress for eternity because it is something that cannot, by definition, be completed. All explanations must have an ultimate terminus, an uncaused First Cause.

One of the greatest mathematicians of the 20th century, David Hilbert, provided a mathematical proof of the impossibility of an infinite regress, concluding his famous paper on the subject saying: "Our principal result is that the infinite is nowhere to be found in reality. It neither exists in nature nor provides a legitimate basis for rational thought...The role that remains for the infinite is solely that of an idea."[17] Think of it in simple terms; no matter many zeros are written down from now until the end of time, you can always write another. Because there is no "before" the Big Bang, an infinite regress in the causal chain that produced the universe is impossible; there must be a stopping point. Even if this universe was created by something else in a different timeline, say from the highly speculative idea of a "multiverse" (the notion that there are countless other universes beyond our perception), the logic still holds; the regress must ultimately arrive at the First Cause existing outside of all possible timelines. Something caused the Big Bang, so that "something" must be outside of time and be itself infinite.

If an atheist wants to stop the infinite regress with the universe as a self-creating brute fact, he would have to assert that while it is true that everything manifest in the universe has a cause, the universe itself requires no explanation because it is the uncaused cause of itself. The atheist's universe pulled itself up by its own bootstraps, and did it so astoundingly well that it produced sentient beings capable of probing its secrets. To say that the universe is self-existent is analogous to saying that a man can literally be the father of himself, or perhaps a man can exist without being fathered. On that argument alone, which is the most logical inference to the best explanation: God or nothing? An "omnipotent" but mindless universe takes us back to the pre-1930s world in which the universe was considered a past eternal brute fact, and thus we need not bother ourselves pondering its origin. Such an argument hobbles the entire enterprise of cosmology, which exists precisely to discover how the universe came into being. Rather than claiming that matter and energy somehow always existed and created itself, Nobel Laureate physicist Sir Edmund Whittaker applies Occam's razor: "It is simpler to postulate creation *ex nihilo*—Divine Will constituting Nature from nothingness."[18]

A contingent physical universe exists, and since nothing can be the cause of itself, an explanation for its existence must be found outside the physical; to something that exists non-contingently, namely, God. A moment's reflection will reveal that atheism's notion of a self-creating universe is more miraculous than theism's creation by God because it posits that all energy/matter came from nothing. This not only defies logic; it defies the first law of thermodynamics that decrees that under *natural* conditions, energy cannot be created or destroyed, but once it is in existence it can be transformed. What was this Godless naturalistic nothing that circumvented its own law and created matter/energy from nothing? The atheist has neither a natural nor a supernatural explanation for creation, which underlines the hollowness (literally) of his arguments—nothing created something from nothing for no reason!

Endnotes

1. Carter, B., 1974, p. 293.
2. Leslie, J. 1989.
3. Tipler, F. 1988, p. 28.
4. All cited in Susskind, L. 2005, p. 353.
5. Carter, B., 1974, p. 294.
6. Corey, M. 2001, p. 47.
7. Barnes, L. 2012, p. 562.
8. Barrow, J. & Tipler, F. 1986, p. 23.
9. Pagels, H. 1985, p. 38.
10. Planinić, J. 2010, p. 47.
11. In Isaacson, W. 2007 p. 46.
12. Lipton, P., 2000, p. 185.

13. Meyer, S., 1999, p. 27.
14. In Howard, D. 2005, p. 34.
15. Hawking, S. 2018, p. 38.
16. Craig, W., 2008, p. 39.
17. In Maurin, A., 2013.
18. In Heeren, F., 2000, p. 121.

CHAPTER FOUR

The Big Bang:
The Event that Traumatized Physics

"Probably every physicist would believe in a creation if the Bible had not unfortunately said something about it many years ago and made it seem old-fashioned."
George Thomson, Nobel Laureate physicist

In the Beginning

Genesis *1:1* tells us that: "In the beginning God created the heavens and the earth." God's creation was *creatio ex nihilo* ("creation from nothing"). While these exact words are not found in the Bible, there are a number of passages from which their meaning is inferred. For example, Psalm 33:6 says: "By the word of the Lord the heavens were made, their starry host by the breath of his mouth," and John 1:1 "In the beginning was the Word, and the Word was with God, and the Word was God." Creation from nothing was deemed absurd by scientists since nothing (no-thing) comes from nothing, although some have done an about-turn on this recently to affirm that the universe created itself from nothing. We have seen that before the 20th century most scientists believed that the universe had no beginning, that it was static, eternal, and infinite in space and matter. It was simply an inexplicable brute fact of existence that needed no jump start, and certainly not from the supernatural. This model of the universe was known as the "steady-state" model and was scientifically satisfying because it meant that scientists did not have to get into the messy metaphysical questions about the universe's origin and cause.

The steady-state view re-emerged in 13th-century Europe with the discovery of Aristotle's ancient writings. Aristotle believed that everything that exists must have a cause, but the universe was past eternal. Aristotle's universe was organized out of chaos by the Demiurge, a Platonic god subordinate to God, whom Plato called "the One." The great Isaac Newton accepted a past eternal universe for religious reasons, even though it conflicted with Genesis. Astrophysicist Guillermo Gonzales and philosopher of science Jay Richards note that Newton "viewed the universe as the 'divine sensorium'—the medium through which God acted on the world. To be adequate to this task, the cosmos had to be infinite." He held this view because: "For him, an infinite universe answered the question of why gravity did not cause the constituents of the universe to collapse in on one another."[1]

The eternally existing steady-state universe began to unravel with the 1915 publication of Einstein's general theory of relativity. Einstein was unsettled to find that his equations predicted the expansion of the universe, which did not fit the current scientific orthodoxy. He "corrected" his equations by adding what he termed the "cosmological constant" (Λ; the Greek letter *lamda*) representing a repulsive force to counter gravity's attraction, thus leaving the universe static, or so he thought. He later called this the greatest blunder of his life, but his equations turned out to be right after all—the universe had a beginning and was expanding. Dark energy is now considered the repulsive force countering the attraction of gravity.

It was Belgian priest and mathematician-physicist, Georges Lemaître, who noted in the early 1920s that all was not right with the static universe. Like Newton, he reasoned that in a state of past eternity, gravity would have long ago pulled all matter in the universe together into one huge mass. Lemaître drew the conclusion that to avoid this crunch the universe had to be expanding, and if it was expanding, it had to do so from a finite point in time. Simon Appolloni explains the reasoning: "Lemaitre concluded that the universe had to be anything but static, and reasoned it must be expanding; what is more, all matter would stay separated [from the gravitational pull] as the expansion force slightly exceeded the gravitational force."[2]

Lemaître reasoned that if the universe was expanding, rewinding the cosmic clock should arrive at a point when all matter was condensed into a single entity, which he called the primeval atom. Modern physicists call this the singularity; a "point" of almost infinite density and temperature. Everything that exists in the universe, every last atom, every physical force, and space/time itself was contained in this super-dense concentration of energy. Stranger yet, this singularity was not some tiny dot hanging around in space somewhere because there was no "somewhere" for it to be, nor was it hanging around waiting to pop into existence because there was no time before it popped. Because this sounded all too weird, most physicists dismissed Lemaître's reasoning; after all, he was a man of God.

In 1929-1930, American astronomer Edwin Hubble, working at the Mount Wilson Observatory in California, provided observational evidence for the expanding universe. Hubble found that all galaxies are moving away from each other and that the farther away they were the faster they are moving. This is determined by the wavelength spectrum of stars, with galaxies moving away from us becoming redder (more "red-shifted") as their light wavelength is stretched. This is known as the Doppler Effect and is seen in all wavelengths. We experience it most obviously with sound, but it is equally true of color. We experience it every time we hear the sirens of emergency vehicles. As they come closer to us the sound waves are compressed, and as they recede, they are stretched as the siren's pitch decreases. The inescapable conclusion from these observations was that some gigantic event caused the universe to expand with unfathomable force some 13.8 billion years ago, give or take a few million years.

This event is known as the Big Bang, the event that brought matter/energy, time, and space into being in a split-second flash. The Big Bang was not a literal explosion, but rather an expansion of the universe creating space and time as it expanded at an unimaginable rate. The expansion is not an expansion into previously unoccupied space such as a balloon being blown up or a bomb being detonated; these things are embedded in three-dimensional space with centers and edges into which they expand. The universe has no center or edge to expand into. Rather, it creates space as it expands. As physicists like to put it, the Big Bang was not an explosion *in* space but an explosion *of* space. Literal explosions throw matter apart in all directions, after which it falls back randomly under the influence of gravity. They never result in matter clumping together in orderly patterns such as galaxies, stars, and planets. The attractive force of gravity pulling matter in had to be exquisitely calibrated to the "explosive" force driving it forward. How exquisite was this balance? Physicist Paul Davies informs us that if the rate of expansion from the beginning differed by more than 10^{-18} seconds we wouldn't be here. In his words: "The explosive vigour of the universe is thus matched with almost unbelievable accuracy to its gravitational power. The big bang was not evidently, any old bang, but an explosion of exquisitely arranged magnitude."[3]

Scientists estimated that in 2013 the observable universe stretches in all directions from us for 2.7×10^{23} miles.[4] That's 270 sextillions, or about a billion trillion miles, and it is still expanding ever faster. The universe had a beginning and a cause after all, and that cause has to be an entity that transcends time, space, and matter/energy since these things did not exist before the creation. Contrary to Hawking's time/cause confusion, astronomer George Greenstein says: "As we survey all the evidence, the thought insistently arises that some supernatural agency, or rather Agency, must be involved. Is it possible that suddenly, without intending to, we have stumbled upon scientific proof of the existence of a Supreme Being?"[5s]

Interestingly, the Big Bang idea is much more ancient than most suppose. The English theologian and mathematician Robert Grosseteste, perhaps the smartest man you never heard of, wrote a mathematical treatise called *De Luce* (*On Light*) in 1225 in which he explored the nature of matter and the universe. Bower and his colleagues inform us that: "Four centuries before Isaac Newton proposed gravity and seven centuries before the Big Bang theory, Grosseteste describes the birth of the Universe in an explosion and the crystallization of matter to form stars and planets in a set of nested spheres around Earth."[6] Grosseteste's dual role as scientist and theologian (and Bishop of Lincoln) mocks the claim that science and religion are in conflict.

The Kalam Cosmological Argument

The kalam cosmological argument is the modern form of a medieval Islamic argument for the existence of God. It relies on deductive logic, asserting that the universe had a beginning in finite time against the Aristotelian notion the universe is past eternal. The modern champion of the argument is philosopher of science William Lane Craig, who offers the kalam argument in the form of the following syllogism (with my comments):[7]

1. ***Whatever begins to exist has a cause.*** The major premise comports with everyday experience; every physical object or system, or change in an object or system, has a cause preceding the effect. As self-evident as this is, there are objections to it. In what Craig described as the "Gettysburg Address of Atheism," he quotes philosopher Quentin Smith's claim that: "the most reasonable belief is that we came from nothing, by nothing, and for nothing."[8] Smith knew that the only alternative to a beginning is a past eternal universe, and wrote these words long after it was known that past infinity is not realizable. Physicist Sir Fred Hoyle, a strong atheist, tried unsuccessfully to fit infinity into a mathematical model to avoid a beginning, but it was a mathematical equation—Dirac's quantum mechanical equation describing the fine balance between subatomic particles in nucleosynthesis—that made him question his atheism. He did not embrace theism, but rather a "super-intellect." In 1994 he wrote: "It is because of this incredible chain of subtlety that I doubt the nineteenth-century denial of a purposive universe."[9]

2. ***The universe began to exist.*** Although the minor premise logically follows from the major premise, it ran counter to the notion of a past eternal universe. While no scientist doubts that all effects have causes, some require the universe to be the sole exception because positing a cause has theistic implications. To be consistent with a purely naturalistic perspective, we would have to say that the universe somehow created itself, but nothing can be the cause of itself since it would have to have existed prior to itself to do so, which is absurd. Astrophysicist Alexander Viilenkin said at a 2011 conference on The State of the Universe that: "All the evidence we have says that the universe had a beginning."[10]

The second law of thermodynamics argues against an infinite past. The second law states that entropy (disorder) in an isolated system (a system without an external source of energy) always increases. Over time all such systems devolve towards thermodynamic equilibrium, which is a state of maximum entropy. If the universe was past-infinite, it would have been in a state of thermodynamic equilibrium a very long time ago. The inimitable Stephen Hawking has noted that the notion of a past eternity runs into very serious difficulty with the second law, and that there had to be a beginning: "Otherwise, the universe would be in a state of complete disorder by now, and everything would be at the same temperature. In an infinite and everlasting universe, every line of sight would end on the surface of a star. This would mean that the night sky would have been as bright as the surface of the Sun."[11]

3. *The universe has a cause*. It was the conclusion of the syllogism that rattled many scientists determined to leave a First Cause off the discussion table. If everything from a paper-clip to a space shuttle requires intelligence, how much more intelligent must the Creator of everything be? It is an inescapable fact that if the universe has a cause, it was an uncaused cause. This is something that made scientists very uncomfortable. William Craig notes that: "A cause of space and time must be an uncaused, beginningless, timeless, spaceless, immaterial personal being endowed with freedom of will and enormous power. And that is a core concept of God." [12]

Early Opposition to a Beginning

In the early 20th century, scientists had an almost religious faith in the materialist laws of physics, which had proved to be beautiful, true, and useful, and could not fathom a beginning of time, space, and matter. Scientists in the officially atheist Soviet Union were fierce critics of the Big Bang because it supported the Genesis account of creation. Marxist philosopher Georges Politzer noted what the Big Bang implied for materialism: "The universe was not a created object. If it were, then it would have to be created instantaneously by God and brought into existence from nothing. To admit Creation, one has to admit, in the first place, the existence of a moment when the universe did not exist, and that something came out of nothingness. This is something to which science cannot accede."[13]

Biologist John Maddox also announced the Big Bang "philosophically unacceptable" because it supports Genesis: "Creationists and those of similar persuasions seeking support for their opinions have ample justification in the doctrine of the Big Bang." [14] It does not matter if the Big Bang is scientifically acceptable if it is unacceptable to atheist philosophy. The deep disquiet the Big Bang generates among atheists is proof enough that it points to Christianity's view of creation. Noting that many scientists have a religious faith in materialism, astrophysicist Robert Jastrow points out: "This religious faith of the scientist is violated by the discovery that the world had a beginning under conditions in which the known laws of physics are not valid, and as a product of forces or circumstances we cannot discover. When that happens, the scientist has lost control. If he really examined the implications, he would be traumatized."[15]

Many scientists were indeed traumatized; even the phrase "Big Bang" was cynically coined by Fred Hoyle. Scientists committed exclusively to materialist philosophy rejected the idea because it led to echoes of the "spooky" Genesis story of divine creation *ex nihilo*. Other scientists who railed against the idea of a universe with a finite past include astronomer Arthur Eddington, who said that "Philosophically, the notion of a beginning is repugnant to me." Eddington was not animated by anti-religious motives because he was a deeply religious Quaker. Appolloni tells us that Eddington "considered that no matter how far science advanced, God's creation remains ultimately mysterious and wonderful. While Eddington seeks truth as a scientist, like Lemaitre, as a Quaker, the pursuit of truth (in all aspects of life whether scientific or religious) will always remain just that, a pursuit, not a realization."[16] This is, of course, true.

Chemist Walter Nernst angrily dismissed a beginning: "To deny the infinite duration of time would be to betray the very foundations of science," while physicist Philip Morrison had more of the attitude of a true scientist by following the data where they lead: "I find it hard to accept the Big Bang Theory. I would like to reject it, but I have to accept the facts." Astronomer Allan Sandage, concluded that "It is such a strange conclusion....it cannot really be true."[17] Sandage, the "Grand old man of cosmology," later became a Christian, noting that "It was my science that drove me to the conclusion that the world is much more complicated than can be explained by science. It was only through the supernatural that I can understand the mystery of existence."[18]

There is little doubt that much of the opposition to the Big Bang was motivated by the idea that a beginning implied a Creator. However, an increasing number of scientists soon accepted it, much to the consternation of Sir Fred, who complained that: "The reason why scientists like the 'big bang' is because they are overshadowed by the Book of Genesis. It is deep within the psyche of most scientists to believe in the first page of Genesis."[19] For Hoyle, Genesis influenced scientists' acceptance of the Big Bang, but it is surely more likely that the science of the Big Bang influenced their acceptance of Genesis. This is supported by the words of famous scientists themselves. Robert Jastrow said, "Now we see how the astronomical evidence supports the Biblical view of the origin of the world. The details differ, but the essential elements in the astronomical and Biblical accounts of Genesis are the same: the chain of events leading to man commenced suddenly and sharply at a definite moment in time, in a flash of light and energy."[20] And in a *New York Times* interview, Nobel Laureate physicist Arno Penzias stated: "The best data we have (concerning the big bang) are exactly what I would have predicted had I nothing to go on but the five books of Moses, the Psalms, the Bible as a whole."[21]

The Cosmic Microwave Background Radiation

Hubble's observations did not conclusively convince all scientists to accept the Big Bang. The clincher for most was the discovery of the cosmic microwave background (CMB) radiation. The CMB was discovered accidentally in 1964 by Arno Penzias and Robert Wilson working at Bell laboratories in New Jersey. They were working with supersensitive radio telescopes designed to detect and measure radio waves from balloon satellites. To do this they had to eliminate all interference, such as radar or broadcasting signals. After weeks of doing everything possible to eliminate interference, they found an annoying hiss that was coming from every direction with equal strength. Penzias and Wilson finally concluded that radiation was coming from outside our galaxy, but they could not explain it.

Princeton physicists had long tried to find the radiation predicted to exist as a remnant of the Big Bang that Penzias and Wilson found accidentally. It had been predicted that the radiation would be microwaves because of a massive redshift all the way back to about 380,000 years after the Big Bang (microwaves are weaker than all other waves in the electromagnetic spectrum except radio waves). Because of this, and because it was coming from everywhere at once, scientists came to realize that: "Not only was it real signal, it was evidence for the big bang itself."[22] According to National Aeronautics and Space Administration (NASA) scientists, the fact that CMB radiation is detected everywhere we look and has a uniform temperature of 2.725° Kelvin above absolute zero to better than one part in a thousand "is one compelling reason to interpret the radiation as remnant heat from the Big Bang; it would be very difficult to imagine a local source of radiation that was this uniform. In fact, many scientists have tried to devise alternative explanations for the source of this radiation, but none have succeeded."[23] You may have witnessed this background radiation yourself. The hissing and "snow" static sometimes seen on your out-of-tune television is the microwave echo of creation.

The Abundance of Light Elements

The third piece of evidence for the Big Bang is the abundance of light elements in the universe. To appreciate this evidence, we have to understand the origin of the chemical elements in the stars. Scientists trace the history of the universe all the way back to Planck time, which is an astounding 10^{-43} seconds after the Big Bang. At this time, the fundamental forces of nature—gravity, electromagnetism, the strong and weak nuclear forces—were one. The universe was so hot (about 80 million trillion, trillion, degrees Fahrenheit) at this point that atoms could not form. At 10^{-35} seconds, the universe had expanded "from something smaller than a proton to something the size of a grapefruit, and the strong and electroweak forces were unified."[24] With further cooling at 10^{-10} seconds came the unification of the electromagnetic and weak forces, and at 10^{-5} seconds, quarks, the first "matter," appeared. At first, quarks and anti-quarks zipped around in unbound states, but with a little more cooling they were able to combine and form protons and neutrons.

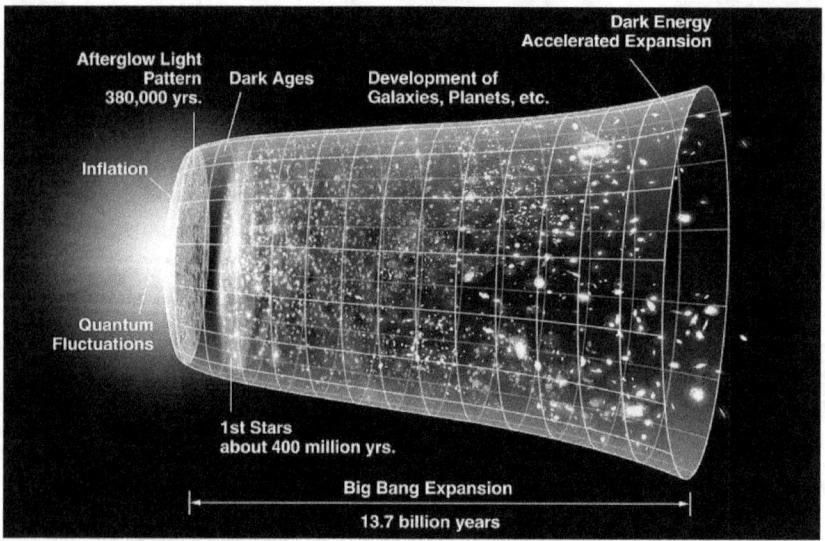

Figure 4.1: The Timeline for the Expansion of Space from the Big Bang

Protons and neutrons are matter that has mirror images called anti-matter with the same mass. Matter has negatively charged electrons, and anti-matter has positively charged positrons. As these particles whizzed around in the hellish maelstrom, they annihilated each other in a flash of radiation, with both kinds spontaneously arising from that same radiation. The laws of physics tell us that we should have expected equal amounts of matter and anti-matter, but because there were slightly more matter particles than antimatter particles (1 in 100,000,000), matter prevailed. If it did not, the universe would be nothing but pure energy.

At three minutes (a huge time jump on this scale) protons and neutron were able to form stable nuclei. If they had formed before this cooler time, violent collisions would have immediately torn them apart. It was hundreds of thousands of years later when the temperature was cool enough for electrons to attach themselves to protons and neutrons to form atoms. These atoms were hydrogen, helium, and a tiny amount of lithium, which have 1, 2, and 3 protons in their nuclei, respectively. The abundance of these light elements is consistent with their creation in a Big Bang nucleosynthesis (the process of creating new and more complex atoms from simpler preexisting ones). The bounty of helium and deuterium (an isotope of hydrogen—heavy hydrogen) is particularly important because it is much more abundant than could have been produced by stellar nucleosynthesis (making elements in the stars). Stars destroy deuterium, so deuterium synthesis could have only occurred in Big Bang nucleosynthesis.

The 75/23 hydrogen/ helium ratio is taken to be the ratio which existed at the time when the deuteron (a particle consisting of a proton and a neutron, but no electron) which, as an atom, is deuterium, became stable, thus halting the decay of free neutrons with the expansion and cooling of the universe.[25] Heavier elements had to wait for the formation of stars from hydrogen and helium gases because they require the extreme temperatures and pressures found within stars and cooked up in the process of stellar nucleosynthesis. This process produces elements up to iron; all elements heavier than iron are formed in the massive energy released by supernovae explosions in the process of supernovae nucleosynthesis (more on this in the next chapter).

Cosmologists call the time before the formation of atoms the cosmological "dark ages" because there was literally no light. Light is composed of photons and is part of the electromagnetic spectrum. The immense heat of the Big Bang created a soupy mix of electrically-charged subatomic particles—protons, neutrons, and electrons—that had not yet combined to form atoms. Photons interact strongly with charged matter, and only travel a short distance before being scattered as a dense fog scatters the light from a car's headlights. When the protons and neutrons were able to capture electrons to form atoms, photons were freed to travel through space, leaving behind the CMB radiation. With this decoupling of matter and radiation we get light, and thus cosmologists have been able to see back into the universe to about 380,000 years after the Big Bang.[26] Figure 4.1 from NASA of the timeline for the expansion of space and the formation of galaxies, stars, and planets.

A Little Bit about Probability and Really Big Numbers

Because we will be mentioning a lot of ridiculously unlikely events and some really big numbers, it is a good idea to briefly provide a little refresher course on probability. A great deal of scientific data evaluation involves calculating probabilities. We cannot always say that given X, Y *will* occur; we say that given X, Y has a certain *probability* of occurring. As noted earlier, when scientists make scientific observations, they are like jury members in a criminal trial instructed not to convict the offender unless convinced that he is guilty beyond a reasonable doubt. The judicial system assumes innocence until the facts indicate otherwise. Likewise, scientists have a presumption of "innocence"— that X does *not* cause Y. Scientists will not reject an assumption that X does *not* cause Y unless probability calculations tell them that they can do so "beyond a reasonable doubt." A very liberal probability would be 0.01, meaning that an observation (X *does* cause Y) we could get the observed result by chance 1 out of 100 times, and a more conservative one is 0.001, or one chance in a thousand. Rare things obviously happen; we just have to determine how likely or unlikely they are to happen.

What is the point at which something improbable becomes impossible? Mathematician William Dembski has computed the absolute limit of probability using three estimates from astrophysics: (A) the estimated number of atoms in the known universe (10^{80}), (B) Planck time (10^{45}), and (C) the number of seconds since the Big Bang at the time of his calculations (10^{25}). Planck time sets an absolute limit on the rate at which the properties of elementary particles can transition from one state to another. Dembski concludes: "If we now assume that any specification of an event within the known physical universe requires at least one elementary particle to specify it and that such specifications cannot be generated any faster than the Planck time, then these cosmological constraints imply that the total number of specified events throughout cosmic history cannot exceed 10^{80} x 10^{45} x 10^{25} = 10^{150}".[27] This is by far the most conservative estimate of the probability boundary. It completely exhausts all probability resources since it includes the product of all atoms in the universe, all seconds since the universe began, and the fastest possible time in which an event can occur.

Exponential numbers such as 10^{150} are a lot larger than they seem. Dembski's probability boundary is 1 followed by 150 zeros. To put it in perspective, one million is 10^6, a billion is 10^9, and a trillion is 10^{12}, or 1 followed by just 12 zeros, and remember scientists are willing to use 10^3 (1 in 1,000) as a conservative level to rule out chance (while not denying that chance remains a possible explanation for an observation). Numbers larger than a trillion are difficult for us to imagine, but astrophysicist Hugh Ross provides us with a nice visual image that helps us to understand the immensity of the number 10^{37}, which is vastly smaller than Dembski's 10^{150}.

Cover the entire North American continent in dimes all the way up to the moon, a height of about 239,000 miles (In comparison, the money to pay for the U.S. federal government debt would cover one square mile less than two feet deep with dimes.). Next, pile dimes from here to the moon on a billion other continents the same size as North America. Paint one dime red and mix it into the billions of piles of dimes. Blindfold a friend and ask him to pick out one dime. The odds that he will pick the red dime are one in 10^{37}. [28]

Endnotes

1. Gonzales, G. and Richards, J., 2004, p.260
2. Appolloni, S., 2011, p.23.
3. Davies, P.1984, p. 184.
4. Scharf, 2014, p. 211.
5. In Strobel, L., 2002, p. 189.
6. McLeish, T. et al, 2014, p.161.
7. Craig, W. 2008.
8. In Strobel, L., 2002, p. 122.
9. In Bussey, P., 2016. p. 70.
10. In Grossman, L., 2012, p.7.
11. Hawking, S. No date (nd), no page (np).
12. In Strobel, 2004, p. 132.
13. In Yahya, H. 1999, p. 19.
14. Maddox, J. 1989, p. 425
15. Jastrow, R., 1981, p. 19.
16. Appolloni, S., 2011, p. 29.
17. All cited by R. Jastrow, 1978, p. 122 and 123.
18. In Strobel, L., 2004, p. 84.
19. In Wallace, P. 2016, p. 101.
20. Jastrow, R., 1981, p. 19.
21. In Schaefer, H., 2003, p. 49.
22. Trefil, J. and Hazen, R., 2007, p. 318.
23. National Aeronautics and Space Administration, np.
24. Trefil, J. and Hazen, R., 2007, p. 321.
25. Bromm, V., and Larson, R., 2004.
26. Ibid.
27. Dembski, W. 2004, pp. 84-84.
28. Ross, H. 1993, p. 115.

Chapter Five

Fine-Tuning and Stellar Alchemy

> "If you think strongly enough you will be forced by science to the belief in God, which is the foundation of all Religion. You will find science not antagonistic, but helpful to Religion."
> Lord William Kelvin:
> physicist; father of thermodynamics

The Fine-Tuning of the Four Fundamental Forces

We live in a life-friendly planet, but the conditions for the existence of such a planet, never mind for life on it, are so highly improbable that it leaves physicists in awe. Each of the four fundamental forces of nature are so fine-tuned that even the slightest change in any of their values and the universe would not exist. Stephen Hawking and Leonard Mlodinow note that: "The emergence of the complex structures capable of supporting intelligent observers seems to be very fragile. The laws of nature form a system that is extremely fine-tuned, and very little in physical law can be altered without destroying the possibility of the development of life as we know it. Were it not for a series of startling coincidences in the precise details of physical law, it seems, humans and similar life-forms would never have come into being."[1] Each of the fundamental forces except gravity has a known carrier particle, although there is a hypothetical carrier associated with gravity called a "graviton." These particles are fundamental particles that cannot be broken down, not composite particles such as protons and neutrons. Figure 5.1 shows the carrier particles and differentiates between matter particles and force-carrying particles.

Gravity

Einstein's general relativity theory tells us that gravity is not a force between masses; rather it is the curving of spacetime in the presence of mass. It is included as a fundamental force because the warping of spacetime accelerates objects as if acted on by a force. Gravity causes smaller bodies to orbit larger ones, such as the Earth to orbit the Sun and the Moon to orbit the Earth as the smaller body "falls" into the bend of spacetime caused by the larger body. Gravity sculpted the universe by gathering the material of the Big Bang and forming it into stars and planets. Once formed, the continued existence of stars becomes a balancing act between gravity pushing in and the pressure from the explosive energy produced by nuclear fusion, but gravity will be the ultimate victor.[2] Gravity is thus very powerful at the level of big things like stars, but it is by far the weakest of the four forces. According to physicist Robert Krebs, gravity is approximately 10^{38} times weaker than the strong force, 10^{36} times weaker than the electromagnetic force, and 10^{29} times weaker than the weak force.[3] It is only because the multiple trillions upon trillions of particles in large bodies add up that gravity has the power that it does.

The law of gravity states that its strength increases proportional to the masses involved, and decreases with the square of their distance apart. If gravity had been slightly weaker by the smallest degree at the moment of creation, it would not have been able to pull matter together to form stars and planets. If it had been slightly stronger to the same degree, it would have pulled matter back into a big crunch long before stars and planets were able to form.[4] To help us to understand the extraordinary fine-tuning of gravity, physicist Robin Collins asks us to imagine a dial governing gravity's setting broken down into one-inch increments that stretches right across the universe. This would be more inches than all the grains of sand on Earth. He noted that if we moved the setting just one inch out of those unimaginable trillions from its precise setting, "That small adjustment of the dial would increase gravity by a billion fold."[5]

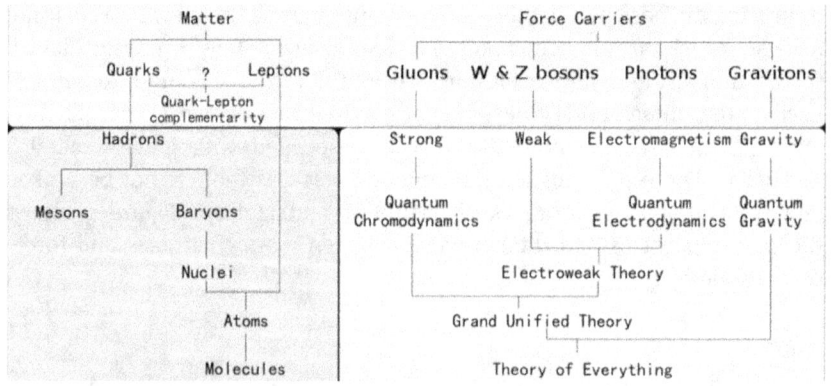

Figure 5.1: Matter and the Carriers of the Fundamental Forces

Gravity is engaged in a cosmic tug-of-war with the repulsive energy of Einstein's cosmological constant. We call this energy "dark" or "vacuum" energy. The conventional wisdom used to be that the cosmological constant was zero or even slightly negative, but we now know that the expansion rate of the universe is actually increasing, which can only mean that the cosmological constant is positive, and that dark energy is pushing the universe apart. No one understands what dark energy actually is or how it works, but one hypothesis comes from Einstein's relativity theory that predicts that space possesses its own energy. NASA Science explains: "Because this energy is a property of space itself, it would not be diluted as space expands. As more space comes into existence, more of this energy-of-space would appear. As a result, this form of energy would cause the universe to expand faster and faster."[6]

In this expanding universe, matter is not packed as closely together as it once was. This means that the gravitational pull is weakened, allowing dark energy to play a more dominant role. Gonzalez and Richards state that "There is only one 'special time' in the history of the universe when the vacuum and matter energies are the same, and we're living near it. If the vacuum energy had become prominent a few billion years earlier than it did in our universe there would have been no galaxies. If it had overtaken gravity a little earlier still, there would have been no individual stars"[7] The vacuum field is extremely weak, but the early universe had to have a value large enough to allow it to expand against gravity's pull: "These particle fields require an extraordinary degree of fine-tuning—at least 10^{-53} to get such a small, positive, non-zero, value for the vacuum energy."[8]

Electromagnetism

The electromagnetic force is the combination of all electrical and magnetic forces and is the best understood of the fundamental forces. The electromagnetic force has a very large range, with its strength proportional to the inverse square of the distance. Unlike gravity which only attracts, the electromagnetic force both attracts and repulses. It is the force responsible for generating visible light, as well as radiation in other wavebands such as gamma rays, X-rays, microwaves, radar, and radio waves; all carried by units (or "quanta") of light called virtual photons. The electromagnetic force is the force that makes chemical bonding possible and gives matter its strength, shape, and hardness. Just as gravity holds everything together on a cosmic scale, the electromagnetic force holds everything together between the cosmic and sub-atomic scales. It holds electrons in their orbit around the nucleus, and if electromagnetic bonding in the nuclei was the slightest bit weaker, they could not be held in orbit; if slightly greater, electrons could not bond with electrons of other atoms to form molecules. Any change that rules out atom formation also rules out molecule formation, and thus life. Its force holds atoms and molecules together by the action of its attraction and repulsive charges. It is so powerful that the contributions of the other forces as determinants of atomic and molecular structures are negligible by comparison.

As small as the relative contributions of the other forces may be, without them the electromagnetic force would be useless. Paul Davies notes that if the ratio of the nuclear strong force (discussed below) to the electromagnetic force had been different by 1 part in 10^{16} the stars could not have formed. He also tells us that if the ratio of the electromagnetic force to the gravitational force were increased by one part in 10^{40} only small stars can exist, and if it were decreased by the same amount there would be only large stars. "You must have both large and small stars in the universe: the large ones produce elements in their thermonuclear furnaces; and it is only the small ones that burn long enough to sustain a planet with life."[9]

The Strong Force

The strong nuclear force is the most powerful of the four forces, although it has the shortest interaction distances. The strong force is carried by a type of boson called gluons, so called because they "glue" the constituent protons and neutrons of atoms together. It is the force that binds quarks together to form the protons and neutrons in the nuclei of atoms. Each atom contains a number of positively charged protons. We know that positively charged objects (think magnets) brought closely together repel one another by the action of the electromagnetic force. Protons that must have a way of sticking together—of combating the repelling force—or we would have no elements heavier than hydrogen. It is the strong force that overcomes the proton's natural "shyness" to mate with others. This is a very good thing because it is the strong nuclear force that powers the stars. It does this by crushing hydrogen atoms so tightly that their nuclei overcome their natural repulsion and fuse together, resulting in massive amounts of energy being released to keep stars alive. This force, along with the weak force, is called "nuclear" because their activity is confined to the nuclei of atoms.

This release of massive amounts of energy from nuclear fusion is best known from the awesome power of thermonuclear bombs. The theoretical basis of the bomb is the fact that the combined mass of protons and neutrons is slightly less than the sum of their individual masses. This phenomenon exists because when protons and neutrons come together, a small portion of their mass is converted to energy (see the example in Figure 5.2). However, that small amount of energy is multiplied trillions of times in any amount of matter, as $E = MC^2$ tells us. That is, the energy in a given amount of matter is determined by the speed of light squared; a truly massive amount. Physicist Martin Rees informs us that the mass converted to energy is only .007 of the particle's masses, but if it was .006, a proton would not bond to a neutron to make helium and the universe would consist only of hydrogen. On the other hand, if it was .008, there would be ready and rapid fusion, and no hydrogen would have survived.[10]

The Weak Force

While gravity, electromagnetism, and the strong force hold things together, the weak nuclear force makes atoms come apart by nuclear decay. Decay is vital for building different chemical elements. During what is called beta decay, a neutron is replaced by a proton or a proton by a neutron, with an electron being ejected from the nucleus. Interaction between subatomic particles is facilitated by the weak force, which is carried by bosons called W (named from the Weak force), which has either a positive or a negative charge, and Z (so-called because it has Zero charge). The stars could not exist without this radioactive decay process. It is this force that drives the fusion of hydrogen protons and neutrons to form deuterium, and the energy generated from nuclear fusion is the source of the heat we get from the sun. The tiniest increase in the strength of the weak force would have driven the hydrogen-to-deuterium process faster, making stars use up their energy faster than their planets could cool, and thus life could not develop. A weaker weak force may have been too feeble to do much fusing at all, and all we may have in the universe is hydrogen.

As weak as it is, the weak force plays a crucial role in life. We have seen that the heavier elements necessary for life are formed in giant stars and spewed into space in supernovae explosions. Supernovae explosions fuel the cosmic cycle by pollinating new stars formed from its gasses containing the heavy elements. Such explosions would not occur if the weak force was not exquisitely calibrated. As Paul Davies explains: "If the weak interactions were slightly weaker, the neutrinos [hard to detect subatomic particles lacking an electric charge and produced by the decay of radioactive elements], would not be able to exert enough pressure on the outer envelope of the star to cause the supernova explosion. On the other hand, if it were slightly stronger, the neutrinos would be trapped inside the core, and rendered impotent." [11]

The fine-tuning of the relationships among subatomic particles is noted by Hawking and Mlodinow: "If protons were 0.2 percent heavier, they would decay into neutrons, destabilizing atoms. If the sum of the types of quark that make up a proton were changed by as little as 10 percent, there would be far fewer of the stable atom nuclei of which we are made; in fact, the summed quark masses seem roughly optimized for the existence of the largest number of stable nuclei."[12] Additional fine-tuning involves the proton-to-electron mass ratio. The mass of a neutron is slightly more than the combined masses of a proton, an electron and a neutrino. If neutrons were less massive by even the slightest amount, they could not decay without energy input. "If its mass were lower by 1%, then isolated protons would decay instead of neutrons, and very few atoms heavier than lithium could form."[13]

Carbon: The Indispensable, Improbable, Scaffolding of Life

As Carl Sagan once famously remarked, we humans are literally "made of star stuff." Six of the 118 elements in the periodic table—carbon, hydrogen, nitrogen, oxygen, phosphorous and sulfur (the so-called "CHNOPS" elements)—make up about 97 percent of the elements in the human body. Of these six, carbon stands head and shoulders above the rest because carbon atoms form the backbone of almost all the biological molecules in our bodies except water. Carbon has 6 protons and 6 neutrons (collectively called nucleons) in its nucleus and 6 electrons swirling around it. The carbon atom has two electrons in its inner shell and four valence electrons in its outer shell. Valence electrons are the electrons that determine the number of other atoms with which it can form covalent bonds (sharing of electrons between atoms). Carbon has room to form 4 bonds with other elements, which will fill its outermost shell with the 8 electrons needed for stability. Carbon forms an unparalleled number of compounds, and is one of the few elements that can bond with itself, making long chains of molecules.

All elements in the column that carbon occupies in the periodic table possess four valence electrons and can form a variety of bonds, but none can do so with the ease and stability of carbon. Silicon, the element right below carbon in the table, forms a large number of molecules, but unlike double-bonded carbon, double-bonded silicon is unstable and quickly becomes two single-bonded silicon atoms again. Also unlike silicon, the carbon in our bodies contributes crucially to plant life by interacting with oxygen to produce carbon dioxide (a gas formed by a carbon atom covalently double-bonded to two oxygen atoms). Humans exhale carbon dioxide after inhaled oxygen reacts with carbon during respiration. Plant life thrives on carbon dioxide and plants provide us with oxygen in return. When silicon reacts with oxygen it produces quartz, which is a solid. We don't exhale quartz bricks, and plant life can't use it if we did. In short, no better element exists for the chemistry of life than carbon.

Carbon is forged in the stars, but only a series of wildly improbable happenings make the manufacture of this crucial element possible. Carbon is made when three "alpha" particles (the nuclei of helium) fuse their combined 12 nucleons (6 protons, 6 neutrons) to form carbon-12 (^{12}C). However, there is a seemingly insuperable problem here because, "as soon as ^{12}C is synthesized from helium, it absorbs another α [alpha] particle and becomes ^{16}O [oxygen] leaving no carbon. The reaction forming ^{12}C was much slower than the reaction that destroys it. If so, argued Hoyle, life should not exist!"[14] But life does exist, so Fred Hoyle reasoned from this fact that there must be a way to make carbon that avoids its instantaneous destruction. Physicists John Gribbin and Martin Rees note that Hoyle reasoned from the fact we exist that carbon must have an energy level at 7.6 MeV, and that was precisely the level that experiments found it to be: "As far as we know, this is the only genuine anthropic principle prediction; all the rest are

Fine-Tuning and Stellar Alchemy

'predictions' that might have been made in advance of the observations, if anyone had had the genius to make them...There is no better evidence to support the argument that the Universe has been designed for our benefit—tailor-made for man."[15]

"MeV" denotes "mega (one million) electro-volts." Electro-volts are units of energy that provide particle acceleration. A simple dictionary definition of eV is the amount of energy an electron gains after being accelerated by one volt of electricity. The energy level of 7.6 MeV is Hoyle's prediction of the resonance state of carbon-12, known as the Hoyle state. The Hoyle stare describes the state of carbon immediately after the triple alpha reaction (described later) and returning to its ground state (the state in which its electrons are in their lowest possible energy state).

Resonance is important to understanding carbon production. Although the concept seems daunting, we use the language of resonance in everyday language: "I feel her vibes;" "I'm in tune with you"; "Sally and I are on the same wavelength;" "That song really resonates with me." We get the idea from this that resonance is somehow the matching of energy levels, and that this is achieved by one thing "exciting" another. Resonance in physics means the matching of energy achieved by some outside source exciting the internal motion of something else. By "exciting" we mean producing oscillation or vibration.[16] Every object in the universe has its own natural frequency of vibration. A common classroom demonstration of resonance involves two identical tuning forks mounted on sound boxes. The first tuning fork is struck ("excitation") with some object and begins vibrating at its natural frequency. These vibrations set the air inside the second sound box vibrating at the same frequency because of the sound waves impinging on it. The professor then grabs the prongs of the first fork to prevent it vibrating further, but the same sound is heard from the fork that wasn't hit. Since the incoming sound waves hitting the second fork share the same natural frequency as the first, the second fork begins vibrating at its natural frequency—its internal motion was excited by an outside source—the vibration from the first tuning fork. This is resonance.

Nuclear reactions are either helped or hindered by the resonances of the maelstrom of reacting nuclei within stars. Particles must collide with enough energy to overcome the electromagnetic repulsion of the positively changed protons. As Michael Cory put it: "If the sympathetic viberational frequency of one of the reactants happens to 'resonate' with the energy of an incoming nucleus, the two nuclei are more likely to fuse together; if no resonance at all occur they are more likely to 'bounce' off one another and remain separate."[17] To get the energy needed to get these resonances just right, we need an ageing giant star massive enough for the force of gravity to start the contraction of its core. As the star contracts, higher and higher densities and temperatures are produced, and a more powerful type of nuclear reaction occurs. This allows for an abundance of precious carbon to be made and spewed out into space when the star goes supernova.[18]

Carbon is made in such a star by a process known as the triple-alpha process, which has to take place in an incredibly short period of time. The triple-alpha process involves two steps as shown in Figure 5.2. First, two helium particles (^{4}He) combine to form beryllium (^{8}Be) and emit excess energy in the form of a gamma ray. Recall from our discussion of the strong force that when particles come together a small portion of their mass is converted to energy as the strong force pushing them together overcomes the electromagnetic force holding them apart. This is the gamma ray emitted from the ^{4}He/^{8}Be collision. Second, another ^{4}He must fuse with ^{8}Be, making 12 nucleons and hence ^{12}C.

Sounds simple enough, except for the fact that beryllium has a lifetime of approximately 10^{-17} seconds (one ten-thousandth of a trillionth of a second) before it decays. Nucleons must thus come together and bond within an unimaginably small period of time. This does not necessarily produce carbon, but rather the excited Hoyle State, which is a resonance state: "The resonance allows approximately four out of ten thousand decays to produce the ground state of carbon-12."[19] Note that ^{8}Be is an isotope of beryllium and is not the same as the stable soft metal beryllium isotope found on Earth, which has one more neutron (^{9}Be).

What happens to the newly formed ^{12}C if another alpha particle bumps into it to produce oxygen (^{16}O)? We need oxygen, but we don't want it at the expense of carbon. The avoidance of such an expense also has to do with energy levels. The Hoyle state of carbon is key; the slightest change in either the electromagnetic or the strong force would change the energy levels and have a catastrophic effect on the ability of a star to make carbon. Sylvia Ekström and her colleagues note that: "The requirement that some ^{12}C and ^{16}O be present at the end of the helium burning phase allows for permille [a sign indicating parts per thousand] limits on the change in the nuclear interaction and limits of the order of 10^{-5}."[20] What that means is that a change of just 1 part in 100,000 in the electromagnetic coupling constant (a number that determines the strength of the force exerted in an interaction) would result in the inability of stars to synthesize both carbon and oxygen. If the energy state was a little less there would be an abundance of carbon, but helium would burn into carbon much earlier and the star would not be hot enough to produce sufficient oxygen. Cassé explains that: "It turns out that the sum of the mass energies of carbon and helium is just 1% above an energy level of oxygen-16 [this ensures that resonance does not occur]. But this 1% difference is not enough for all the carbon to disappear in the stellar crucible, thereby destroying any chance of life at a later date."[21].

Cassé describes the energy levels of carbon and oxygen and the fine-tuning of the forces as coincidences: "Apparently, our existence depends on a series of coincidences."[22] Physicist George Greenstein also puts it all down to "lucky breaks": "Other nuclear reactions do not proceed by such a remarkable chain of lucky breaks...It is like discovering deep and complex resonances between a car, a bicycle, and a truck. Why should such disparate structures mesh together so perfectly? Upon this our existence, and that of every life form in the universe, depends."[23] Perhaps only a physicist can really appreciate the fine-tuning required to manufacture carbon. Fred Hoyle marveled at the "miraculous" relationship between the energy levels of carbon and oxygen in the *Annual Review of Astronomy and Astrophysics*, avoiding the "lucky breaks" mantra:

"If you wanted to produce carbon and oxygen in roughly equal quantities by stellar nucleosynthesis, these are the two levels you would have to fix, and your fixing would have to be just where these levels are actually found to be." He then tries to understand how it could have happened: "*A commonsense interpretation of the facts suggests that a super-intellect has monkeyed with physics, as well as with chemistry and biology, and that there are no blind forces worth speaking about in nature.* The numbers one calculates from the facts seem to me so overwhelming as to put this conclusion almost beyond question [my emphasis]."[24]

Figure 5.2: The Triple-Alpha Process

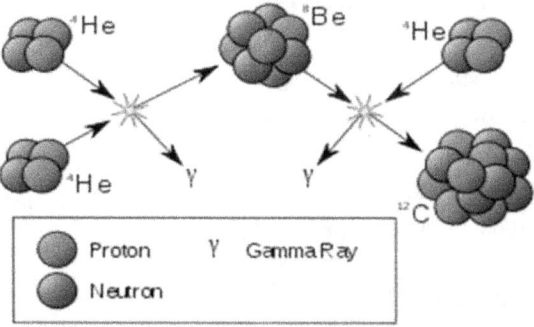

Who else but Almighty God could be the super-intellect that "monkeyed" with the laws of physics (carbon's improbable fusion), chemistry (its amazing bonding features), and biology (its basis for life)? And which of the interacting "coincidences" out of the thousands found so far will be the final straw to lead the atheist scientist to God? Hoyle himself said that he believed that any scientist who examined the evidence would "draw the inference that the laws of nuclear physics have been deliberately designed with regard to the consequences they produce inside the stars. If this is so, then my apparently random quirks have become part of a deep-laid scheme. If not, then we are back again to a monstrous sequence of accidents."[25] Yet, there are many who continue to prefer to place their faith in a "monstrous sequence of accidents" than in a Divine Creator.

Endnotes

1. Hawking, S. and Mlodinow, L. 2010, pp. 160-161.
2. Bussy, P. 2016.
3. Krebs, R., 2006, p.133.
4. Gonzalez, G. and Richards, J., 2004.
5. In Strobel, L., 2002, p. 161.
6. NASA, 2018.
7. Gonzales and Richards, 2004, p. 205.
8. Ibid., p. 205.
9. In Lennox, J., 2009, p. 70.
10. In Lemley, B., 2000, p.64).
11. Davies, P. 1982, p.68.
12. Hawking, S. and Mlodinow, L. 2010, p. 160.
13. Borwein, J, and Bailey, D. 2014.
14. Shaviv, G. 2015, p. 311.
15. Gribbin, J. and Rees, M. 1989, p. 247.
16. Cassé, M. 2003, p. 123.
17. Corey, M. 2001, p. 102.
18. Cassé, M. 2003.
19. Hjorth-Jensen, M. 2011, p. 38.
20. Ekström, S, et al, 2010, p. 1.
21. Cassé, M. 2003, p. 143.
22. Ibid., p .143.
23. Greenstein, G. 1988, pp 43-44.
24. Hoyle, F. 1982, p. 16.
25. In Holder, R., 2013, p. 48

CHAPTER SIX

The Universe Engineered "Just Right" for Us

> "Why do I believe in God? As a physicist, I look at nature from a particular perspective. I see an orderly, beautiful universe in which nearly all physical phenomena can be understood from a few simple mathematical equations."
> William Phillips, Nobel Laureate physicist

Location, Location, Location

Real estate agents tell us that a basic principle of their profession is that location is the major determinant of the desirability of a property. How would you fine-tune your choice to build a home on the safest street, in the safest neighborhood in the safest country in the world if money were no object? You would certainly like to be in an affluent democratic country with a peaceful reputation and a low crime rate, such as Switzerland. After deciding on a country, you further refine your tuning to a location in which neighbors are helpful, the weather is good with lots of opportunities for recreation, and you settle on building a villa overlooking Lake Lugano. You want to install the best protective devices money can buy, and although you don't want to get too close to people from other areas of town who may pose danger to you, you don't want to live too far away from the goods and services they provide either. If you were likewise asked to locate the most advantageous galaxy, star, and planet location where life itself is possible, you could do no better than the Milky Way, the Sun, and the Earth, because as far as we know, these are the only locations that fit the bill. We are in the right place with just the right kind of star, just the right distance from it, with a stabilizing moon, and enjoy helpful neighboring planets.

Before you can choose the location of your villa it is necessary to have land available. For any land to be available at all, we must have a planet, and for that a solar system, galaxy, and universe, so the planning for your villa really occurred billions of years ago with the creation of the universe. In planning your villa, the architect makes a few simple land surveys and room dimension calculations, which he can recalculate and amend if necessary. But the Grand

Architect of the universe used far more complex mathematics to plan a universe, and He did not have to go back to the drawing board. Cosmologists increasingly recognize the unbelievable mathematical precision by which the universe had to unfold for humanity to exist. Physicist Freeman Dyson phrased it in strongly anthropic terms: "As we look out into the Universe and identify the many accidents of physics and astronomy that have worked together to our benefit, it almost seems as if the Universe must in some sense have known that we were coming."[1]

Many people have wondered why, if God intended to build a home on Earth for us and the universe "knew that we were coming," it was necessary to wait almost 14 billion years and to salt the universe with billions of galaxies and stars—it seems such a waste of time and material. Think of it in terms of your decision to build a villa. You first had to find a reputable contractor capable of doing the job. The contractor then had to obtain building permits and employ a surveyor and architect to work with you. He then had to have the necessary building materials transported to the site and hire skilled journeymen to put them together. That's just for one villa on a planet where men and materials already exist; getting those men and the materials they use, not to mention the planet itself, from literally nothing is another matter. For one thing, it requires a galaxy 9 or 10 billion years old to get sufficient abundance of life-sustaining heavy elements.[2] From our human perspective this is an immense period of time, but for a timeless God existing in the eternal Now, it may be just the blink of an eye.

Now that the decision to build has been made, we consider the familiar adage known as Murphy's Law, which states that anything that can go wrong, will. Your car, your house, your relationships, and everything else, requires you to devote energy to maintain because there are many ways they can go wrong, and only one way to go right. The early universe is as subject to this as anything else; there were almost infinitely more ways for it to have gone wrong at its inception, and only one way for it to go right if intelligent life was to emerge on Earth. Unlike human beings, the universe could not learn from a series of mistakes (putting off the oil change or house painting, or forgetting a spouse's birthday, will teach you important lessons you won't forget); it had to get all the fundamental constants right out of the multiple trillions of alternative values on the first try with all parameters needed for life factored in at the beginning.

Getting it Started in Phase Space

The most remarkable cosmological fine-tuning of all is getting the whole thing started in phase-space. Phase space is dynamic multidimensional space in which all possible states of a system are represented, with each of these possible states corresponding to one unique point. The issue of the state of the universe at the moment of the Big Bang is intimately connected to the second law of thermodynamics and its principle of entropy, which is the really big brother of Murphy's Law. Recall that entropy is the degree of thermodynamic disorder, which is always increasing in a closed system. Given this, there had to be an immense degree of order at the Big Bang because a universe capable of supporting life must begin with the lowest possible entropy.

Mathematical physicist Sir Roger Penrose asks us to imagine all the possible ways that the universe might have started off in phase-space, and the probability that the Creator could hit the exact point in to create a life-producing universe. He calculated the probability of the initial entropy conditions of the Big Bang by calculating the maximum entropy of the universe. This figure is the logarithm of the total phase-space volume of all possible beginnings of the universe, or 10^{123}. Because logarithms and exponents are inverse functions, the total phase-space volume is $10^{10^{(123)}}$. Penrose asks: "How big was the original phase-space volume W [W = original phase-space volume] that the Creator had to aim for in order to provide a universe compatible with the second law of thermodynamics and with what we now observe?" He then remarks on two ways to estimate this figure and writes: "Either way, the ratio of V [total phase-space volume available] to W will be, closely V/W = $10^{10^{(123)}}$."[3]

Penrose notes that this number could not be written down if we had every elementary particle in the universe to write a zero on, and it is basically the ratio of probabilities of ending up in a particular macroscopic state if you chose a microscopic initial state of the system at random. Penrose's calculations present problems for atheists who have wrestled with the initial entropy problem themselves, as evidenced by a *Journal of High Energy Physics* paper by three physicists titled "Disturbing implications of a cosmological constant." They noted that it is an unshakable given that the universe could only make sense if it began in a state of minimum entropy, and added: "there is no universally accepted explanation of how the universe got into such a special state...Far from providing a solution to the problem, we will be led to a disturbing crisis."[4] What is this "disturbing crisis" they found after examining all naturalistic explanations for such exquisite fine-tuning and finding them wanting? According to the authors, it is no less than forcing cosmologists to think the unthinkable: "Another possibility is an unknown agent intervened in the evolution, and for reasons of its own restarted the universe in the state of low entropy characterizing inflation."[5] The "unknown agent" they fear to name is God.

Matter Density and Galaxy Formation

With the universe in motion, the first galaxies are thought to have formed about 400 million to one-billion years after the Big-Bang. At his point, the universe was a near homogeneous mass of matter and energy of immense density. We know that if you crush enough matter together it is destined to collapse into a black hole, so why didn't it? For matter to coalesce into galaxies there must be some contrast or "roughness" in the smooth homogeneity of the distribution of matter to enable it to collapse under the pull of gravity to form galaxies. The formation of galaxies thus depends crucially on matter density variation from one location in the universe to another, but this variation must be very small. Astrophysicist Abraham Loeb tells us that: "The Universe we live in started with primordial density perturbations of a fractional amplitude~10^{-5}" [6]

We see these perturbations in the discoveries of the Wilkinson Microwave Anisotropy Probe (WMAP), which was a space-based mission lasting from 2001 to 2010 charged with measuring temperature differences in the cosmic microwave background across the visible universe. "Anisotropy" refers to small temperature fluctuations in the background radiation, and is the opposite of isotropy, or universal homogeneity. As noted earlier by Loeb, this anisotropy is fine-tuned to about one part in 100,000. If it had been "significantly smaller, the early universe would have been too smooth for stars and galaxies to have formed" ... and "galaxies would have been denser, resulting in numerous stellar collisions, so that stable, long-lived stars with planetary systems would have been very rare."[7]

The contrast and density of matter had to be just right from the moment of the Big Bang. Too much and the gravitational pull would be greater than the expansive force causing it all to collapse back on itself; too little and the gravitational pull would be insufficient for matter to accrete. This is known as the energy density of matter (p). There is a critical value (p_{crit}) of the energy density that prevents gravity from overcoming the force of expansion and pulling all matter into a big crunch. The value of p has to be microscopically close to p_{crit} to avoid this, and in fact it had to vary by less than one part in 10^{60} from the very beginning of creation. Paul Davies expresses his amazement of this fine-tuning: "We know of no reason why p is not a purely arbitrary number...to choose p so close to p_{crit}, fine-tuned to such stunning accuracy, is surely one of the great mysteries of cosmology."[8] NASA scientists tell us that: "The value of the critical density is very small: it corresponds to roughly 6 hydrogen atoms per cubic meter [that's just over 35 cubic feet], an astonishingly good vacuum by terrestrial standards!" [9]

If that doesn't serve as a useful gauge of the mind-boggling fine-tuning required, here's another. Various estimates of the number of grains of sand on Earth hover around 10 sextillion, that's 10,000,000,000,000,000,000,000, or 10^{22}. This is 1 followed by 22 zeros whereas 10^{60} is 1 followed by 60 zeros. Thus, if we say that if the critical value varied by less than one part in 10^{60} from the very beginning of creation we wouldn't be here, it is crudely analogous to saying remove or add one or two grains of sand to the Earth and we wouldn't exist!

The Geography of the Universe and the Cosmological Constant

The density of matter in the universe affects the geometry of space-time, with the critical value of 10^{60} being the requirement for a flat universe. Only in a flat universe is the energy of matter balanced by the energy of the gravity that mass creates. This feat is like balancing a pencil on its sharpened tip for 13.8 billion years or so. But what do we mean by a flat universe; don't we live in a 3D world with an up and a down? When physicists say the universe is flat, they do not mean it in the same sense that a piece of paper on your desk is flat. They mean it in terms of the geometry of the universe. Figure 6.1 illustrates that how the geometry of the universe is determined by its density (Ω, the Greek letter "omega"); that is, if it is less than 1, equal to 1, or greater than 1. In a spherical universe with greater than critical density ($\Omega>1$) Euclidean geometry breaks down; the three angles of a triangle no longer equal 180 degrees and lines that start out parallel will eventually meet. A universe with such density will curve space-time in on itself and gravity will collapse it in a "big crunch." This was the basis for the now discredited oscillating universe in which the universe was said to undergo endless cycles of big bangs and big crunches, but the universe has passed the threshold where gravity can reverse its expansion. In a saddle-shaped (hyperbolic) universe ($\Omega<1$) there is insufficient mass to cause the expansion of the universe to stop and it will expand forever. Lines that start off parallel in such a universe would eventually diverge. In a flat universe with exactly the critical density ($\Omega=1$) parallel lines will never meet and the angles of a triangle will always add up to 180 degrees.[10]

At the early phases of the universe, matter was clumped closer together and therefore more likely to be pulled together by the grip of gravity. The explosive force of the Big Bang was enough to balance out gravity for a long time, but this is slowly dissipating, thus requiring another force to prevent a "big crunch." A project known as the Supernova Cosmology Project began in 1998 expecting to measure the deceleration of the universe but found that it was accelerating instead.[11] Gravity needed to dominate during the period of matter accretion into galaxies, stars, and planets, but for some reason known only to God, dark energy now rules the roost. As we saw in the previous chapter, Einstein's cosmological constant (the energy built into the vacuum of space—dark energy) has taken on the job of keeping the universe expanding.

There is much about the amazing precision of the cosmological constant that puzzles the best minds in physics. Alejandro Jenkins and Gilad Perez remark that "the most serious fine-tuning problem in theoretical physics: the smallness of the 'cosmological constant,' thanks to which our universe neither recollapsed into nothingness a fraction of a second after the big bang, nor was ripped apart by an exponentially accelerating expansion."[12] Physicists Livio and Rees inform us that anthropic reasoning is becoming seriously discussed in physics and may have predictive power for sorting out certain cosmological phenomena such as the mystery of the cosmological constant. They ask: "Why is the force so small? If there was an inflationary era with a large cosmic repulsion, how could that force have been switched off (or somehow have been neutralized) with such amazing precision? In our present universe, Λ ["lambda," the symbol for the cosmological constant] is lower by a factor of about 10^{120} than the value that seems natural to theorists."[13]

Livio and Rees mean that the cosmological constant is 10^{120} orders of magnitude smaller than physicists expected from their calculations, and that makes it "unnatural." The "switching off" or "neutralization" of repulsion they refer to must therefore be unbelievably fine-tuned to 120 decimal places from the very beginning of the universe. Livio and Rees go on to note that: "If Λ were larger, then the acceleration would have overwhelmed gravity before galaxies had a chance to form."[14] Nobel Laureate physicist Steven Weinberg's understanding of the razor's edge balance between dark energy and gravity lead him to exclaim that: "This is the one fine-tuning that seems to be extreme, far beyond what you could imagine just having to accept as a mere accident."[15] John Donoghue takes this huge improbability differently, using it to argue that the cosmological constant as proof of a multiverse: "In particular the exponential fine-tuning needed for the cosmological constant seems to have no technically natural explanation, and is a strong motivation for a multiverse explanation."[16] If it has no natural explanation, it must have a supernatural explanation, but rather than admit to that, Donoghue resorts to the multiverse lottery to explain the "exponential fine-tuning."

Figure 6.1: Curved, Hyperbolic, and Flat Geographies of the Universe

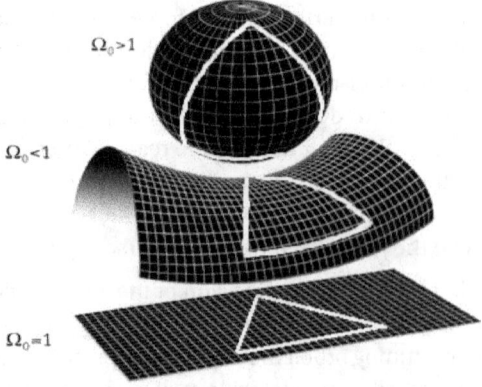

The Higgs Boson and the Standard Model of Physics

On July 4th, 2012, it was announced with great fanfare that the elusive Higgs Boson had finally been detected. Bosons are force carriers and one of two classes of fundamental particles; the others are fermions. Composite particles such as the familiar protons and neutrons are fermions (see Figure 5.1 in the previous chapter). The Higgs boson was first theorized in 1964 when scientists, including Peter Higgs, after whom it is named, were disturbed that their theories predicted a universe without mass, when it obviously does have mass. They theorized that in order for their theories and observations to correspond there must be a quantum field pervading the universe that gives particles their mass, which they later called the Higgs field. It is the Higgs field, rather than the boson itself, that physicists want to study, because it is the field that does the interesting stuff in the universe. However, the only way to understand the field is to find Higgs particles.

Like all quantum fields, the Higgs field is a field of oscillating energy. When particles pass through it, they slow down and gain small quantities of that energy. If they interact with the field long enough, they accumulate energy as mass since energy and mass are different forms of the same thing. There are only two known particles that zip through the Higgs field without interacting and are thus massless: photons that carry the electromagnetic force, and gluons that carry the strong force. Quantum theory informs us that all fields have particles associated with them, so the Higgs field must have its own force-carrying particle. The Higgs boson (which was whimsically nicknamed the "God Particle") is the quantum of energy with which the Higgs field interacts with other particles.[17]

The existence of a field that imparts mass to particles was taken as self-evident, but the search for it had to await the building of the largest machine ever built: The Large Hadon (a hadron is any particle made of quarks, such as a proton) Collider (LHC). The LHC is a 17-mile underground ring of superconducting magnets spanning the borders of Switzerland and France run by CERN (*Conseil Européen pour la Recherche Nucléaire*). It operates year-round sending sub-atomic particles along a number of accelerating structures in opposite directions at within decimal points of the speed of light and smashing them together. The resulting debris from these collisions yields 25 million gigabytes of data per year. The majority of this data is not useful, but Nobel-winning gems like the Higgs boson sometimes emerge. The difficulty in finding the Higgs is that it decays in 10^{-22} seconds, which led William Craig to comment that the reason it is labeled the "God particle" (apart from its obvious publicity value) is because like God, it underlies every physical object that exists, and is very difficult to detect.[18]

The discovery of the Higgs boson capped 50 years of an uninterrupted streak of successes for the Standard Model of particle physics that classifies and describes the behavior of all known elementary particles. Yet, physicist Harry Cliff tells us that physicists regard the Standard Model as highly "unnatural" because of the large number of particles and forces that are precariously balanced such that changing any of the values: "you rapidly find yourself living in a universe without atoms. This spooky fine-tuning worries many physicists, leaving the universe looking as though it has been set up in just the right way for life to exist."[19]

What does Cliff mean by calling the most successful model in science unnatural? Naturalness in physics is basically the prohibition against fine-tuning. If a theory requires the tweaking of its parameters in order to agree with experimental observation, it is considered "unnatural." The Higgs boson has such a lot of unnaturalness about it that Euan McLean says has led some physicists to panic about its "spooky" nature. The problem is that its mass is fine-tuned by multiple trillions of degrees more than the Standard Model predicted. That is, the actual measured mass (m) differs dramatically from its theoretical, or "bare" mass (m_0). As McLean points out: "The universe would be radically different if that value of m_0 was changed even a tiny bit.... It seems like, to generate a universe remotely like the one we live in, nature needs to decide on a parameter m_0, highly tuned to 33 decimal places."[20]

A fine-tuned universe is only "unnatural" and "spooky" if one is a committed ontological materialist. Particle physicist Michael Strauss is not, and asks if the discovery of the Higgs provides us with insights into the existence and nature of God. He marvels at the fact that six physicists sat down in 1964 and came up with mathematical calculations that predicted the Higgs particle should exist, and then 48 years later it was discovered. The fact that we can describe the universe mathematically and can trust math to predict things about the universe, leads Strauss to opine that all explanation other than God are inadequate to explain our incredibly complex universe. He notes of the Higgs particle that: "Though it may not be properly 'The God Particle,' the mathematical description and complexity of our universe, along with its actual existence, gives a clear indication of a true deity who has designed and created what we now have the privilege to observe and study."[21]

Notwithstanding Strauss' analysis, an article by journalist Jeffrey Tayler in the left-wing atheist magazine, *Salon*, used the discovery of the Higgs as an opportunity to mock God and His followers: "Seriously though, it insults our intelligence to be enjoined to believe, now that we have split the atom, discovered the Higgs Boson, and sent a probe to Pluto, in the veracity of a supernatural account of the origins of our cosmos."[22] Tayler does not recognize that it is the material cause (energy and matter) of science that is natural, but the efficient cause (the Being that created it) is supernatural. He is oblivious to the fact that the scientists who actually did much of the heavy lifting in nuclear fission and fusion such as Lise Meitner, who pioneered it, and Nobel Laureate Ernest Walton, who was the first person to split the atom, belie his implied and deeply arrogant "I'm too bright to believe in God" claim. Both Meitner and Walton were committed Christians, and Walton wrote of his view of science and God: "We must pay God the compliment of studying His work of art and this should apply to all realms of human thought. A refusal to use our intelligence honestly is an act of contempt for Him who gave us that intelligence."[23] Meitner and Walton must have been pretty stupid according to Tayler, since they were intimately involved with splitting the atom and embraced God all the more for it.

Others have made remarks similar to Tayler's. For instance, physicist Lawrence Krauss has said that the Higgs particle is "more relevant than God" because it could be the quantum burp by which the universe created itself out of nothing. Rather than reiterating what was said in Chapter 2 about such nonsense, I will quote John Lennox's response to Krauss: "The philosopher Ludwig Wittgenstein pointed out that the meaning of a system will not be found within the system. The meaning of the universe will be found where Newton and Clerk Maxwell found it: in God. So what can we say about the Higgs boson? Simply this: God created it, Higgs predicted it, and CERN found it."[24] The astonishing success of physics is remarkable confirmation of the truth of Judeo-Christian theism. It returns to Christianity that which Christianity gave to it—an ever-increasing faith in the intelligibility of God's creation, and in mathematics as "God's cosmic language."

Endnotes

1. Dyson, F. 1979, p. 250.
2. Ross, H., 2016, p. 35.
3. Penrose, R., 2016, pp. 445-446.
4. Dyson, L., Kleban, M., and Susskind, L., 2002. p.1.
5. Ibid. p. 19.
6. Loeb, A., 2010, p. 8.
7. Bailey, D. 2018.
8. Davies, P. 1982, p. 90.
9. National Aeronautics and Space Administration, 2019.
10. Kolb, E. 2018.
11. Perlmutter, S., Aldering, G., Goldhaber, G., et al, 1999.
12. Jenkins, A., & Perez, G. 2010, p. 44.
13. Livio, M., & Rees, M. 2005, p. 1022.

14. Ibid., p 1022.
15. In Folger, 2008.
16. Donoghue, J. 2016, p. 27.
17. Peskin, M. and Schroeder, D. 2018.
18. Craig, W. 2013.
19. Cliff, H. 2013.
20. McLean, E. 2017.
21. Strauss, M. 2017.
22. Tayler, J. 2015, np.
23. In *Coming Untrue* (2015), np.
24. Lennox, J. 2012, np.

CHAPTER SEVEN

The "Just Right" Galaxy and Solar System

> "I want to know how God created this world. I am not interested in this or that phenomenon, in the spectrum of this or that element. I want to know His thoughts, the rest are details."
> Albert Einstein, Nobel Laureate physicist

The Milky Way Galaxy

Now that the landscape has been surveyed we can explore our cosmic neighborhood called the Milky Way Galaxy. Our galaxy's name is derived from Greek mythology. As the story goes, the god Zeus took the infant Heracles, the offspring of one of his dalliances with mortals, for his sleeping wife, Hera, to breastfeed. Hera was ill-disposed to the half-mortal and pushed him away while he was suckling, causing milk to spill into the night sky forming the Milky Way. We know today, of course, that the Milky Way consists of billions of stars of numerous types. The Milky Way is what is known as a spiral galaxy; the other two major types of galaxy are elliptical and irregular which, like ours, are salted with billions of stars.

In the first few minutes of the universe's existence it was composed of a swirling mass of protons and neutrons not yet combined with electrons. These were the primordial building blocks of everything, including the men and materials engaged in building your villa. You can't build your villa from these early particles; you need the heavy elements, and these take time to be fabricated in the stars. Stars are the home of a perpetual battle between gravity and nuclear fusion. The more a star is crushed, the hotter and denser it gets, and the faster the rate of nuclear fusion. As long as the star has fuel it will push back against the crush. The most massive stars burn their fuel supply quickly to provide the energy needed to counteract the gravitational crush, and may go supernova after only 3 to 10 million years. A star with our Sun's mass can keep on fusing its hydrogen for about 10 billion years.

Stars begin their lives by fusing hydrogen into helium, which produces a lot of energy for fusing heavier elements. As hydrogen is used up, fusion slows down, which results in less energy being released. With less energy released, gravity causes the star's core to contract, which raises the core temperature to the point where helium fusion can begin. When the helium has been used up, the core contracts and becomes hotter, allowing for the fusion of heavier elements. Each successive cycle is shorter, and the heavier the element the less fusion energy it produces. The most massive stars (at least 8 times as massive as the Sun) can fuse elements up to iron, at which point the energy required to fuse iron is more than the energy the star gets back from fusing it. With no energy being released, the star immediately begins to collapse under the crush of gravity and it goes supernova, spewing its elements out into space.

But iron is only the 26th element in the periodic table, so how do we get all the others such as nickel, copper, lead, and gold? These, and all other elements past iron, are made in the mighty energy of supernovae explosions. When a star is about to go supernova, the heat of the core increases so quickly that iron is smashed apart into helium, which, in turn, is shattered into protons and neutrons.[1] The tremendous energy released by a supernova explosion releases a superabundance of free subatomic particles cascading from the collapsing core. This energy drives massive fusion reactions among these particles that forge the remaining elements found on Earth. Your villa thus got built because some stars died a few billion years ago.

The Milky Way galaxy is in the one or two percent of galaxies massive enough to have gravity strong enough to attract enough hydrogen and helium to construct the heavier elements needed for life to exist. But our galaxy also has several unsavory neighborhoods containing densely packed collections of millions of ancient stars revolving around the galactic core called globular clusters. Earth-like planets cannot exist in globular clusters because their stars are poor in the heavier elements necessary to build them. Additionally, the gravitational pull of the myriad of stars in a cluster results in highly elliptical orbits, which would plunge planets both into each other and/or into extremes of heat and cold. Hugh Ross likens elliptical galaxies to dilapidated neighborhoods containing limited resources for building planets. Star formation ceases too early in elliptical galaxies to produce a sufficient amount of the heavy elements, and stars in them are so crowded that long-term and stable planetary orbits are impossible. Ross says that the situation is just as bad irregular galaxies: "Irregular galaxies...are chaotic in appearance without either spiral arms or a nuclear bulge. Large irregular galaxies possess active nuclei, which spew deadly radiation. Small irregular galaxies lack the quantities of heavy element that life requires. All irregular galaxies manifest chaotic stellar orbits, which can disrupt or bring bright young ultraviolet-emitting stars into the vicinity of a life-sustaining planet."[2]

The Galactic Habitable Zone

Our galactic neighborhood is in what is known as the Galactic Habitable Zone (GHZ), the safe zone of the city. The Milky Way is about 180,000 light-years across, and our relatively safe zone is about 27,000 light-years from its center, and tens of thousands of light-years away from the outer rim of the galaxy's spiral arm. Remember, a light-year is the distance light travels in a year at 186,282 miles per second, so 27,000 light-years is a *very* long way. Our solar system orbits the center of the galaxy about once every 200 to 250 million years, traveling at about 500,000 miles an hour; a really wild ride.[3]

According to Gonzales, Brownlee, and Ward: "The boundaries of the galactic habitable zone are set by two requirements: the availability of material to build a habitable planet and adequate seclusion from cosmic threats."[4] The center of the galaxy is a very dangerous place full of exploding supernovae and a gigantic black hole, so best not to build there. The GHZ band is far enough from the center to avoid the effects of deadly radiation of exploding stars or the possibility of drifting to close to the black hole and getting sucked in. However, just as we don't want a villa too far from the stores, our planet must be close enough to benefit from the heavy elements that supernovae provide. If we were located further out in the spiral arm of the galaxy, there would not be enough heavy elements to build Earth-like planets and we would be exposed to hazardous giant gas clouds the spiral arms often visit.[5]

Our solar system exists as part of the galactic spiral called the Orion Arm. Orion is located between two other spiral arms—Sagittarius and Perseus—and wraps around the galaxy at, or very close to, the corotation circle. This is a very good thing because being in the corotation circle means that we are moving at the same speed as the spiral arms. Moving at identical speeds means that our solar system will remain between the other two spiral arms and will not visit them. Most other stars in the Milky Way periodically wander into and out of their spiral arms, which disrupts planetary orbits due to gravitational interactions, and such disruptions would be deadly for complex life.

Stars in the galaxy's outer spiral arms are what astronomers call metal-poor (poor in the heavy elements, which inhibits planet formation) population II stars, "and the few planets that exist there lack the requisites for life."[6] Fortunately, we are far away from the spiral arms, and orbit the center in an almost perfect circle (very low eccentricity). Most other solar systems do not evidence this orbital pattern, and thus very few of our neighbors enjoy the same level of safety. As Plaxo and Gross write about our "just right" orbit: "Such low eccentricity orbits are, however, relatively, rare, [they tend to be elliptical rather than circular] and the majority of Sun-like stars currently in our neighborhood spend a significant fraction of each galactic orbit far too close to the galactic center for comfort...less than 5% of all stars lie in the life-supporting zone."[7] Just because they exist in the life-supporting zone, however, does not mean that they have life-sustaining planets. A myriad of other physical, chemical, and biological conditions must be satisfied for the emergence of complex intelligent life.

Unlike our Sun, which is solitary, most stars (about 85%) are locked with one or more other stars resulting in wild gravitation pulls that make stable planetary orbits impossible in such systems, and thus life-supporting zones in other galaxies are improbable.[8] Most observable galaxies in the universe fall into the elliptical category and have stars with relatively random orbits (since many are conjoined). Gonzales and Richards describe them thus: "Elliptical galaxies, which have less gas and dust, contain stars with a wide range of orbits—most highly inclined and eccentric."[9] Eccentric orbits would take planets into the danger zone at the center of their galaxies that are littered with supernovae explosions and black holes. The situation in irregular galaxies appears even worse.

Thus, for a variety of reasons elliptical and irregular galaxies cannot support complex life, but not any spiral galaxy will do either. Hugh Ross notes that such a galaxy can be neither too big nor too small. Galaxies more massive than ours spawn massive back holes with enough mass "to ignite potent relativistic jets of deadly radiation" and their immense gravitation forces chaotic "mergers with multiple smaller galaxies." Galaxies that are both too large and too small run into the problem of the corotation radius, or "the precise distance from the galactic center at which at which stars rotate around the center at the same rate as the galactic arms."[10] The problem for larger galaxies is that rotation is far too long for the heavier elements to reach any planets that may otherwise have potential for life. The problem for smaller galaxies is that the corotation radius is too short for life because planets will be too close to the galactic core's deadly radiation. As Gonzales, Brownlee, and Ward remark: "We live in prime real estate."[11]

The Circumstellar Habitable Zone

If the solar system is in a prime neighborhood, so is Earth's location within the solar system. We live in what is called the Circumstellar Habitable Zone (CHZ). The CHZ is a band of space around the Sun that is hospitable to life because planets residing there can potentially hold water. Depending what parameters are included, the CHZ band is either between .85 and 1.75 astronomical units (AUs). One AU equals the distance between the Earth and the Sun, or 92,955,807 miles = 1.0 AU. Venus (barely) and Mars (at the far end) also lie within the CHZ, but neither has surface water.

The Earth orbits the Sun well within the CHZ. The Earth's orbit is very special in terms of its orbital eccentricity; that is, the value from which it departs from circularity. Eccentricity values of orbits range from zero (perfectly circular orbit at constant speed) to 1. Our eccentricity value is a mere 0.0167, which makes our weather patterns stable.[12] The closer a planet gets to an eccentricity of 1, its orbit becomes increasingly longer, and it will eventually escape the parent body's gravity and fly off into deep space. This is because all bodies of matter in space want to keep moving forward in a straight line, not in circles. Without the Sun pulling us toward it, the Earth would zoom away into the cold depths of space. Orbits are the result of this tug-of-war between planets and their parent body. The parent body's gravity wants to pull a planet in, but a planet wants to move straight ahead. Orbits are thus the result of a precise balance between the forward motion of a body and the pull of gravity on it from another body.

If we were orbiting much closer to the Sun, we may be caught in a tidal lock. Tidal locking means that a body's rotational period equals its orbital period around its parent body. This occurs because the gravitational pull of a parent body slows rotation of objects orbiting it until one side always faces the parent body and the other always faces away. Mercury is tidally locked to the Sun, and the Moon is tidally locked to the Earth. Tidal locking results in an extremely hot surface on the side facing its star, and extreme cold on the side facing away from it, and both extremes are hostile to life. Tidal locking is a major problem for planets orbiting stars with low luminosity. Such planets have to hug their stars more closely to capture its heat, but in doing so they eventually become tidally locked.[13] Unlike most planets, the Earth's low eccentricity keeps it in the CHZ permanently. Venus's orbit is more circular than the Earth's, but it is close to being tidally locked. It has such a slow rotation that its day of 243 Earth days is longer than its year of 225 Earth days. That is, it rotates on its axis slower than it orbits the Sun. Earth's 24-hour spin rate keeps us from being both too long in the heat of the day or the cold of the night.[14]

The Sun

Let us take a look at our marvelous star we call the Sun. It is not just any old star but rather one—and perhaps the only one—that has all the necessary characteristics to make complex life possible. The Sun is a relatively young star, forming only about 4.6 billion years ago. It formed from an immense cloud of gas called a nebular cloud composed mostly of hydrogen. It would have also contained the heavier elements from the first generation of stars formed after the Big Bang which subsequently became supernovae. The Sun formed at just the right time to benefit from the inheritance of the heavy elements bequeathed to it by these early supernovae. At some point in our neighborhood, the gas cloud began to spiral around its center until gravity overcame the gas pressure and fused hydrogen into helium to ignite our Sun into the flaming ball of plasma it is today. The material not used for the Sun (less than 1%) coalesced into planets and moons, and the solar system was born.

According to NASA, the Sun's diameter is 864,400 miles, which is about 109 times the diameter of Earth, and has a core temperature of about 27 million degrees Fahrenheit and a surface temperature of about 10,000 degrees

Fahrenheit. It is composed mostly of hydrogen (about 76%) and helium (about 22%), with traces of other elements with heavier nuclei such as lithium, oxygen, carbon, and iron. Fortunately, the Sun is not hot enough to fuse the lighter elements into iron. The small amount of iron it has comes from supernovae. That small amount, however, (about 0.1% of the Sun's mass) is 330 times the mass of the Earth since the Sun is 333,000 times the mass of the Earth. The Sun spins on its axis at about 4,400 miles per hour compared to the Earth's approximate 1,000 miles per hour at the equator. At its equator, the Sun takes 27 Earth days to make one complete rotation. Because it is plasma rather than solid, it rotates much slower at its poles, taking about 35 Earth days to complete the rotation. [15]

The Sun's photons are manufactured in its core by thermonuclear fusion as hydrogen is fused into helium. When this occurs, highly energetic gamma rays, the most energetic type of photon, are emitted. Since they are photons, they travel at the speed of light. Given the Sun's radius, we might expect them to reach the surface (the photosphere) in a couple of seconds, and then speed on to Earth. Thankfully, this does not happen, because if it did life would not be possible. In fact, they take an average of 170,000 years (depending on assumptions, some say up to a million years) to reach the surface, time mostly spent in the Sun's radiative zone. In this super-dense zone, gamma-rays bounce around wildly like the ball in a pinball machine. When they hit a proton, they are absorbed and reemitted. After many trillion such encounters, they eventually become less energetic X-rays and UV rays. Once they escape the radiative zone, the photons enter the cooler and much less dense convective zone. Here they are carried up toward the surface by hot gas, like bubbles in a pan of boiling water. This journey takes only a few days, and then these ancient photons are free to finally zip unimpeded to Earth in about eight minutes as visible light—that tiny sliver of the electromagnetic spectrum.[16]

We have already seen that our Sun is something of an oddity in that it is only one of the 15 percent of stars that is not partnered with other stars, but it has other properties that make it ripe for life on a nearby planet called Earth. Sunlight is the energy plants use to make food from carbon dioxide and water in the process of photosynthesis. Animals use plants for their energy, and humans eat these plants and animals to get theirs, so in a sense we are eating recycled photons. We are able to breathe because oxygen is a byproduct of photosynthesis, and the byproduct of our breathing is the carbon dioxide that plants need to live. When plants die and are compressed deep within the earth over millions of years, they become the fossil fuels we need to keep the economy humming, so our cars and gas stoves are also using recycled photons.

Some of the Sun's energy must be returned to space to keep Earth from overheating. In fact, the Earth sends a little more heat out into space that it as it receives from the Sun because the Earth generates its own heat from radioactive decay. About a third of the Sun's energy is reflected by clouds, water, and snow. What is not reflected is absorbed and used or sent back into space. "How is this balance maintained?" asks a University of California, Davis, online course on the electromagnetic spectrum, and replies that: "Earth warms up to exactly the temperature that is necessary to re-radiate *exactly the right amount of energy*"[my emphasis].[17] The professor conducting the course appears to accept this as another fortuitous accident.

Wind is another source of energy produced by the Sun. Wind is caused by differences in atmospheric pressure, which, in turn, is caused by differences in solar heat in different areas of the planet. Land and bodies of water are heated at different rates because land and water absorb or reflect sunlight differently. This uneven heating results in changes in the atmosphere as hot air rises and cool air moves in to replace it. The flow directly moves around the high/low pressure systems because of the rotation of the Earth. The Earth's rotation causes the wind to deflect counterclockwise in the northern hemisphere and clockwise to in the southern hemisphere (this deflection is called the Coriolis Effect). Wind plays a vital role in keeping the air, land, and sea fresh. It drives the ocean currents and aids plant life to disperse their seeds. Wind-driven turbines are also playing an increasing role in providing us with clean and renewable energy.

Stars are categorized according to their mass and luminosity. Luminosity is a star's intrinsic brightness and energy output, which depends on its mass and temperature, and our Sun's is "just right" for life. The Sun is one of the nine percent of the most massive stars in our galaxy and is highly stable. Despite its massive size relative to other stars in our galaxy, relative to other stars in the universe it is classified as a "yellow dwarf." There are stars that are hundreds of times more massive than our Sun in the universe, but as noted previously, more massive stars have a shorter lifetime and get more luminous more quickly, thus precluding any chance of obtaining a life-sustaining planet around them. If a star is less massive than our Sun, a planet must orbit closer to receive its energy, which will eventually lead to deadly tidal locking.[18]

The Sun's luminosity varies only by one-tenth of a percent, which is very important because more variety would lead to wild climate changes on Earth. Many stars frequently undergo large increases in luminosity as they burn their hydrogen into helium, thus releasing immense additional heat energy radiating toward bodies orbiting them. We are fortunate that our star is as massive as it is, but it cannot be so massive that it burns out before the planet has time to develop and to manufacture an oxygen atmosphere and the water. At the other end of the spectrum, low mass stars are more likely to tidally lock any planets it may have, and their unstable luminosity would sterilize life on an Earth-like planet due to regular large and deadly stellar flares.[19] We are indeed blessed.

Gonzales and Richards write: "Taken together, then, the anomalies [mass, luminosity, stability] suggest that the Sun is atypical in ways that enhance Earth's habitability for technological life."[20]

The Moon

The role of the Sun for life is obvious, the role of our Moon is less so. The Moon is anomalously large in comparison to its parent planet at almost one-third the size of the Earth. But it is not just this ratio that is odd; a recent in-depth analysis of our Moon and moons found orbiting other galactic planets revealed that ours may be extremely rare throughout the known universe.[21] The best evidence we have indicates that the Moon was formed about 4.5 billion years ago by a collision at an oblique angle of a Mars-size object, dubbed Theia, with the embryonic Earth, which was then still mostly molten rock. This "big whack," as astronomers have dubbed it, threw off a massive amount of material that gravity eventually gathered into the bright body in the night sky that poets rhapsodize about. The Moon does more for us than planting romantic notions in the minds of lovers and poets. The gravitational pull of the newly formed Moon slowed the Earth's rotation, slowly lengthening our day from about 6 hours to the current 24 hours. Rotation affects wind, and a rotational period of 6 hours would have meant constant cyclonic winds raging around the Earth. [22]

The Earth's axis (the degree to which it tilts toward or away from the Sun) varies little from its present angle of 23.5 degrees, but without the moon's gravity it would vary much more, and chaotically so. If the tilt of the Earth was around 60 degrees, it would mean that all of the Northern hemisphere would lean toward the scorching heat of the Sun and would experience perpetual summer light for half of the year. Likewise, the Southern hemisphere would experience perpetual darkness and freezing cold in a six-month winter. Complex life could not exist under those conditions. On the other hand, a tilt of much less than the 23.5 degrees would prevent the distribution of wind patterns and hence the distribution of rain around the world. As it is, our stable tilt gives us our predicable four seasons and wind patterns. Ward and Brownlee explain the Moon's gravitational effect has kept this angle almost constant for hundreds of millions of years. If the Moon was not present, the Earth's tilt angle would wobble due to the gravitational pulls of the Sun and Jupiter. They further add: "The monthly motion of our large Moon damps any tendency for the tilt axis to change. If the Moon were smaller or more distant, or if Jupiter were larger or closer, or if Earth were closer to or further from the sun, the Moon's stabilizing influence would be less effective. Without a large moon, Earth's spin axis might vary as much as 90 degrees."[23] Thus, the Moon literally makes life possible on Earth.

The gravitational pull of the Moon causes a slight bulge in the Earth that tidally moves the oceans about. Tides help to clean and oxygenate the oceans, and bring them vital nutrients from land erosion. Tidal currents mix arctic water, which is unable to absorb much solar energy, with water from warmer regions, balancing planetary temperatures and making for a more predictable and habitable climate. If there were no Moon, tides would be solely due to the Sun's gravitational pull and thus much smaller, which would have inhibited life. Physicist Joseph Spradley provides us with insight into the importance of the Moon for life on Earth and its theistic implications. He notes that the uniqueness of the Earth-Moon system violates materialism's faith that life should be commonplace in the universe. "For Christians, it supports the belief that God can work through natural and seemingly random processes to achieve his purposes in creation. It encourages a new appreciation for the special gift of life and an environment suitable for its survival." He then notes that it is reminiscent of Psalm 8:3–4, "When I look at your heavens, the work of your fingers, the moon and the stars, which you have set in place, what is man that you are mindful of him, and the son of man that you care for him?"[24]

Jupiter, Saturn, and Mars

If few people have thought about the Moon's vital contribution to life on Earth, even fewer have considered the role of the gas giant, Jupiter. Jupiter is 10 times larger than the Earth and 300 times its mass (mass refers to the amount of matter contained an object whereas size refers to its dimensions), and is considered both a maker and destroyer of planets. Astrophysicists suggest that without Jupiter the Earth might not exist as the life-friendly planet that it is. According to Konstantin Batygin and Greg Laughlin, our solar system is highly unusual. Noting that the most common mode of planetary formation generates planets with much greater masses than the Earth and are tightly packed around their parent star (closer than Mercury in our system) with short orbital periods, they set out to develop a model to explain why solar system is apparently unique. Jupiter was the first planet to form and was able to dominate the planet-building process thereafter. Where the rocky planets are today, there were nascent planets destined to be bigger and gassier (like Jupiter) than the Earth, and uninhabitable.

Batygin and Laughlin's model posits that Jupiter migrated in and then back out of the inner solar system early in its history. The gravitational interactions during its long journey caused planets to crash into each other with the debris either falling into the Sun, or leaving the remnants to form the asteroid belt located between Mars and Jupiter. Jupiter itself would have fallen into the Sun if it had maintained its inner trajectory, but was rescued by the formation of Saturn. As the two planets came closer together, it caused a gravitational interaction (orbital resonance) that fixed their current orbits. This explains our "oddball" solar system and how the events in the model caused the debris to coalesce into the small rocky planets, including Earth. The authors conclude: "Most dramatically, our work implies that the majority of Earth-mass planets are strongly enriched with volatile elements [substances that evaporate readily] and are uninhabitable."[25] In other words, these planets lack many things that make the Earth habitable such as its solid surface, its water, and its optimal atmospheric pressure.

Jupiter and Saturn (indirectly as the savior of Jupiter), have contributed to the Earth's formation as well as to its habitability. Some of the debris (asteroids) of the destroyed planets was flung off into space when it came too close to Jupiter's orbit. Millions of these asteroids and meteorites pummeled the early Earth over an estimated period of about 400 million years in what is known as "the period of heavy bombardment," seeding it with vital life-essential elements. We needed an asteroid belt neither too large nor too small. A much smaller belt would have failed to bring the essential element to Earth; a larger one would have continued pummeling Earth for many more millions of years. There are far fewer asteroids today, but we need to be protected from them, and the respective masses of Mars and Jupiter, the two planets marking the boundary of the asteroid belt, are most helpful in this regard. If Mars was more massive, as all models of planetary formation say that it should be, its gravity would sling more of these things our way but instead it takes a number of hits for us. Jupiter is an even more efficient guard. Its powerful gravity functions as a giant vacuum cleaner sucking up asteroids that have been bumped out of orbit. Jupiter's clean-up role was in evidence when Comet Shoemaker-Levy 9 impacted it in 1994. Had Jupiter's gravity not sucked up the comet, it may have entered our neighborhood. The solar system's bully has become Earth's bodyguard.

British geneticist J. B. S. Haldane's once quipped that "the Universe is not only queerer than we suppose, but queerer than we *can* suppose." So many astronomically improbable events came together to form our livable planet that it is quite beyond human comprehension. Physicist Donald Page, a former student of, and collaborator with, Stephen Hawking, calculated the odds against our universe randomly taking a form suitable for life as one in 10^{124}.[26] The series of ridiculously improbable things that have led to our inhabitable Earth leads physicists to the notion of unnaturalness. Theoretical physicist Nathan Seiberg expresses surprise at what he sees as this emerging unnatural view of the universe with all its outrageous "coincidences." He notes that: "Ten or 20 years ago, I was a firm believer in naturalness. Now I'm not so sure. My hope is there's still something we haven't thought about, some other mechanism that would explain all these things. But I don't see what it could be."[27] Scientists wedded to ontological materialism will remain puzzled as long as they refuse to understand the reason the universe seems so unnatural and "queerer than we *can* suppose" is because it was created by a supernatural Mind powerful beyond human comprehension.

Endnotes

1. Cassé, M. 2003, p. 147.
2. Ross, H. 2016, p. 31.
3. Astronomy Essentials, 2015.
4. Gonzales, G., Brownlee, D. and Ward, P., 2001, p. 62.
5. Ibid., p. 209,
6. Plaxo, J. and Gross, M., 2006, p. 35.
7. Ibid., p. 36.
8. Ibid., p. 34.
9. Gonzales, G. and Richards, J., 2004, p. 151.
10. Ross, H., 2016, pp, 32-33.
11. Gonzales, G., Brownlee, D. and Ward, P., 2001, p. 67.
12. Berger, A., Loutre, M., and Mélice, J. 2006.
13. Barnes, R. 2017.
14. Kopparapu, R., Ramirez, R., SchottelKotte, J. et al, 2014.
15. National Aeronautics and Space Administration, 2017.
16. Trefil, J. and Hazen, R. 2007.
17. University of California, Davis (nd).
18. Barnes, R, 2017.
19. Tarter, J., Backus, P., Mancinelli, R., et al, 2007.
20. Gonzales, G. and Richards, J., 2004, p.137.
21. Vieru, T. 2011.
22. Ward, P. and Brownlee, D., 2000.
23. Ward, P. and Brownlee, D., 2000, p., 223.
24. Spradley, J., 2010, p. 273.
25. Batygin, K., & Laughlin, G., 2015, p. 4217.
26. In Zacharias, R. 2008, p.35.
27. In Wolchover, N. 2013, np.

CHAPTER EIGHT

Earth:
Our Privileged and Improbable Home I

> "As we learn more about our world, the probability of its having resulted by chance processes becomes more and more remote, so that few indeed are the scientific men of today who will defend an atheistic attitude."
> Arthur Compton, Nobel Laureate physicist

Our Pale Blue Dot

We finally arrive at our own rocky "pale blue dot," as astronomer Carl Sagan dubbed the Earth on viewing photographs taken by Voyager 1 from 3.6 billion miles away. Our prize patch of cosmic real estate was formed from the left-over materials from the creation of the Sun. Solar winds blew away most of the lighter elements (hydrogen and helium) from our planet, leaving behind the heavier elements needed to form rocky planets. If we had been further away from the Sun, the solar winds would have been too weak to have this effect, and the light elements would have coalesced into a gas giant like Jupiter. We were in the right place at the right time to obtain the "habitable trinity" for life: an atmosphere, an ocean, and a landmass. These are the minimum requirements for life, each of which have their own minimum requirements.

Earth is often referred to as the "Goldilocks" planet because it seems to be the only place in the universe that is "just right" for intelligent life. Guillermo Gonzales and Jay Richards ask us to imagine if we had detailed knowledge about extrasolar planets and as a result we found the probability of getting a live-sustaining Earth by chance was 10^{180}: "This would mean that, even in a universe with 10^{11} stars per galaxy and 10^{11} galaxies, totaling 11^{22} available attempts, the chances of getting one such system would still be one chance in 10^{158}".[1] This calculation is far beyond William Dembski's probability boundary, which you will recall is one chance in 10^{150}. There is a host of interacting and interdependent things that must be very finely calibrated for intelligent life to exist, beginning with our planet's just-right mass.

The "Just Right" Planetary Mass

The first essential thing for a planet's habitability is its mass, and just like Goldilocks' chair, the Earth's mass is just right; not too big and not too small. Of the thousands of planets discovered so far outside our solar system, the smallest one is 5.5 times more massive than the Earth. Such huge mass means a gravitational crush too big for the formation of a landmass. Any land that may form on such a planet would quickly become eroded and become a giant water world if situated in its system's CHZ.[2] However, a planet must have sufficient mass to hold on to its atmosphere and greenhouse gases. This allows for surface water and planetary warming, but not so much that it leads to a runaway greenhouse effect and the evaporation of all water. Furthermore, the right surface pressure and temperature must stay in the same range for billions of years to prevent atmospheric escape into space. Gravity must be sufficiently strong to prevent this, but not so strong that the planet cannot rid itself of the thick hydrogen-rich atmosphere that existed on the early Earth, as it did on all planets. Astrophysicist Francois Forget notes the importance of this just-right mass: "with a solid body slightly more massive than the earth, a potential 'super-Earth' may ultimately remain like Neptune, with a massive H^2-He [hydrogen-helium] envelope that would prevent water from being liquid by keeping the surface pressure too high."[3]

The Earth's atmosphere provides protection against deadly solar radiation in the form of the ozone layer. About 90 percent of our ozone exists in the stratosphere, a layer 6 to 31 miles above the Earth's surface, although it is present in small concentrations throughout the atmosphere. Ozone (trioxygen) is formed when ultraviolet rays split oxygen atoms and three of them become covalently bonded. The amount of ozone has to be finely tuned. Too much would hinder human respiration and reduce crop yields; too little would increase the ultraviolet radiation reaching us, leading to increased health risks and damaged crops. Physician Arthur Brown sees God's hand in the creation of the ozone layer: "The Ozone layer is a mighty proof of the Creator's forethought. Could anyone possibly attribute this device to a chance evolutionary process? A wall which prevents death to every living thing, just the right thickness, and exactly the right defense, gives every evidence of a plan."[4]

Before any of the Sun's particle radiation (a broad span of the electromagnetic spectrum, not just ultraviolet rays) reach Earth, a significant amount has been deflected by the magnetic shield that surrounds the Earth and extends about 370,000 miles out into space. Only a few particles penetrate the shield, and these are channeled toward the poles, producing the beautiful auroras. In Earth's early life, the Sun was far more excitable, throwing more UV rays out into space. Without the shield the atmosphere would have been stripped. If you saw the movie *The Core*, you will know that the magnetic field is generated by a pool of molten metal and a solid iron core at the center of the Earth. The solid inner core is two-thirds of the size of the Moon and is as hot as the Sun's surface. The gravitational pressure—which is really crushing at close to 4,000 miles below the Earth's surface—prevents the core from becoming liquid despite the high temperature. Its surrounding 1,243-mile outer core of iron, nickel, and some other metals is kept fluid because there is lower gravitational pressure there. Scientists at the Institute of Physics tell us how the core produces our magnetic shield:

> Differences in temperature, pressure and composition within the outer core cause convection currents in the molten metal as cool, dense matter sinks whilst warm, less dense matter rises. The Coriolis force, resulting from the Earth's spin, also causes swirling whirlpools. This flow of liquid iron generates electric currents, which in turn produce magnetic fields. Charged metals passing through these fields go on to create electric currents of their own, and so the cycle continues. This self-sustaining loop is known as the geodynamo. The spiraling caused by the Coriolis force means that separate magnetic fields created are roughly aligned in the same direction, their combined effect adding up to produce one vast magnetic field engulfing the planet.[5]

The Core movie showed some of the initial effects when the iron core stopped rotating. Scientists in the movie determined that within a year the magnetic field will have dissipated, leaving the Earth victim to deadly solar radiation. There was a lot of really bad science in the movie, but they got that one right; without the rotating core we would be toast.

The Earth's atmosphere and magnetic shield protects us from the Sun's damaging rays, but the first line of defense from deadlier rays from outside our solar system is the Sun itself. These galactic rays are extremely high-energy particles originating from supernovae explosions. Their energy is great enough to easily penetrate the Earth's protective atmosphere and magnetic shield, so without the Sun's protection life would be impossible. The protection provided by the Sun is an interplanetary magnetic field known as the heliosphere. Our Sun is a magnetic star whose plasma atmosphere (mostly electrons, protons and alpha particles—helium nuclei) blows from its surface and cause huge solar winds, and: "Embedded in the solar wind is the Sun's magnetic field wounded up in a spiral and transported with the solar wind into space forming the heliospheric magnetic field."[6] The heliospheric magnetic field deflects about 90 percent of galactic rays away from Earth, and our own atmosphere and magnetic shield takes care of the almost all rest to the point that they present almost zero harm to life.

Plate Tectonics

The Earth is a dynamic system like a river; always changing yet staying the same. Looking at a modern map of the world, we note that the bulge of South America can fit into the southwestern concave of Africa. These areas were once joined, as was the rest of the world's landmass, in one massive super-continent that slowly broke into the continents that we now recognize. The mechanism driving Earth's geological-geographic evolution is plate tectonics; the movement of plates of the outer shell of the Earth drifting on top of a thick fluid mantle beneath the Earth's surface. The continents move on gigantic plates pushing against one another, jostling the Earth and occasionally causing earthquakes. The Earth's lithosphere, consisting of the crust and upper mantle, is crisscrossed by a jigsaw of large and small rigid plates in motion relative to adjacent plates. The motion of these plates causes deformation of the land by subduction ("taken beneath"), a process that occurs when one plate sinks beneath another. This movement is generated by convection currents carrying heat from the Earth's interior to its surface. The Earth's inherent heat comes partly from what is left over from its early molten state, and partly from the decay of radioactive elements in the core and mantle.[7] Geophysics has demonstrated that the composition of these radioactive elements are exquisitely fine-tuned for long-term stable plate tectonics and Earth's habitability.[8]

Over thousands of years, the movement of continental plates has pushed the Earth's crust up to form magnificent mountain chains. But mountains are not just to look at. Mountains are vital for life because without them there would be no rivers, and thus precious little fresh water to be found. Mountains capture and hold snow during the winter, and release it with the summer melt, sending water cascading down the rock face to feed agricultural and industrial activity. The hydroelectric power from water rushing down mountains generates renewable energy that promotes economic development. We seldom think of what mountains gift us, but: "Mountains are crucial for a global green economy. Providing 60–80% of the world's freshwater resources for domestic, agricultural, and industrial consumption, mountains are a critical driver of food security and clean energy. Mountains also supply important minerals and genetic resources for major food crops"[9]

When we look at the majesty of a mountain range, we may think that its mountains have always existed, but mountains that existed early in the Earth's history have eroded away long ago. Geologists using a simple formula calculate that a typical mountain mass of 1.2 miles in height and 2.5 square miles wide, and note this mass would have completely disappeared in 123 million years; a very short time in geological terms.[10] But we see mountains everywhere on the planet instead of a flat, lifeless water world. Some gigantic force has to be continuously creating them, and plate tectonics is that force. Thus, while tectonic activity in the form of earthquakes and volcanos can be destructive, the activity is necessary for life.

Both plate tectonics and the magnetic field depend on convection, a process by which the less dense material of the Earth rises and more dense material sinks. Rocks, water, and air become less dense as temperature increases, and so they rise. When they become colder, they become denser and sink. It is this convective motion that, over vast amounts of time, breaks the tough lithosphere into plates. A planet lacking convective activity lacks plate tectonics and its benefits. Just as large-mass planets harbor too much gravity for mountains and continents to form, small-mass planets lack plate tectonics, which results in the same outcome. A planet with low mass cools early from its initial molten state, and its crust solidifies forming a "stagnant lid" (no movement of its crust); too much mass and plate tectonics may be too mobile. The Earth's plate tectonics have just the right amount of vigor, because it "falls within a zone of transition between 'hard' stagnant lid and mobile plate regimes."[11] Figure 8.1 illustrates plate tectonics and volcanic activity.

The "big whack" collision that formed the moon helped to create both the magnetic field and plate tectonics. The collision generated such intense heat that liquid iron sank to Earth's center, providing the mechanism for generating magnetism. It also removed a large amount of Earth's crust, which is important because "a thicker crust may have prevented plate tectonics."[12] Without plate tectonics constantly pushing material upwards and maintaining volcanic activity the Earth would be a lifeless water world. Water is great, and is crucial for life, but we don't owe our existence to water alone: "The Earth is not unique because if its oceans. Any planet in the right part of the habitable zone will have those. What is unique about the Earth is that it has LAND. If the moon had not carried away most of the crust, there would be no ocean basins, no land, and no chance for life to evolve on land." [13]

Note that Genesis 1:9-10 described first the appearance of water on the planet, and then the land: "And God said, Let the waters under the heaven be gathered together unto one place, and let the dry land appear: and it was so. And God called the dry land Earth; and the gathering together of the waters called the Seas: and God saw that it was good." It wasn't until 1963 that science conclusively accepted the "outlandish" idea that the Earth's interior actually moved to "let the dry land appear." Wallace Pratt, perhaps the 20th century's most prominent geologist, has noted the accuracy of Genesis given its intended audience: "If I as a geologist were called upon to explain briefly our modern ideas of the origin of the earth and the development of life on it to a simple, pastoral people, such as the tribes to whom the Book of Genesis was addressed, I could hardly do better than follow rather closely much of the language of the first chapter of Genesis."[14] Pratt further noted that the sequence of events according to the science of geology: first there was water, then the emergence of land, then marine life, followed by land animals, is the same ordering revealed in Genesis.

Figure 8.1: Illustrating the Process of Plate tectonics and Volcanic Activity

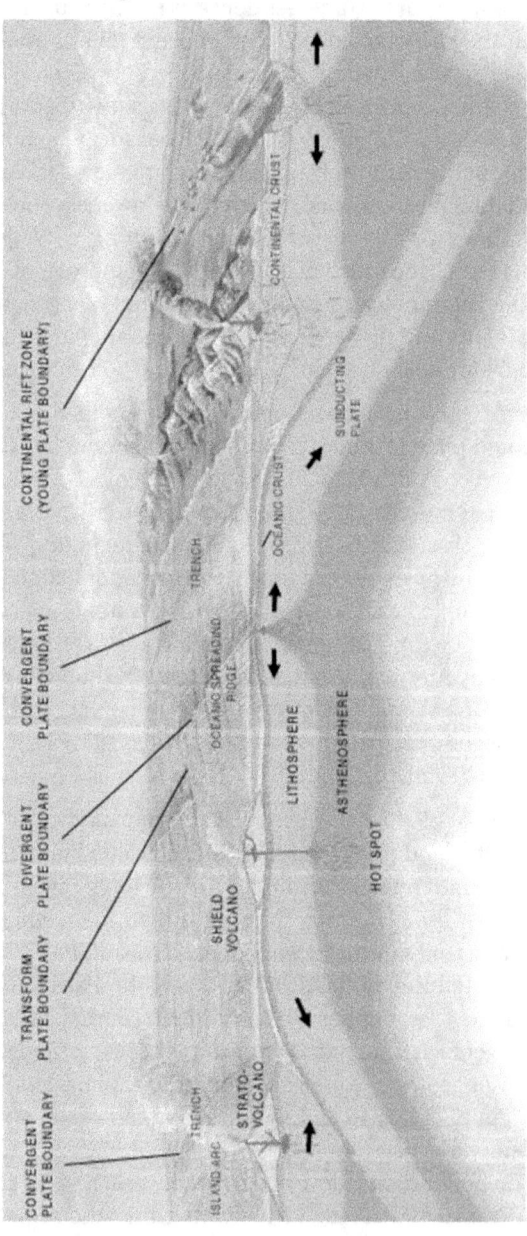

Volcanoes and Photosynthesis

It may seem strange to pair something as destructive as volcanoes with something as life-giving as photosynthesis, but it really isn't. Job 28:2 tells us that "The earth, though out of it comes forth bread, is in fiery upheaval underneath." The Earth we live on is indeed a fiery cauldron, but its volcanoes play an important role for the planet's habitability. While it is true that volcanoes can be very destructive, they are also essential for life because they indirectly "bring forth bread." When volcanoes erupt, lava, and the particulate matter released into the atmosphere, contains many vital minerals, such as phosphorus. Phosphorus is an essential component of life because it supplies part of the sugar-phosphate backbone of DNA and RNA. It is also an essential ingredient of adenosine triphosphate (ATP) that provides the energy needed for cellular activity, and is the nutrient that regulates the rate of photosynthesis. Volcanic gases are the source of most of Earth's water and most of the atmospheric gases the Earth needs. Because these gases created more than 80 percent of the planet's surface and laid the foundations for life, Maya Wei-Haas has described volcanoes as Earth's geologic architects: "Their explosive force crafts mountains as well as craters. Lava rivers spread into bleak landscapes. But as time ticks by, the elements break down these volcanic rocks, liberating nutrients from their stony prisons and creating remarkably fertile soils that have allowed civilizations to flourish."[15]

But that's not all; volcanism also moves carbon in and out of the Earth's interior, thus regulating the amount of carbon dioxide in the atmosphere (and thus the carbon dioxide/oxygen ratio). Without this process, too much carbon dioxide (a greenhouse gas) would become trapped in the atmosphere, and the Earth would heat up so much that the Earth would become a hot, dead planet. Too little carbon dioxide and the Earth's heat would escape, and it would become a cold, dead, planet. The Earth has many similar cycles by which various elements are used, changed, used in their changed form, and then changed back and reused again. The constituents of these finely-calibrated cycles are never lost entirely but only changed back and forth.

Plants are the first link in the food chain; we eat plants, and we eat the animals that eat the plants. Plants are living things that grow and reproduce, so they also need nourishment. The amazing thing about plants is that unlike humans and animals, they make their own food internally from non-living materials in the marvelous process of photosynthesis. The ingredients from which plants make their food are atmospheric carbon dioxide, sunlight, and water. Carbon dioxide and sunlight enter plants through microscopic holes on the underside of the leaf called stomata, which have cells that capture photons. Water enters the plant through its roots and makes its way to the leaves where photosynthesis takes place.

Within each cell on the plant's leaves are structures called chloroplasts that contain a chemical called chlorophyll. Chlorophyll transfers the sun's energy to a photon reaction center where it is converted into the chemical energy used to split water molecules into hydrogen and oxygen. After the hydrogen-oxygen split, hydrogen is combined with carbon dioxide and used by the plant to produce starch and sucrose. The oxygen is released into the atmosphere through the stomata as a waste product. Biochemists Gust, Moore, and Moore describe how vital for life this process is: "Photosynthesis is the largest-scale, best-tested method for solar energy harvesting on the planet, and it is responsible not only for the energy stored in coal, petroleum, and natural gas but also for the energy that powers most of the biological world, and for the earth's oxygenated atmosphere."[16] Photosynthesis is vital for life, but for it to endure it requires demanding constraints that must fall within the right ranges of light intensity, temperature, humidity, carbon dioxide concentration, seasonal stability, water, and minerals.

There is a lot of physics involved in photosynthesis. Sunlight is composed of photons, and it is their electromagnetic energy that causes photosynthetic chemical reactions. A plant's antennae gather photons and transfer their excitation energy to reaction centers. These centers contain mechanisms that regulate the rate of energy delivery long enough to resist natural recombination and perform the necessary work. Photosynthesis needs a resisting mechanism because photons energetic enough to break water apart are also energetic enough to break apart most other biological molecules. The resisting mechanism relies on quantum mechanics to transport energy in the most efficient way possible according to the temperature and moisture in the plants' environment. In the weird world of quantum mechanics, there is a principle called superposition. Superposition means that a quantum particle/wave is everywhere at once and takes every possible path to its destination, and the shortest path it will take in photosynthesis depends on its local environment.[17]

But how do volcanoes contribute to photosynthetic activity? Plants need sunlight to photosynthesize, but most plants thrive better in softer, shaded, sunlight than in its direct glare. This is where the volcanic eruptions come in. Such eruptions are not as rare as you may think. There are at least 20 volcanoes actively erupting at any given moment, and they may be thought of as the Earth's efforts to cool itself.[18] Scientists looking at the effects of the 1991 Mount Pinatubo volcanic eruption in the Philippines found that global terrestrial photosynthesis was increased by this eruption, and that it resulted in a 0.5 degree reduction of average global temperature. Volcanic ash and sulfur dioxide decrease temperatures by producing sulfur-rich particles called aerosols. Photosynthesis increased because these minute particles in the atmosphere scatter and absorb sunlight. As noted, most plants grow better in diffuse light than in direct light, and the increased particulates scattered the light globally for two or three years. The researchers concluded: "We estimated that this increase in diffuse radiation alone enhanced noontime photosynthesis of a deciduous forest [Harvard Forest located in Petersham, Massachusetts] by 23% in 1992 and 8% in 1993 under cloudless conditions."[19]

As with all living things, plants respire (breathe) constantly because their cells need energy to stay alive. To be precise, plants don't breathe like animals, rather, they respire by diffusion of carbon dioxide from regions of high concentration to regions of lower concentration. Plants only photosynthesize in the sunlight, but respiration occurs constantly. Plant respiration occurs internally by burning sugar and oxygen, which produces energy and carbon dioxide that is released as a by-product, just as oxygen is released as a by-product of photosynthesis. During daylight, photosynthesis and respiration occur simultaneously, so overall the amount of oxygen produced by plants greatly exceeds that of carbon dioxide. This is an incredibly complex process involving many enzymes that effectuate biochemical reactions in cycles.

Evolutionary biologists take it for granted that photosynthesis evolved in tiny steps, although it is acknowledged that: "We know very little about the earliest origins of photosynthesis. There have been numerous suggestions as to where and how the process originated, but there is no direct evidence to support any of the possible origins."[20] It seems that all the structures, physics, and biochemistry involved in the many steps of photosynthesis and respiration must be present simultaneously if the process is to occur at all. Just the assembly of chlorophyll takes 17 enzymes, which require many base pairs (the "letters" in the genetic code) to line up correctly. "Why would evolution produce a series of enzymes that only generate useless intermediates until all of the enzymes needed for the end product have evolved?" asks plant geneticist Rick Swindell. He answers in the mathematics of probability:

> if groups [of bases] of 1,000 recombined at a rate of a billion per second (10^9 tries) for 30 billion years (10^{18} seconds), with the number of bases being equal to the number of electrons that could fit with no space between them into a universe of 5-billion-year radius (10^{130}). This would yield 10^{157} total tries, an inconceivably huge number, utterly and absolutely trivial in comparison with the number of tries needed to have any chance of generating one small gene."[21]

Plant and animal life thus share a symbiotic relationship with one another in this wonderful world of ours, and it is inconceivable to think that there is no design behind the millions of finely-tuned mechanisms from the Big Bang onwards necessary for this symbiosis to exist. The immense complexity of the process of photosynthesis can be gauged by the fact that numerous chemists, biologists, and physicists are at work trying to duplicate God's ingenuity. Sunlight is by far the most usable form or renewable energy that we have, and if any company could learn to harvest it the way plants do it would be worth billions. The development of artificial photosynthesis was first put forward as an alternative to fossil fuels in 1912, and scientists are still working feverishly to duplicate what humble plants do all the time. If scientists are eventually able to produce a cheap and efficient method of artificial photosynthesis it would be a great boon to humankind, providing civilization with a constant source of cheap energy that will last as long as the Sun shines.

Endnotes

1. Gonzales, G. and Richards, J., 2004, p. 327.
2. Beaulieu et al, 2006.
3. Forget, F., 2013, p. 180.
4. In Hick, J. 1963, p. 25.
5. Physics.org (nd, np).
6. Ferreira, S. and Potgieter, M. 2004, p. 115.
7. Trefil, J and Hazen, R., 2007, p. 362.
8. Jellinek, A. and Jackson, M. 2015.
9. Kohler T; Pratt J; Debarbieux B; Balsiger J., et al. 2012, p. 7.
10. Trefil, J and Hazen, R., 2007, p. 355.
11. Valencia, D., O'Connell, R. and Sasselov, D., 2007, p. 47.
12. Gonzales, G. and Richards, J. 2004, p.6.
13. Hoffman, N., 2001.
14. In Copithorne, W. 1971, p. 14.
15. Wei-Haas, M. 2018, np.
16. Gust, D., Moore, T., and Moore, A. 2009, p. 1891.
17. Collini, E., Wong, C., Wilk, K., et al, 2010, np.
18. Wei-Haas, M. 2018, np.
19. Gu, L., Baldocchi, D., Wofsy, S., et al. 2003, p. 2035.
20. Blankenship, R. 2010, p. 434.
21. Swindell, R. 2003, p. 79.

CHAPTER NINE

Earth:
Our Privileged and Improbable Home II

> "This sense of wonder leads most scientists to a Superior Being – *der Alte*, the Old One, as Einstein affectionately called the Deity – a Superior Intelligence, the Lord of all Creation and Natural Law."
> Abdus Salam, Nobel Laureate physicist.

Weird and Wonderful Water

Perhaps because of its abundance (70% to 75% of the Earth's surface is covered by it), few things are more unappreciated than water. Yet water is the weirdest liquid on the planet, and the more scientists learn about it the weirder it seems to be, because it bends all the rules of physics and chemistry. The first interesting thing to note is that the two elements that make up the water—a great fire extinguisher—are combustible. Hydrogen is extremely flammable, and oxygen sustains fire. The reason that water douses a fire rather than feeds it is that water is already "burnt." When hydrogen and oxygen combine to form a water molecule, energy in the form of heat is released. A water molecule is at a lower energy configuration than its separate hydrogen and oxygen atoms because, as we have seen, the process of atomic fusion decreases energy. Like ashes in your fire pit, the hydrogen/oxygen combination has given off as much energy as possible in the reaction and cannot "burn" any further.[1]

Water makes up about 60 percent of a human body, and it is essential for the body's metabolism, temperature regulation, and for the flushing of its toxins. In liquid form, it is the solvent required for biochemical reactions. Water is not just a passive solvent dissolving and transporting chemicals to and from a body's cells, however. It has an active role in the stability, flexibility and the structure of proteins, nucleic acids, and DNA, and without it there is no chemistry of life. Bellissent-Funel and colleagues remark on the many functions of water in the cell: "Structurally, water participates chemically in the catalytic function of proteins and nucleic acids and physically in the collapse of the protein chain during folding through hydrophobic [tending to repel water] collapse and mediates binding through the hydrogen bond in complex formation. Water is a partner that slaves the dynamics of proteins, and water interaction with proteins affect their dynamics."[2]

As simple as its H_2O structure is, it is one of the strangest molecules known to science. Because life depends on its anomalous properties; if it behaved "normally" we wouldn't be here. H_2O is a polar molecule with a tiny positive charge at the hydrogen pole and a tiny negative charge on the oxygen pole. The molecule has a tetrahedral geometry (having four triangular faces like a triangular pyramid); that is, an oxygen atom located at the center, a hydrogen atom in each of two corners, with loose electron pairs in the other two corners. As a gas, H_2O is lighter than almost any other gas; as a liquid it is much denser, and as a solid, it is much lighter than chemists are led to expect based on the nature of other molecules.

Chemists graph the boiling points of molecules by their atomic weights and find that they all behave as expected—except water. If water followed the rules, its atomic weight leads chemists to expect that it would boil at 200°C less than its actual boiling point of 100°C; an anomaly that Michael Cory maintains: "seems to point decisively in the direction of Intelligent Design."[3] Corey quotes Harvard biochemist Lawrence Henderson on the tailor-made properties of water and carbon, both essential for life: "The chance that this unique ensemble of properties should occur by 'accident' is almost infinitely small. The chance that each of the unit properties of the ensemble, by itself and in cooperation with others, should 'accidentally' contribute a maximum increment is also almost infinitely small."[4]

Water's weirdness does not end there. In almost all substances, atoms and molecules huddle closer together as they get colder and eventually solidify. Water follows this rule until it attains its greatest density at just over 39°F, at which point it begins to sink to the bottom of seas, lakes, and ponds. Then, defying all the rules that molecules other substances follow, water molecules move farther apart rather than coming together as the temperature drops below the level of its maximum density and become ice, which floats to the top. Hugh Ross tells us of the importance of the fine-tuning of water's polarity: "if greater: heat of fusion and vaporization would be too great for life to exist; if smaller the heat of fusion and vaporization would be too small for life's existence; liquid water would become too inferior a solvent for life chemistry to proceed; ice would not float, leading to a runaway freeze-up."[5] So we should thank God that water does not behave "naturally" and is really weird stuff.

Water is crucial for maintaining our planetary atmosphere by both heating and cooling it. The Sun bombards the Earth with a tremendous amount of heat, the majority of which is absorbed by the ocean like a massive, heat-retaining solar panel. Water has a very high heating capacity, which means that it takes more energy to raise its temperature than it does for other liquids. Water can soak up a lot of energy before it changes from liquid to gas, which enables the ocean's water to store heat during the day keeping us from getting too hot, and releases it at night preventing us from getting too cold. The Sun's radiation is unevenly distributed on the surface of the Earth, but our oceans even thing out a little. The equatorial areas get the lion's share of heat, but the oceans distribute excess heat to colder areas via ocean currents caused by the Earth's rotation, winds, and tides. Like a gigantic conveyer belt, rotation moves heated water from the hottest areas of the planet to the coldest areas, and in return, the hotter areas get cold water from the coldest. Without this temperature regulation it would be super-hot at the equator and super-frigid toward the poles, thus reducing the Earth's habitable landmass.

As wonderful as this font of life is, water itself needs protection. An atmosphere at the right temperature supplying the right gravitational pressure on a planet's surface is needed to keep water from boiling away into space. Through its role in temperature regulation, plate tectonics is important in sustaining the conditions required for surface water to exist over billions of years.[6] But since, like every other solid body in the universe, Earth began as an inhospitable ball of white-hot molten rock, where did all our water come from? Scientists admit that water's origin on Earth is still something of a mystery, but according to Plaxco and Gross: "Jupiter's massive gravitational effects perturbed the orbits of icy, volatile-rich planetesimals (asteroids and icy comets) from the outer solar system and 'tossed' them into the inner Solar System, where they collided with—and provided the volatile [chemical elements and compounds] inventory of–the rocky inner planets."[7] The ice in the first planetesimals hitting the young volcanic Earth it would have instantly turned to steam, but as the Earth cooled below 100°C about 3.9 billion years ago it could hold onto a large amount of liquid water.[8]

Other scientists maintain that less than 50 percent of the Earth's water could have come from planetesimals, and that probably most of the Earth's water has been here from the beginning, albeit not in ready-made form. The current notion is that when Earth was forming, hydrogen from the solar nebula was incorporated into its interior. Some of this hydrogen combined with oxygen—the most abundant element in the Earth's crust—during such things as volcanic eruptions and/or chemical reactions in the mantle to help create Earth's abundant water supply.[9] Indeed, using data from hundreds of seismographs over a number of years, scientists have discovered a vast reservoir of water three times the volume of all the oceans deep beneath the Earth's surface.[10]

Water is everywhere in the cosmos; after all, hydrogen is the most abundant element in the universe, and oxygen is the third most abundant, so: "As long as the supply of hydrogen can be sustained, one can speculate that water formed from this process could be a contributor to the origin of water during Earth's early accretion. Water formed in the mantle can reach the surface via multiple ways, for example, carried by magma in the form of volcanic activities."[11] The abundance of water in the universe does not mean that Earth-like planets are abundant, however. Geologists Jan Zalasiewicz and Mark Williams marvel at the uniqueness of our planet: "The more we learn about how Earth acquired and retained its water, the more it seems the situation was incredibly fortuitous…Even in a water-filled cosmos, Earth might still be one of a kind amid water worlds far weirder—and more hostile to life—than our own. We might be in possession of an extraordinarily precious, rare jewel: our oceans."[12]

The fine-tuning of water for life is even evident at the level of quantum mechanics. In an article about water's strange properties, Lisa Grossman states: "Water's life-giving properties exist on a knife-edge. It turns out that life as we know it relies on a fortuitous, but incredibly delicate, balance of quantum forces." Water is held together by weak hydrogen bonds whose lengths keep changing, as we expect because of the Heisenberg uncertainty principle. In this context the uncertainty principle says that molecules cannot have a definite position with respect to others. This indefiniteness should destabilize the network structure of water and would remove many of water's marvelous life-sustaining properties were it not for a second quantum effect that cancels the effect of the first. Evidently, the uncertainty principle affects each water molecule's bond length by strengthening the attraction between them, thus keeping the network intact. Grossman concludes: "We are used to the idea that the cosmos's physical constants are fine-tuned for life. Now it seems water's quantum forces can be added to this 'just right' list."[13]

Water has been recycling on the earth for millions of years moving nutrients, pathogens, and sediment in and out of aquatic environments in a process called the hydrologic cycle. This mechanism is a perfect recycling method of distributing fresh, clean water around the planet to plants, animals, and us. We have to begin the cycle somewhere, so we begin with the power of the Sun's heat lifting water as gas from the surface through evaporation. These gas molecules are too small and weak to bring with them to the clouds any of the contaminants muddying the waters, so they are pure H_2O. Clouds form from these molecules, and when the air cools to the point that it cannot support water vapor, droplets of rain, hail, sleet, or snow become precipitation. When these droplets fall to the ground, we have our water supply replenished, which is then ready to evaporate again, keeping the cycle going via endlessly different paths that water can take on its pilgrimage of life.

Soil

If water is unappreciated, the stuff under our feet, dirt and soil (dirt is dead soil), are even less so. Soil is composed of inorganic solids such as clay, silt, sand, water, gas, and organic matter such as plant, animal, and microbial residues in various states of decomposition. Live microbes and worms create pockets in the soil for air and water to penetrate. The Earth's soil was produced over hundreds of millions of years by physical, chemical, and biological weathering of rocks. Physical weathering occurs by gradual breakdown of rocks through collisions, and through cracking by ice and frost. Chemical breakdown occurs when the chemical properties of the minerals within rocks interact with air, water, or other chemicals, and biological weathering is the result of plant roots growing on rocks and splitting them. The forces of weather, geological movement, volcanos, chemistry, and life forms have thus combined to grind the rocky layer of the Earth into smaller and finer grains which has been infused with nutrients provided by dead plant and microscopic animal life. Without soil we would not be able to grow food, and animals would have nothing to eat. Soil also performs other services that contribute to a healthy planet, such as nutrient cycling and water filtration.

Plants grown in the soil help to regulate the atmosphere by absorbing and storing carbon dioxide and other greenhouse gases from the atmosphere. For plants to exist and provide these services they need soil; to get soil we need a dynamic planet with the right atmosphere, water, and plate tectonics, as well as the early appearance of single-celled bacteria and archaea (similar to bacteria but a little more complex). Our planet's long history was not a waste of time; rather, it required a vast amount of time to ready it for human habitation. Only a dynamic planet has soil, and it is the interaction of the tectonic activity, weather, and hydrologic systems combined with the early appearance of microorganisms could make it possible.[14] The coming together of so many highly improbable things renders it almost impossible not to see Divine creation at work. It certainly took an inordinately long time to get from hydrogen to intelligent human beings, but time is a human construct which means nothing to a timeless God. God commands us to take care of the soil and the bounty it produces in Genesis 2:15: "The LORD God took the man and put him in the Garden of Eden to work it and take care of it." It takes a very long time to make living soil, but it is fragile and can easily turn to dead dirt if neglected.

The Nitrogen Cycle

We have noted that photosynthesis makes food and oxygen for us in the carbon dioxide/oxygen cycle. The nitrogen cycle just as amazing, because of all Earth's cycling of elements the nitrogen cycle may be the most complex. It is so complex because it takes a tremendous amount of energy to turn relatively unreactive nitrogen gas into a usable form so that it can be incorporated into our cells in reactive form. Nitrogen is important because it makes the things that make us—amino acids, proteins, and the nucleic acids for DNA and RNA. Because of its unique structure, no other element could possibly be substituted to make these important molecules. Nitrogen makes up 78 percent of the gas in the atmosphere; oxygen 21 percent, and the other one-percent is made up of other gases. Although there is abundant nitrogen in the atmosphere, it is not available to us directly. The only way we can get nitrogen is from the plants we eat or from the meat of animals that have eaten plants that absorb nitrogen-containing nitrates in the soil.

Figure 9.1: The Nitrogen Cycle

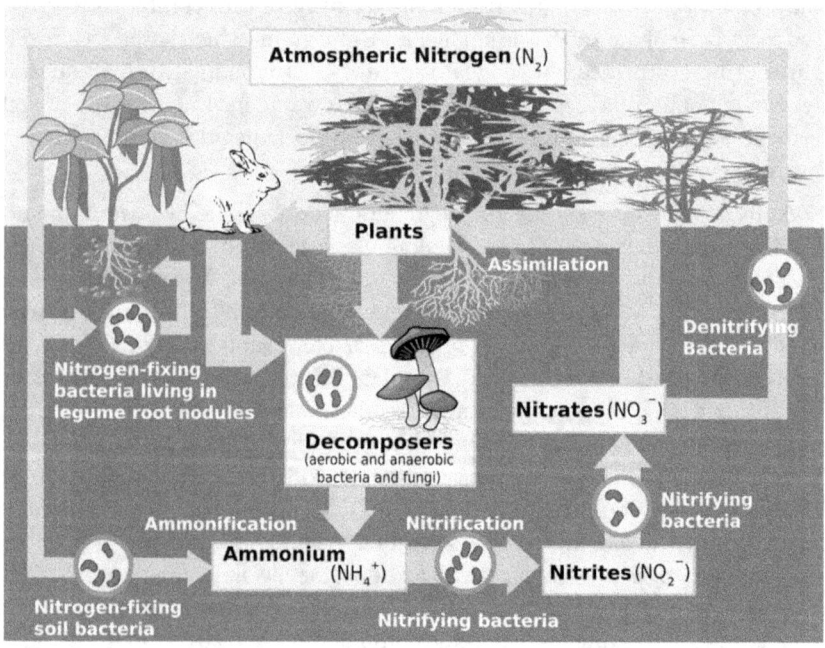

We cannot absorb nitrogen directly because nitrogen molecules consist of two tightly bonded atoms that do not readily interact, and the human body's chemistry does not provide sufficient energy to break them apart. The conversion of nitrogen gas into a usable form is accomplished by both biological and physical means. A process called nitrogen fixation is needed to change atmospheric nitrogen into forms that can be used by plant life (see Figure 9.1). Most nitrogen fixation is done by bacteria that possess an enzyme that combines nitrogen with hydrogen to make ammonia, and then nitrites. The other way to separate nitrogen's tight bonds is the awesome power of lightning. Lightning has enough energy (a lightning bolt is about 50,000°F) to separate the nitrogen atoms floating in the air. When they are separated, some of the free atoms combine with oxygen to form nitrogen dioxide. Nitrogen dioxide dissolves in water, which creates nitric acid, which then creates nitrates. Nitrates are powerful natural fertilizers that mix with the rain and fall to earth, watering and fertilizing the soil at the same time. No wonder farmers love lightning storms and their nitrogen-charged raindrops! When plants and animals die, their nitrogen compounds are broken down by bacteria as plants and animals decay, which transforms nitrates back to nitrogen. This nitrogen is then released into the soil and back into the atmosphere, completing the cycle.[15]

The Gaia Hypothesis

I have endeavored to show in very simplified way how the Earth has been able to successfully and efficiently recycle all its toxic (yet necessary in the right doses) and non-toxic products that we need for many millions of years, and has managed to retain its delicate balance of oxygen, hydrogen, nitrogen, methane, and traces of some other elements. The feedback systems of the Earth are so finely-tuned that atmospheric chemist James Lovelock proposed his famous Gaia hypothesis to try to account for it. The Gaia hypothesis proposes that the Earth functions as a self-regulating system, similar to a living organism, and that the organic and inorganic matter of the planet are a mutually sustaining integrated whole. The Earth's crust, its atmosphere, and oceans are regulated by the behavior of the biota—living organism—just as the biota is regulated and sustained by soil, air, and water. Without the presence of plants breathing in carbon dioxide, Earth's atmosphere would be about 95 percent carbon dioxide and only trace amounts of oxygen rather than 21 percent oxygen and a mere 0.04 percent carbon dioxide, and the Earth would be uninhabitable. In Lovelock's own words:

> the Gaia hypothesis says that the temperature, oxidation state, acidity, and certain aspects of the rocks and waters are at any time kept constant, and that this homeostasis is maintained by active feedback processes operated automatically and unconsciously by the biota. Solar energy sustains comfortable conditions for life. The conditions are only constant in the short term and evolve in synchrony with the changing needs of the biota as it evolves. Life and its environment are so closely coupled that evolution concerns Gaia, not the organisms or the environment taken separately.[16]

The notion that organic life affects the inorganic environment, and the inorganic environment affects organic life, was too much for many scientists because it smacked of teleology—X exists for the *purpose* of sustaining Y. Such end-directed thinking is anathema to atheistic scientists who emphasize that there is no purpose in anything at all. The notion that everything in nature works together harmoniously for the good of the whole, might get folks thinking in terms of why, and that might lead them to think of God. In order to achieve this wonderful equilibrium, living organisms need to be fine-tuned for positive feedback to sustain the environment that sustains them. Atheists expect us to believe that the spectacular maintenance of the ecosystem is just accidental. It requires an awful lot of faith to be an atheist, doesn't it?

Are we Alone?

John Haught writes of this fine-tuning: "So impressive is the still accumulating information about the many emergent levels of the world's fine-tuning for life that some scientists can hardly suppress a suspicion that something momentous, perhaps even purposive, is afoot in the cosmos."[17] The more we find out about fine-tuning the deeper the mysteries have become, and the more implausible naturalistic explanations seem to be. If all the parameters of the universe are so exquisitely fine-tuned, then they require a fine-tuner. Thus, by inference to the best explanation, a theistic account of a life-sustaining universe is logically preferable to the atheistic "happy accident" account. Astronomer Robert Jastrow's account of how science caught up with theology on the matter of creation is both interesting and poetic: "For the scientist who has lived by his faith in the power of reason, the story ends like a bad dream. He has scaled the mountain of ignorance; he is about to conquer the highest peak; as he pulls himself over the final rock, he is greeted by a band of theologians who have been sitting there for centuries."[18]

With all the incredibly precise fine-tuning in evidence, perhaps we are tempted again to ask the question that every philosopher of the heavens has asked: "Are we alone?" Is it inexcusably egocentric to claim that Earth is the only planet in the cosmos with intelligent life? For an ontological materialist, it is. The narrative of materialism has long been guided by the notion that there is nothing special about our planet. This narrative invokes the Copernican Principle, which is bogus, because all Copernicus did was to remove Earth from the center of the cosmos. In doing so he humbled Earth a little, but he did not humble creation or the uniqueness of human beings made in God's image. It really doesn't matter anyway if we are not alone; the Creator could have literally salted the universe with intelligent life if that was His desire. His existence hardly rests on an affirmative answer to the question of Earth's uniqueness.

But returning to the question anyway; there are far more parameters necessary for a life-bearing planet than I have addressed, and the more of these there are, the more unlikely it is that another planet bearing intelligent life exists. Hugh Ross computes the probability of a planet falling within necessary parameters by chance at less than 1 in 10^{215} and states that: "fewer than a trillionth of a trillionth of a percent of all stars will have a planet capable of sustaining advanced life. Considering that the observable universe contains less than a trillion galaxies, each averaging a hundred billion stars, we can see that not even one planet would be expected, by natural processes alone, to possess the necessary conditions to sustain life."[19]

Ross is an astrophysicist, but also a strong Christian, so perhaps his calculations are unconsciously biased. However, many other astrophysicists—theists, deists, agnostics, and atheists—have reached a similar conclusion. After exploring the thousands of parameters that have to be just right for complex life, astrophysicist John Gribbin asks if it is likely to exist elsewhere in the universes and answers: "Almost certainly no, given the chain of circumstances that led to our existence."[20] Then we have Stuart Taylor's address upon being awarded the prestigious Leonard Award for outstanding contributions to planetary science. Taylor looked at the staggering improbabilities of a myriad of things from the formation of the Milky Way to intelligent life on Earth and concluded: "When the remote chances of developing a habitable planet are added to the chances of both high intelligence and a technically advanced civilization, the odds of finding 'little green men' elsewhere in the universe decline to zero."[21] Biologist Francisco Ayala is also skeptical. He calculated that the probability of intelligent life existing on another planet is $10^{-1,000,000}$![22] Finally, astrobiologists Plaxco and Gross weighed the probability of intelligent life on other planets and concluded: "The range of values in Drake's parameters [an equation for estimating the probability of intelligent life outside the Earth] could adopt is so great, that despite the huge numbers of stars in the Universe, current scientific knowledge is entirely consistent with N=1. That is, Fermi [Enrico Fermi, the Italian-America Nobel Prize-winning physicist] was right, and we are alone."[23]

Endnotes

1. Sharma, S., and Ghoshal, S. 2015.
2. Bellissent-Funel, M., et al, 2016, p.7673.
3. Corey, M. 2001, p. 120.
4. In Montefiore, H. 2016, p. 90.
5. Ross, H., 1993, pp.112-113.
6. Cockell, C., et al, 2016.
7. Plaxo, K. and Gross, M., 2006, p. 53.
8. Trefil, J. and Hazen, R, 2007, p. 379.
9. Wu, J., et al, 2018.

10. Coghlan, A. 2014.
11. Coghlan, A. 2017.
12. Zalasiewicz, J. & Williams, M. 2014.
13. Grossman, L. 2011, p. 14.
14. Ross, H., 2016.
15. Fowler, D., et al, 2013.
16. Lovelock, J. 1995, p. 19.
17. Haught, J. 2008, p. 38.
18. Jastrow, R. 1992, p. 107.
19. Ross, H. 1994, pp. 169-170.
20. Gribbin, J. 2018, p. 99.
21. Taylor, S. 1998, p. 327.
22. In Tipler, F. 2003, p. 142.
23. Plaxo, K. and Gross, M., 2006, p. 247.

Chapter Ten

Losing God in the Multiverse

> "When these scientists talk about the multiverse, that's actually their way of talking about theology! It's their way of doing metaphysics without using the G—word!"
> Robin Collins, philosopher of science.

If You Don't Want God, Better Get a Multiverse

Just as many scientists were uncomfortable with the theistic implications of the Big Bang in the early 20th century, there are some today that are uncomfortable with the theistic implications of fine-tuning. They would like to avoid this fine-tuning with its outrageously large improbabilities that suggests, as Hoyle remarked, some "super intellect has monkeyed with the physics, as well as the chemistry and biology." Famous Stanford University physicist Andrei Linde notes this sublime fine-tuning and arrives at a remarkable anthropic-like conclusion. He notes that physicists dislike coincidences, are uneasy with the teleological notion that human life seems central to the universe, and that scientific discoveries are forcing them to confront anthropic ideas. Lind is not uncomfortable with anthropic reasoning himself: "Life, it seems, is not an incidental component of the universe, burped up out of a random chemical brew on a lonely planet to endure for a few fleeting ticks of the cosmic clock. In some strange sense, it appears that we are not adapted to the universe; the universe is adapted to us."[1]

Rather than wrestling with the challenges fine-tuning poses, some physicists have instinctively turned to the extravagant speculation that our universe is but one of trillions of other universes in a multiverse in which every possible combination of physical laws exists. Because we cannot, even in principle, observe these universes, they have invented one metaphysical entity that we can never know to get rid of another we can. The crux of the matter, according to cosmologist Bernard Carr, is: "If there is only one universe you might have to have a fine-tuner. If you don't want God, you'd better have a multiverse."[2] There are many scientists, theists and atheists alike, who believe that the universe's exquisite calibration means it either must have a fine-tuner or else we must have a multiverse of untold trillions of universes. Multiverse proponents believe that we are here because we just happened to have won the cosmic lottery. I am not suggesting that all physicists attracted to the multiverse concept are animated by atheism. I'm only saying that the concept is most attractive to atheists to counter the shock of the remarkably "unnatural" fine-tuning that science has discovered.

Physicist Alan Lightman is one who says we must choose; noting that intelligent design does not appeal to atheistic scientists wedded to ontological materialism: "The multiverse offers an explanation of the fine-tuning conundrum that does not require the presence of a Designer."[3] He adds that our universe only appears fine-tuned because we are here to observe it, period. Of course, we are here to observe it, but the goal is to try to understand it at both a physical and metaphysical level, not to say that we cannot, and then turn to speculations outside our universe. Lightman is an enthusiastic supporter of the multiverse hypothesis, and in an ultra-materialist mode, he writes: "Not only *must* we accept that the basic properties of our universe are accidental and incalculable. In addition, we *must* believe in the existence of many other universes. But we have no conceivable way of observing these other universes and cannot prove their existence. Thus, to explain what we see in the world and in our mental deductions, we *must* believe in what we cannot prove" [my emphasis].[4]

Sounding like materialism's pope speaking *ex cathedra*, Lightman concedes that the multiverse hypothesis cannot be tested, but also that all devout materialists *must* take the existence of trillions of unseeable universes as a matter of faith. He echoes Richard Lewontin's view noted in Chapter 2 that scientists are forced by their *a priori* commitment to a materialist worldview to reject anything supernatural. Of course, it is understandable that physicists want to explain how the universe works without reference to metaphysics, but they also believe that it is legitimate to inquire into all aspects of human reality. While they do not want to erect "No Trespassing" signs around any area of intellectual inquiry, they are content to leave non-materialist inquiry to the philosophers when confronted by the mysterious nature of God. Yet, many philosophers, physicists, and mathematicians view multiverse proponents as sneaking metaphysics in through the back door with their non-empirical speculations about the multiverse.

Beating the Odds with the Multiverse

What is this multiverse that Lightman says must believe in? There are a number of different models, all based on finding a non-design explanation for cosmic fine-tuning. The argument from design is basically that the probability of functional higher-order complexities arising from random step-wise interactions of simpler constituent parts is exceedingly low. The probability is enormously higher that such a process will produce only non-functional combinations. We have a finite universe capable of sustaining life, albeit one that is a probabilistic impossible given the trillions of non-life-sustaining alternate values the universal constants could have taken from the Big Bang onwards. Recall Roger Penrose's calculations of the precision of the "Creator's aim" necessary to create a universe consistent with the second law of thermodynamics. And that was just to get the whole show started.

The multiverse hypothesis allows the design argument to be rejected because, given an infinite number of universes and an infinite amount of time, we could insert an infinite number of probabilities into our equations and the impossible becomes probable, and the probable becomes inevitable. Hypothesize sufficient universes and you will beat the odds against finding one with its physical constants fine-tuned to such an incomprehensible degree as ours is. In an infinity of universes, at least one should contain all the "coincidences" that lead to intelligent life. This looks very much like a "multiverse-of-the-gaps" maneuver, because if you buy all the lottery tickets you are bound to win the lottery. Be that as it may, the multiverse hypothesis is the latest weapon in the atheist armory, and so it is necessary for Christians to engage it.

Multiverse Models

There is a variety of multiverse models, but those proposed by Max Tegmark are the best known. Tegmark is a brilliant and imaginative physicist who has spent decades exploring multiverse possibilities, has arranged his models in hierarchical fashion into different "levels" such that each subsequent level encompasses and expands on the level below. Each of these levels increasingly reveal how far some scientists will go to deny the Creator.

Tegmark describes Level I as: "A generic prediction of cosmological inflation in an infinite 'ergodic' space, which contains Hubble volumes realizing all initial conditions—including an identical copy of you about $10^{10(29)}$ m [meters] away."[5] "Ergodic" refers to dynamic processes which, given sufficient time, will impinge on all points in a given space. In simpler terms, it means if you observe randomly determined processes long enough you will see every possible outcome. Thus, an infinite ergodic universe must contain Hubble volumes in which all possible initial laws of physics obtain. The Hubble volume is the observable volume of the universe, which extends from Earth to the maximum distance that light traveled since the universe became transparent (because the universe is constantly expanding, light from the most distant regions will never reach us). According to Tegmark, given an almost infinite number of Hubble volume having the same laws and constants as ours, there are universes with similar, and even identical, configurations. Because they are identical, each of us will have identical twins residing in them, but we will never be able to see these universes or Skype our twin(s) since they are beyond our Hubble volume.

The essence of Tegmark's Level II is the assumption that "different regions of space can exhibit different effective laws of physics and open up infinite possibilities."[6] As opposed to universes in Level I, where each universe has a different initial distribution of matter but the same laws of physics, Level II assumes that different regions of the multiverse exhibit different laws of physics. Given this, Tegmark assumes there are infinite developmental possibilities for these universes. This model posits that the Big Bang was just one of an infinite number of spacetime universes arising within a larger system, like bubbles popping into existence as we run water in a bathtub, and our universe is just one of the bubbles. A constant series of big bangs implies the idea of a past eternity, but universes stretching infinitely into the past are not possible: "Here we have addressed three scenarios which seemed to offer a way to avoid a beginning, and have found that none of them can actually be eternal in the past."[7] A past eternity would have taken everything into thermodynamic equilibrium long before today. This point has been made before, but it bears repeating when discussing multiverse scenarios.

Level III invokes quantum mechanical universes, with quantum events unfolding in every possible way. In the world of quantum mechanics, you can have a simultaneous "event" and a "non-event." As was noted when we discussed photosynthesis, this is called superposition. Schrödinger's famous thought experiment in which a cat was placed in a box for an hour with a radioactive atom and a vial of poison illustrates this. If the atom decayed within the hour, it would trigger the release of the poison and the cat would die. If it did not decay, the cat would live, thus producing a cat that is both dead and alive in quantum superposition. Since quantum events can only be predicted probabilistically, we can only know the cat's fate (and thus the atom's) by opening the box after the allotted time and making the observation.

Tegmark argues that quantum superpositions are not confined to the micro world. He says that we, and everything else, are made of atoms, and if atoms can be in superposition (in two places at once), then so can we (don't let your boss know this!). Tegmark informs us that: "The only difference between Level I and Level III is where your doppelgangers reside. In Level I they live elsewhere in good old three-dimensional space. In Level III they live on another quantum branch in infinite-dimensional Hilbert space [a mathematical concept used to infer dimensions beyond the three spatial dimensions of everyday reality]"[8]

Tegmark revs up the speculation throttle to the max with Level IV. This level assumes that multiple universes are made up of all mathematical structures we can conceive of governed by different equations from those that govern our universe. Level IV is Tegmark's favored level because he argues that any conceivable universe is subsumed within it, and therefore there can be no fifth level. He explains, that this level "can be viewed as a form of radical Platonism, asserting that the mathematical structures in Plato's realm of ideas...exist 'out there' in a physical sense, casting the so-called modal realism theory [a metaphysical theory claiming that "possible worlds" are just as real as the world we inhabit] in mathematical terms akin to what Barrow refers to as 'π in the sky.'"[9]

For Tegmark, as for Plato, mathematics is the ultimate reality. The empirical things they describe are merely imperfect copies of their real form, which is found only within their mathematical description. He really believes this, as his remarks in a *Scientific American* article attest: "I argue that it means that our universe isn't just described by math, but that it is math in the sense that we're all parts of a giant mathematical object, which in turn is part of a multiverse so huge that it makes the other multiverses debated in recent years seem puny in comparison."[10] What Tegmark proposes as tests for Level IV ideas is to get physicists and mathematicians to dream up more mathematical structures. However, testing theories in science always rests on the firm ground of induction from experimental data to theory; adjusting the theory as the data warrants. Mathematics is a deductive enterprise with its own notions of truth-seeking separate from the inductive empirical sciences in which experimentation and observation rule.

M-Theory

The mathematical basis for the multiverse idea is the enormously complex M-theory, which has existed on the fringes of physics since 1968 as different versions of string theory. String theories are attempts to unify general relativity theory with quantum mechanics by smoothing out the mathematical inconsistencies between them. There are many physicists, however, who think the worlds of micro and macro physics are irreconcilable. Relativity and quantum mechanics work perfectly in the own domains, but when they come into contact, they clash violently: "When the equations of general relativity commingle with those of quantum mechanics, the results are disastrous."[11]

M-theory asserts that the fundamental constituents of physical reality are not the particles of standard physics such as quarks, but rather even tinier filaments of energy called "strings." These are strings of vibrating energy in a quantum field that give rise to all particles and forces in the universe. In M-theory, the standard subatomic particles are different vibrations of a fundamental string, some of which are useful, such as those in our universe, and others, such as those in most other universes, are not. Strings are said to vibrate in 11 dimensions—10 spatial dimensions, plus time, that are 'folded" in on one another. All these postulated extra dimensions are curled up in "internal space" in trillions of possible ways, each of which are assumed to be able to describe phenomena within their own restricted range.[12]

Depending on your perspective, M-theory is either the folly or the excitement of science. This is evidenced by the number of possible solutions to its equations, which may be as many as 10^{500}, and "Each solution represents a unique way to describe the universe. This meant that almost any experimental result would be consistent with string theory; the theory could never be proved right or wrong."[13] If there is a remote possibility of such an experiment, and if it should prove one's favored model wrong, the equations can be adjusted to accommodate the findings.

In their book *The Grand Design*, Hawking and Mlodinow argue that while our universe is exquisitely fine-tuned, it is simply blind luck, and we are the lucky winner in the ultimate Powerball lottery. They inform us that: "People are still trying to decipher the nature of M-theory, but that may not be possible."[14] Despite saying that the deciphering nature of M-theory "may not be possible," they continue to write as though it's not only possible, but has been done and dusted. They posit a multiverse governed by the "laws" of M-theory as existing in internal curled spaces: "The laws of M-theory, therefore, allow for different universes with different apparent laws, depending on how internal space is curled. M-theory has solutions that allow for many different internal spaces, perhaps as many as 10^{500}, which means that it allows for 10^{500} different universes, each with its own laws."[15]

The number of universes predicted by M-theory is incredibly large (remember that the best estimate of the number of atoms in the known universe is 10^{80}). However, 10^{500} is not enough to effectively address fine-tuning for intelligent life. If we take the product of all the probabilities starting with the "Creator's aim" in phase space and all the ratios, constants, forces, biological and chemical interactions that need to be precisely the way they are to allow human life, it would vastly exceed 10^{500}. Thus, even such a number of throws of the die cannot account for the fine-tuning "problem," but perhaps M-theorists can tweak the equations and add a zero or two to the exponent.

M-theorists appear to ascribe intelligence and agency to mathematical equations, since they seem to believe that mathematics can bring universes into existence. Witness Hawking and Mlodinow's bizarre notion (akin to Tegmark's) that the universe owes its existence to nothing but mathematical laws: "Because there is a law like gravity, the universe can and will create itself from nothing... Spontaneous creation is the reason there is something rather than nothing, why the universe exists, why we exist. It is not necessary to invoke God to light the blue touch paper and set the universe going."[16] So, we are back to saying that the universe created itself from nothing—something that did not exist managed to create itself. But this mysterious nothing is not nothing, since they say it is gravity. But gravity is a product of locally warped spacetime, and spacetime warps because of matter, so the concept is quite meaningless without the existence of spacetime and mass. Because Hawking and Mlodinow insist that gravity existed prior to the existence of the material universe, they should explain how this is possible, but do not.

Gravity was born simultaneously with spacetime and matter, but to be fair, Hawking and Mlodinow say that the law of gravity, and not gravity itself, that existed before spacetime and matter. But this is even worse. Senior NASA astronomer Seth Shostak ridicules the notion of gravity as the cause of the universe as "Plan B." He notes that the assumption that it is all about gravity: "begs the question, 'who designed gravity?' Isn't it remarkable that this gentle force seems so perfectly suited to the job of assembling a grand and habitable universe? And indeed, even leaving gravity aside, there are many other physical parameters that seem to be nicely adjusted for our presence...Depending on your personal philosophies, you can either credit this custom fitting to the intentions of God, or go for Plan B."[17]

By setting the laws of nature against God who created them and opting for Plan B, John Lennox accuses Hawking and Mlodinow of committing "a classical category error by confusing two entirely different things: physical law and personal agency."[18] Laws are mathematical models that describe the behavior of forces or things that are observed; they do not possess agency to bring those forces or things into existence. An abstraction such as the law of gravity has never created a concrete reality. John Lennox nicely put it in a YouTube debate: "The laws of arithmetic tell me that 2 + 2 = 4, but that has never put £4 in my pocket."

Be that as it may, the law Hawking and Mlodinow say created trillions of universes from nothing is gravity, which is something, not nothing. When asked to where gravity came from, or as Shostak asks, "who designed gravity?", Hawking answered: "M-theory."[19] So gravity was created by math equations coming to life—can you feel Newton and Einstein rolling over in their graves? Tim Radford, science editor of *The Guardian*, captures the God-like nature with which M-theory has been endowed by its proponents: "M-theory invokes something different [from other theories of science]: a prime mover, a begetter, a creative force that is everywhere and nowhere. This force cannot be identified by instruments or examined by comprehensible mathematical prediction, and yet it contains all possibilities. It incorporates omnipresence, omniscience and omnipotence, and it's a big mystery. Remind you of Anybody?"[20]

Theories and Hypotheses in Science

Scientists earn their living seeking new knowledge in an orderly way by first making themselves masters of what is already known. What is already known is organized in a systematic way by fitting the facts into harmonious patterns called theories. A visual example of a scientific theory seen on the walls of university classrooms around the world is the periodic table of the elements. This powerful icon of science rests on the atomic theory of matter. Chemists knew about the basic properties of many of the elements displayed on these charts for centuries, but their relationships were not known until Dmitri Mendeleev placed them into a logical order in 1869. Knowledge of protons and neutrons did not exist in Mendeleev's time, and atoms themselves were only dimly understood, but he realized that the properties of elements were related in a periodic way, and arranged the 63 known elements at the time in groups with similar properties into horizontal columns. The modern periodic table is arranged vertically by increasing atomic numbers according to the number of protons an atom contains, rather than horizontally.

As well as being looking backwards to fit known facts into coherent patterns, theories must also be forward-looking, telling us where we might look to fill gaps in our knowledge. Mendeleev's table had a number of gaps, which indicated that there had to be other elements that fit the properties of others in its group. Chemists have done so, and the table has been adjusted to fit an additional 55 elements since discovered in nature or synthesized in labs. Good scientific theories are thus always open to adjustment as new facts are discovered. Filling the gaps in our knowledge takes the form of hypotheses. These are statements logically deduced from theory about relationships among factors we expect to find based on the logic of our theories. Theories provided the raw material (the ideas) for generating hypotheses, and hypotheses support or fail to support theories by exposing them to empirical testing.

Science is sometimes conducted in an intellectual chaotic environment analogous to a homicide detective's new case. A detective confronted with a number of facts about a murder must fit them together to tell their story. At this point, the detective is like Mendeleev with a bunch of chemical facts in need of an ordered explanation. Using years of experience, training, and good common sense, the detective constructs a theory linking those facts together so that they begin to make sense. An initial theory derived from the available facts then guides the detective in the search for additional facts in a series of "*if* this is true, *then* this should be true" statements (hypotheses). There may be many false starts as our detective misinterprets some facts, fails to uncover others, and considers some to be relevant when they are not. Good detectives, like good scientists, adjust their "whodunit" theory as new facts warrant. Detectives fixated on a particular suspect will do the opposite—adjust the facts to fit their theory. When detectives do this, the crime goes unsolved or the wrong person may be indicted. Even brilliant scientists have been known to have such an emotional attachment to their favored theory that nothing will convince them to the contrary. Witness the genius of Sir Fred Hoyle, who went to his grave in 2001 still denying the reality of the Big Bang.

Is M-Theory a Theory According to the Criteria of Science?

A good scientific theory is considered to possess the following characteristics:

(1). *Predictive Accuracy:* A theory must not only be backward-looking to harmoniously fit known facts together, it must also be forward-looking pointing to ways of finding new facts.

(2). *Predictive Scope:* The scope or range of the theory refers to how much of the empirical world falls under its explanatory umbrella. As the predictive scope of a theory widens it gets more complicated.

(3). *Simplicity:* If two competing theories are equal in terms of the first two criteria, then the less complicated one with the fewest assumptions is considered more "elegant."

(4). *Falsifiability:* A theory is never proven true, but it must be falsifiable. If a theory is stated in a way that no amount of evidence could possibly falsify it, it is not a scientific theory.

Given these universally accepted criteria for judging theories, does M-theory fit the bill? Given its lack of empirical support, it is better characterized as speculation than a theory. Nevertheless, Hawking and Mlodinow inform us with certainty as if it had mountains of empirical support that: "M-theory predicts that a great many universes were created out of nothing. Their creation does not require the intervention of some supernatural being or god. Rather, these multiple universes arise naturally from physical laws."[21] This implies that M-theory possesses both predictive accuracy and predictive scope. But M-theory has not provided one scrap of empirical evidence; it is a gun that's never been fired, so we cannot gauge its accuracy. We cannot apply the term "simplicity" to M-theory because of its esoteric nature. As for the fourth criterion; there is no way, even in principle, that the theory could be falsified empirically.

M-theory could be supported if we could detect a string. As noted in our discussion of the Higgs boson, hypothesized sub-atomic particles are discovered by smashing protons together at near the speed of light and analyzing the debris. The smaller a hypothesized particle, the more the energy needed to detect it, but even the 17-mile Large Hadron Collider is not sufficient to detect a string. There is a formula for calculating the radius of the curvature needed for a collider given the energy of a particle, as well as for the strength required of the magnetic field of the bending magnets that keep particles traveling on a circular trajectory. Physicist Frank Heile has done the math, and shows that the radius would have to be 517 light-years, which means that the diameter of such a collider tunnel would be 1,034 light-years, and that the magnetic power required would be quadrillions of times more powerful than the Earth's magnetic shield.[22] M-theory thus seems to be forever condemned to a purely mathematical existence.

In sum, the multiverse is a highly speculative scenario—albeit one with mathematical underpinnings—which is not empirically testable, even in principle. Yet it is attractive to physicists of an atheistic bent who believe that the multiverse is the only way that they can escape God. "God or the multiverse" is their war cry, but we will see in the next chapter that things are not that simple. Many physicists, atheist and agnostic ones too, dismiss the whole multiverse-string theory-M-theory menagerie entirely, with some claiming it threatens the scientific status of physics. Theologians are not concerned that the multiverse is a threat to belief in God, however.

Endnotes

1. In Folger, T., 2008, np.
2. Ibid., np.
3. Lightman, A., 2011, p. 38.
4. Ibid., p. 40.
5. Tegmark, M., 2009, p.1.
6. Ibid., p. 1.
7. Mithani, A., & Vilenkin, A., 2012, p. 6.
8. Tegmark, M., 2009, p. 8.
9. Ibid., p. 12.
10. Tegmark, M., 2014, np.
11. Greene, B. 2004, p. 15.
12. Hawking, S. and Mlodinow, L., 2010.
13. Folger, T., 2008, np.
14. Hawking, S. and Mlodinow, L., 2010, p. 117.
15. Ibid., p. 118.
16. Ibid., p. 180.
17. Shostak, S. 2011, np.
18. Lennox, J., 2011, p. 36.
19. Ibid., p. 39.
20. Radford, T., 2010, np.
21. Hawking, S. and Mlodinow, L., 2010, pp, 8-9.
22. Heile, F., 2016, np.

Chapter Eleven
Finding God in the Multiverse

> "One way to learn the mind of the Creator is to study His creation. We must pay God the compliment of studying His work of art and this should apply to all realms of human thought."
> Ernest Walton, Nobel Laureate physicist

Mathematics in Science

I do not want to create the impression that the multiverse is just some weird idea atheists have dreamt up to explain God away, even though it has been used for that purpose. There are practicing Jewish and Christian scientists who pursue M-theory and who do not think that it rules out God at all. Indeed, they believe that if the multiverse is a reality it further reveals His creative power. Many such scientists are using mathematical models that they hope might lead to the unity of the macro and micro worlds of physics, and not to multiverse speculations. These models have simply led to the idea of a multiverse for some. It would be preferable if these physicists confined their efforts to trying to better understand this universe than to speculate about untold numbers of others, but curiosity, no matter in what legitimate direction it takes us, should never be discouraged. It is in this spirit that we further examine the multiverse and its critics, beginning with a discussion of mathematics, the sole tool of M-theorists.

M-theory is a purely abstract mathematical theory with a ridiculously incredible number of possible solutions. This is not to criticize mathematics in any way because it is the sturdy backbone of all science. Mathematics provides science with remarkably accurate quantitative information about natural phenomena. The unreasonable effectiveness of mathematics has given it a mystical aura. As Roger Penrose said of it: "There is something absolute and 'God-given' about mathematical truth."[1] Indeed, a number of mathematicians and physicists find themselves in agreement with Nobel Laureate physicist Paul Dirac who opined: "God is a mathematician of a very high order and He used advanced mathematics in constructing the universe."[2]

Mathematics represents natural reality in abstract symbols, and is amazingly successful in doing so. Natural phenomena such as gravity and electricity were known well before they were described by math, but we learned a lot more about them, and to use them for our benefit, when they were so described. Mathematics may describe something esoteric and beyond the senses, such as the bending of light by gravity, as predicted by Einstein's general theory of relativity. This was thought to be impossible because photons are massless and therefore could not bend, but Einstein knew they could because gravity bends spacetime. Arthur Eddington empirically proved Einstein right by observing the bending of light in a solar eclipse. Then there is Higgs boson, uncannily predicted by math almost 50 years before it was found. Numerous other examples justify mathematical models as a faithful representation of empirical reality.

However, M-theorists want to decouple mathematics from empirical validation and to quarantine the problem of empirical verification behind a wall of equations. They claim that the validity of their mathematical models depends on the beauty or elegance of their equations, and not at all on empirical findings. The universe is comprehensible through mathematics, but because it is *describable* in mathematical terms does not mean that the world *is* mathematics. Yet in the rarefied corridors where M-theorists live, they seem to have reached precisely this conclusion, with some, such as Max Tegmark, openly declaring it to be literally true. Arguing against Tegmark's notion that any universe that is mathematical describable is physically real, Don Page notes that, "different mathematical structures can be contradictory, and contradictory ones cannot co-exist. For example, one structure could assert that spacetime exists somewhere and another that it does not exist at all…these two structures cannot both describe reality."[3] M-theorists are aware of this, but have created such a fetish out of mathematical beauty, that their operative equation is evidently "beauty = truth." If M-theorists invent an imaginary world expressed in beautiful mathematic rules, they can use these rules to gather further mathematical evidence about that imaginary world. From such a base, they can shape a theory that is internally consistent to observers who understand the mathematics involved, and can conclude that because the theory is beautiful it must describe a reality existing somewhere.

Mathematics and Imagination

One of the mathematical creations of Stephen Hawking's prodigious intellect is the notion of imaginary time. He described this by asking us to imagine a straight horizontal line describing the arrow of time proceeding from the Big Bang into the future. This is time as we all know it. Hawking's imaginary time, which he says is as real as real time, is imagined as a vertical line intersecting real time at right angles traveling in a different direction. You may ask from where to where Hawking's timeline is traveling, but he admits he has no idea. This imaginary time is said to have existed before the Big Bang and was always there in a "bent" state. Hawking knows that the laws of physics cannot explain anything before the Big Bang because there was no "before," and that his imaginary time has no practical meaning.

To arrive at his model of imaginary time, Hawking uses imaginary numbers such as the square root of -1. Imaginary numbers do not have a tangible value; you can't use them to figure your grocery bill, but they are real in the sense that they are useful in higher mathematics and in physics. Just as imaginary time is perpendicular to real time, imaginary numbers are perpendicular to the horizontal number line and plotted upward and downward. Hawking thus uses imaginary numbers to invent imaginary time, which I am sure is something much admired by mathematicians. Hawkins makes it clear that he is unconcerned about whether or not extra dimensions such as "bent" time are real as long as they appear in the math: "I have been reluctant to believe in extra dimensions. But as I am a positivist, the question 'Do extra dimensions really exist'?, has no meaning. All one can ask is whether mathematical models with extra dimensions provide a good description of the universe."[4] Although imaginary time has no more meaning in everyday reality than is an imaginary bank account; it is a useful idea for multiverse proponents to bandy about.

The multiverse conjecture involving mathematical representations, elegant as they may be, is a creature of the imagination used to explain away the reality of the fine-tuning of our universe. Mathematician George Ellis believes that "it is only the gaps in current theories that cannot tell us why the fundamental physical constants have the values they do that drive the multiverse theories." He goes on to say that if we could explain them, "the drive for a multiverse explanation would fall away."[5] It is pretty lame to use a multiverse beyond our empirical reach to explain away things in this universe that are, or probably will be, within empirical reach.

The math in M-theory is so daunting that it keeps the critical layperson at arm's length; I for one may as well be looking at hieroglyphics or ancient Chinese poetry. I will thus call on Albert Einstein, who said: "As far as the laws of mathematics refer to reality, they are not certain; and as far as they are certain, they do not refer to reality."[6] Math must conform to reality, not reality to math. Herbert Dingle, a formidable mathematician who engaged in many mathematical controversies with Einstein, tells us how almost anything imaginable can be done with mathematics: "In the language of mathematics we can tell lies as well as truths, and within the scope of mathematics itself there is no possible way of telling one from the other. We can distinguish them only by experience or by reasoning outside the mathematics, applied to the possible relation between the mathematical solution and its physical correlate."[7] Witness how Einstein stopped the universe in its tracks on paper by inserting lambda into his equations, and Hawking's mathematical imaginary time running perpendicular to real time.

In his book, *A brief history of time*, Hawking asks: "What is it that breathes fire into the equations and makes a universe for them to describe? The usual approach of science of constructing a mathematical model cannot answer the questions of why there should be a universe for the model to describe. Why does the universe go to all the bother of existing?"[8] This is a "why" question which physicists are not supposed to ask, but it shows that while equations describe aspects of reality they do not confer reality on it any more than an architect's plans for a building is the building. At a few pages later, Hawking notes that the question he posed earlier is of tremendous importance: "If we find the answer to that, it would be the ultimate triumph of human reason—for then we should know the mind of God."[9]

Hawking and Mlodinow make reference to "model-dependent realism" whereby reality depends on what mathematical model is applied: "Each theory may have its own version of reality, but according to model-dependent realism, that is acceptable so long as the theories agree in their predictions whenever they overlap, that is, whenever they can both be applied."[10] This makes scientific sense when we realize that it is meaningless to talk about "true reality" in any absolute sense; true reality is known to God alone. However, no model of reality *is* reality any more than a map *of* Boston *is* Boston. If model-dependent realism can only be judged with reference to other models that are also model-dependent, how can we be sure of getting anywhere? If all maps were drawn without ever having seen the contours and streets of Boston, one should be wary of any map of Boston. Likewise, the various models drawn by M-theorists are unsupported by the contours and streets of empirical data. This is a shaky foundation to build notions of multiple universes on.

An example of an elegant mathematical proof is Kurt Gödel's ontological proof of God's existence. Gödel was a mathematician of genius who has been described as the finest logician since Aristotle. The ontological argument was first proposed by St. Anselm, 11th-century Archbishop of Canterbury, and expanded on by others over the centuries. Briefly, the argument goes like this: God is that which no greater can be conceived and nothing greater can be imagined. If God exists only as an idea in the mind, something greater than God can be imagined. But we cannot, therefore God exists. The argument is obviously much more involved than this, but this is the gist of it for our present purposes.

In 2014, computer scientists Christoph Benzmüller and Bruno Paleo fed Gödel's proof into high-powered computer programs called "higher-order automated theorem provers" and proved Gödel's proof to be right.[11] Would we expect Hawking to accept this? I doubt it very much. He would say that Gödel's theorem was proved only by the internal consistency of the mathematical assumptions on which it is based, and he would be right. We cannot prove God exists by mathematics any more than we can prove the existence of a multiverse. Mathematicians seek to describe and measure defined features of nature. Since God exists outside nature, He lies outside of definition and measurement. We can describe the movement of the planets, and a million other things with math because they are physical things. They are material and natural while God is immaterial and supernatural, and thus not amenable to empirical measurement. Put in mathematical terms, God is not a *theorem* to be proved; He is a self-evident *axiom* from which ultimately everything must be deduced.

Physicists and Mathematicians Weigh in on M-Theory

Although M-theory has many supporters, it also has many detractors. Roger Penrose describes Hawking and Mlodinow's ideas of the multiverse as "hardly science" and "not even a theory." He added that, M-theory is "a collection of ideas, hopes, and aspirations. The book [*The Grand Design*] is a bit misleading. It gives you this impression of a theory that is going to explain everything; it's nothing of the sort. It's not even a theory."[12] Penrose is not the only eminent physicists to call M-theory unscientific. Nobel Laureate physicist Richard Feynman dismissed string theory, on which M-theory is based, as "crazy," "nonsense," and "the wrong direction" for physics, and another Nobel Laureate physicist, Sheldon Glashow, likened it to a "new version of medieval theology."[13]

In 2006, mathematical physicist Peter Woit wrote a stinging book-length criticism of string theory, likening it to that caricature of learning called postmodernism. The words "not even wrong" in the title of his book signifies a hunch masquerading as a theory since it does not make testable predictions by which it could be falsified. Woit says: "There is a striking analogy between the way superstring theory research is pursued in physics departments and the way postmodern 'theory' has been pursued in humanities departments. In both cases, there are practitioners that revel in the difficulty and obscurity of their research, often being overly impressed with themselves because of this."[14] Woit was even more convinced that string theory is "not even wrong" in 2017. He noted that experimental results from the Large Hadron Collider had not shown any evidence of the extra dimensions string theorists had argued for as predictions of the theory, and remarks: "The internal problems of the theory are even more serious after another decade of research. These include the complexity, ugliness and lack of explanatory power of models designed to connect string theory with known phenomena, as well as the continuing failure to come up with a consistent formulation of the theory."[15]

Ellis and Silk add coal to the fire, noting that the mathematical elegance of M-theory generates untestable hypotheses and ignores empirical science. They conclude that because M-theory is metaphysical: "theoretical physics risks becoming a no-man's-land between mathematics, physics and philosophy that does not truly meet the requirements of any."[16] Ellis is a distinguished professor of mathematics and is not denying the huge utility of math in physics. He is only arguing against using it as a sleight of hand in an attempt to deny the obvious. The bottom line on M-theory, mathematically beautiful as the equations may be, is that it has no chance to ever be experimentally supported. Even if other universes exist with laws permitting spontaneous generation of life, it does not explain the only universe we know to exist.

Ellis and Silk also state that M-theory harms physics when its proponents argue for relaxing the criteria by which a theory is judged useful, and note that among the time-honored criteria for a scientific theory is that it must be falsifiable. Yet because M-theorists are faced with fundamental difficulties in meshing their theories to the observed universe, "some researchers called for a change in how theoretical physics is done. They began to argue—explicitly—that if a theory is sufficiently elegant and explanatory, it need not be tested experimentally, breaking with centuries of philosophical tradition of defining scientific knowledge as empirical."[17]

Tom Hartsfield joins the critical chorus: "The fire igniting critics of string theory is not personal animus or professional jealousy. It's the idea that a single theory has become so entrenched and popular in its field that its failures cannot be addressed truthfully. Now, physicists ask that the rules be bent or changed just to accommodate it. To loosen the principles of our fantastically successful scientific method just to allow for one passing theoretical fad to continue would be a disaster."[18] Likewise, Carlo Rovelli calls M-theory the physics of the "why not:" "Why not another dimension, another field, another universe?" He opines that theoretical physics has had a poor record in the last few decades because it has gotten itself trapped in the idea that it can disregard the content of empirical theories. He says that: "Science does not advance by guessing. It advances by new data or by a deep investigation of the content and the apparent contradictions of previous empirically successful theories...But most current theoretical physics is not of this sort. Why? Largely because of the philosophical superficiality of the current bunch of scientists."[19]

In his final paper titled "A smooth exit from eternal inflation," coauthored with Thomas Hertog before he passed away in 2018, Hawking took a giant step back from his 10^{500} possible universes. He apparently recognized that appealing to them is a cop-out with respect to explaining the fine-tuning of our universe. Hawking and Hertog reduced their multiverse to Tegmark's level I in which all hypothesized universes contain the same laws of physics. In an interview with *The Guardian*, Hertog explained the new theory, which had Hawking sounding like an advocate of the Anthropic Principle. Hertog notes that in the earlier theory there were all sorts of universes exhibiting all sorts of strange characteristics, but it did not address the mystery of our existence in this fine-tuned universe. He continues: "This paper takes one step towards explaining that mysterious fine-tuning. It reduces the multiverse down to a more manageable set of universes which all look alike. Stephen would say that, theoretically, it's almost like the universe had to be like this."[20]

What if the Multiverse Exists?

I could go on for pages quoting eminent scientists criticizing M-theory as a "theory too far," but as far as I am concerned, multiverse physicists would do well to heed Isaac Newton, whose first rule of scientific reasoning is: "Nature does nothing in vain, and more is in vain when less will serve; for Nature is pleased with simplicity, and affects not the pomp of superfluous causes."[21] But what if, against all odds, M-theorists turn out to be right and the exquisite laws of nature in our universe turn out to be just local by-laws and other localities have different ones? What are the implications for belief in God?

We have noted that some scientists say that that we have to make a choice between a multiverse ruled by blind chance or this one designed by God. Richard Dawkins' take on Hawking and Mlodinow's *Grand Design* is that physics has administered the *coup de grace* to God, so there is no choice to be made: the multiverse wins. However, Nobel Laureate physicist Steven Weinberg makes the point that God is not ruled out, but he also insists that we must make a choice: "If you discovered a really impressive fine-tuning...I think you'd really be left with only two explanations: a benevolent designer or a multiverse."[22] Such a choice seems irrational to philosopher Richard Swinburne: "To postulate a trillion, trillion universes, rather than just one God, in order to explain the orderliness of our universe seems the height of irrationality."[23] While I feel compelled to accept Swinburne's conclusion, I believe there is no choice involved at all. If a multiverse exists, then it points to an infinitely creative God; there is no reason why one should preclude the other. According to Robin Collins, a number of ancient Christian theologians such as Nicholas of Cusa and Giordano Bruno, as well as theistic scientists such as Newton and Leibnitz, championed the idea of a plurality of worlds (within one universe): "Indeed, many felt that restricting God to creating one universe was contrary to the omnipotence of God."[24]

Although I cannot fathom all the wonders of the universe (or a multiverse) being attributable in any way to chance, I recall Einstein's criticism of the indeterminacy of quantum mechanics, "God does not play dice with the universe." Niels Bohr's reply is a little lesson in humility: "Albert, stop telling

God what to do." It is no part of finite man to presume to know how an infinite and transcendent God decided to create anything. God can work through seemingly random processes, all of which are astronomically improbable to achieve his purpose. To deny that the Lord cannot "work in mysterious ways His wonders to perform" is to deny His omnipotence and to question His judgement. Let us not forget the words of Isaiah 55:8-9: "For my thoughts are not your thoughts, neither are your ways my ways, saith the Lord. For as the heavens are higher than the earth, so are my ways higher than your ways, and my thoughts than your thoughts." We delude ourselves if we think we know His purpose. As Nobel Laureate chemist Richard Smalley remarks: "The purpose of this universe is something that only God knows for sure, but it is increasingly clear to modern science that the universe was exquisitely fine-tuned to enable human life. We are somehow critically involved in His purpose. Our job is to sense that purpose as best we can, love one another, and help Him get that job done."[25]

Astrophysicist Gerald Cleaver, who is a major player in the M-theory community, does not believe that the theory precludes God: "The bulk universe [the multiverse] is consistent with belief in a God whose nature does not change and whose nature contains the attribute of creating. It yields a picture of an infinite, eternal God, who eternally creates and creates infinitely. This should be no surprise, for those who believe in an eternal, self-consistent God, characterized by all of the classical 'omni' attributes."[26] Likewise, famous string theorist Michio Kaku remarks: "I have concluded that we are in a world made by rules created by an intelligence. Believe me, everything that we call chance today won't make sense anymore. To me it is clear that we exist in a plan which is governed by rules that were created, shaped by a universal intelligence and not by chance."[27] Kaku does not invoke a personal God, but rests content with a Hoyle-like mysterious "super intellect," which he sees as compatible with a multiverse.

If a pre-Big Bang scenario is a reality, it just pushes us further back to the beginning of God's creation, and this does not entail the necessity to make a choice between it and God. Cosmologist Bernard Carr also believes that to posit the necessity of a choice between God and the multiverse is wrong-headed for several reasons, because if God created one universe, he is quite capable of creating many. However, Carr finds it: "not surprising that the multiverse proposal has commended itself to atheists. Indeed, Neil Manson has described the multiverse as 'the last resort for the desperate atheist.' For if ours is the only universe, then one has a problem explaining the fine-tunings and might well be forced into a theological direction."[28]

Even if the multiverse exists, it is more reasonable to suggest that it does so by virtue of God's creative hand than investing God-like power to the mindless laws of gravity. As Keith Ward remarks: "it is logically impossible for a cause to bring about some effect without already being in existence...Between the hypothesis of God and the hypothesis of a cosmic bootstrap, there is no competition. We were always right to think that persons, or universes, who seek to pull themselves up by their bootstraps are forever doomed to failure"[29] John Polkinghorne does not rule out the multiverse, but says: "A possible explanation of equal intellectual respectability–and to my mind greater economy and elegance–would be that this one world is the way it is, because it is the creation of the will of a Creator who purposes that it should be so."[30]

Perhaps God purposely created the universe (or multiverse) in such an incredibly unlikely way so that we never stop looking for His fingerprints. If there is intelligent life on other planets and if there are trillions of other universes, it only adds to the majesty of God. This is so because all the things we currently find utterly improbable or even impossible, may at some distant time be found true. And if they are, we can rejoice that they will be found to be the fruits of the true "grand design" inherent in the natural laws of the universe that He set in motion. God did not create a universe incapable of being described in natural terms, nor one that required Him to twiddle the dials occasionally. God is the Agent that designed it all, and He gave us the intelligence and motivation to figure it all out. God's hand is seen in the secondary causes through *His* laws of nature.

Confronted with all the evidence we have from cosmology, all we can do is reason to the best explanation of why we are here, and that explanation points unerringly to a creator God of this universe—or of multiple others. Frank Tipler, a former atheist, came to the same conclusion after a career spent exploring quantum physics and cosmology: "When I began my career as a cosmologist some twenty years ago, I was a convinced atheist. I never in my wildest dreams imagined that one day I would be writing a book purporting to show that the central claims of Judeo-Christian theology are in fact true, that these claims are straightforward deductions of the laws of physics as we now understand them. I have been forced into these conclusions by the inexorable logic of my own special branch of physics."[31] The multiverse may be "the last resort for the desperate atheist," but if it exists it provides no comfort for them because it still needs the Creator. As long as there are stars in the sky, birds in the trees, love in the heart, and one or a trillion universes, we can know that He exists.

Endnotes

1. Penrose, R., 2016, p. 146.
2. In Varghese, P. p. xviii.
3. Page, D. 2007, p. 424.

4. Hawking, S. 2001, p. 54.
5. Ellis, G., 2011, p.295.
6. Einstein, A., 1923, p. 28.
7. Dingle, H., 1972, pp. 31-32.
8. Hawking, S. 1988, p. 174.
9. Ibid., p. 193.
10. Hawking, S. and Mlodinow, L., 2010, p. 117.
11. Benzmüller, C., and Paleo, B., 2014, np.
12. Penrose, R., 2010, np.
13. Holt, J. 2018, p.220.
14. Woit, P., 2006, p. 207.
15. In Horgan, J. 2017, np.
16. Ellis, G., and Silk, J., 2014, p.321.
17. Ibid., p. 321.
18. Hartsfield, T., 2016, np.
19. In Horgan, J., 2014, np.
20. In Sample, I. 2018, np.
21. Newton, I. 1846, p. 384.
22. Gefter, A. 2008, p. 48.
23. Swinburne, R., 1995, p. 68.
24. Collins, R., 2007, p. 461.
25. In Overman, D. 2008, p.11.
26. Cleaver, G. 2006, p. 7.
27. In McLendon, K. 2017, np.
28. Carr, B., 2013, p. 168.
29. In Lennox, J., 2010, p. 64.
30. Polkinghorne, J., 2007, p.95.
31. Tipler, F., 1994, preface p. i.

Chapter Twelve
DNA: God's Book of Life

> "The God of the Bible is also the God of the Genome. He can be worshiped in the cathedral and in the laboratory."
> Francis Collins: physician and geneticist

Decoding the Book of Life

The human genome is God's construction manual for every living thing, and the more scientists discover how exquisitely fine-tuned and amazingly complex it is, the more they stand in awe. The genome provides the instructions for building proteins, which are the substances that build and maintain us. Scientists have struggled to unlock its secrets ever since the Augustinian abbot, Gregor Mendel, established the early rules of heredity in the 1850s-1860s. Progress has been slow but steady, punctuated by significant advances such as the discovery of the architecture of deoxyribonucleic acid (DNA) in the 1950s (DNA itself had been discovered in the 1870s) and DNA "fingerprinting" in the 1980s. A major advance occurred in 2000 when the $2.7 billion Human Genome Project, headed by Francis Collins, succeeded in sequencing the entire human genome. At the ceremony honoring this amazing scientific feat, President Bill Clinton remarked:

> Today's announcement represents more than just an epoch-making triumph of science and reason. After all, when Galileo discovered he could use the tools of mathematics and mechanics to understand the motion of celestial bodies, he felt, in the words of one eminent researcher, that he had learned the language in which God created the universe. Today we are learning the language in which God created life. We are gaining ever more awe for the complexity, the beauty, the wonder of God's most divine and sacred gift.[1]

Genome sequencing is a kind of translation of an ancient language into English; we can see the letters but the original meaning they might have had involves a lot of conjecture. The task of the Human Genome Project was to read and attempt to decipher this miraculous code. This required the pooled wisdom of more than a thousand scientists in biology, chemistry, physics, engineering, mathematics, and computer science from six nations, yet we are expected to believe this exquisite code requiring such brainpower to read and decipher arose fortuitously from ancient sludge by atoms randomly bumping around in the night. We wouldn't expect such a miracle of a two-page instruction manual of how to assemble a child's bicycle, never mind one on how to assemble beings made in God's image to last three score years and ten. This universal code book with its immense information content that reads, interprets, changes, and edits itself (try getting the bicycle manual to do that!) is a perfect code from any point of view.

Physicist and information specialist Werner Gitt, neurophysiologist Robert Compton, and physicist Jorge Fernandez assert that: "From an engineering point of view, and under the criteria that were considered here, the code system used in living organisms for protein synthesis—the *Quaternary Triplet Code*—is the best of all possible codes considering the four requirements that must be met. This testifies to purposeful design."[2] It is indeed remarkably compelling evidence for the existence of God. Gitt and his colleagues quote Romans 1:20— "For since the creation of the world God's invisible qualities—his eternal power and divine nature—have been clearly seen, being understood from what has been made, so that people are without excuse"—to argue that there is no excuse for denying God because the evidence is just too compelling. Some will not contemplate the evidence and some may not be able to, so these people may be forgiven. But those who contemplate and still reject God will do so willfully.

Genes and the Protein Making Process

The human body consists of trillions of cells, with each one—with some exceptions such as mature blood cells—being a microscopic factory for making proteins. Thousands of proteins are constantly being made for everything you need to keep you alive and kicking, and the information needed to make them are carried on specific segments of DNA called genes. Genes are the blueprint, recipe, library, or construction manual (chose your own metaphor) for life, because they contain the information that instructs cells what proteins to make and when to do it.

DNA consists of two strings of nucleotides tightly wrapped around a protein core called a histone and twisted around each other to form the familiar double helix ladder. Each nucleotide is built from a sugar and phosphate backbone, and a base (the rungs). There are four different bases: adenine (A), thymine (T), cytosine (C), and guanine (G), that bond in specific ways: C can only pair with G, and A can only pair with T (see Figure 12.1). There are approximately 3 billion base pairs in the human genome, and a gene is a group of adjacent base pairs that code for the manufacture of a protein. We have approximately 21,000 genes in the genome, but only about two percent code for proteins; many other segments regulate the behavior of the coding DNA.[3] DNA's information content is enormous; Clark and Pazdernik tell us that: "If the sequence were typed onto paper, at about 3,000 letters per page, it would fill 1 million pages of text."[4] Even more stunning is the fact that there is 6 feet of DNA packed into a cell a mere 1/2,500 inches across, and if the human body's entire DNA were to be unwound and placed end-to-end it would reach to the Sun and back.[5]

We inherit two forms of a gene—one from each parent—called alleles that are located at the same position on a specific chromosome. Alleles may be either dominant or recessive and are polymorphic ("having many forms"). Polymorphisms help to determine variation in traits and behaviors by coding for different levels of a protein product, and often differ by only one nucleotide among thousands. We call this a SNP (single nucleotide polymorphism). In the extremely truncated SNP below, adenine is substituted for thymine, and will produce a slightly different protein; albeit, one serving the same function. The bolded bases comprise the actual protein code, called an exon. Exons are sliced off and leave the nucleus to travel to the protein factory. The other bases—called introns—remain in the nucleus.

    Allele 1 from mother   TCACCTTGGA**A****TGGCCTA**AACGTCTTC
    Allele 2 from father   TCACCTTGGA**T****TGGCCTA**AACGTCTTC

Genes do not determine our behavior or our feelings. Neurotransmitters and other gene products certainly affect how we behave or feel, but they do not *cause* us to behave or feel one way or another, they *facilitate* our behavior and our feelings. The relevant protein products produce *tendencies* or dispositions to respond to the environment in one way rather than in another; they do not determine those responses. We can override behavioral tendencies, and we can change our feelings through force of will. God would never have designed a life-giving system that determined human behavior because He endowed us free will so that we may freely come to know him. Genes serve at our beck and call; we do not serve at theirs.

A heating system serves as a useful analogy for the genome's responsiveness to our needs. The thermostat in your house senses when the ambient temperature is below the desired setting and activates the furnace to restore the temperature to where you want it. The body's afferent nerves that carry nerve impulses from sensory organs to the brain may be thought of as a set of physiological thermostats that sense and transmit information about the state of your internal or external environment. When something is not right (say a cut finger), the "furnace" in the nucleus of the cell kicks on and an enzyme called DNA helicase goes to work unzipping the double-stranded DNA into two single strands. An enzyme called RNA polymerase then binds to the promotor region of a gene to synthesize messenger RNA (mRNA). The process by which the DNA code is converted into a complementary RNA code is called transcription.

Figure 12.1: The Making of a Protein

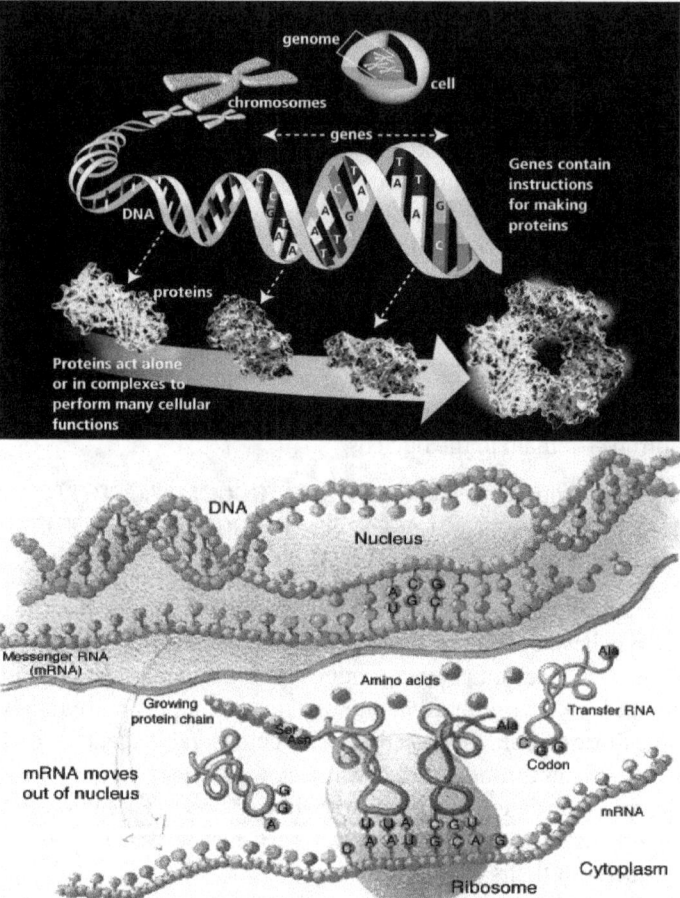

Uracil (U) is substituted for thymine as the base complement of adenine at this time. When the RNA polymerase running up the DNA strand reaches a stop sequence, the mRNA strand is complete, and detaches from the DNA strand. The mRNA then begins its journey to the protein factory where the message is translated and the specified protein is made. Because the environment in the nucleus differs from that of the cell's cytoplasm, they are walled off by a double membrane with channels that allow the passage of mRNA called the nuclear pore complex that recognizes and controls information flow. The mRNA is "tagged" by proteins to direct it toward the particular pore it must use to enter the cell's cytoplasm. After the mRNA strand detaches, the DNA double helix is reconstituted by the billions of free-floating nucleotides in the nucleus at the astounding rate of 50 nucleotides per second. The top of Figure 12.1 illustrates transcription and the bottom half illustrates translation.

The protein-making instructions are transmitted to the cell by mRNA in the form of base triplets (e.g. CAA, AGC, CCU, etc.) called codons. Codons can be thought of as three-letter words corresponding with the words for a particular amino acid, the building blocks of proteins. There are many different amino acids found in nature, but only 20 are used to make the proteins found in humans. Four bases conveyed in units of three means that there are $4 \times 4 \times 4 = 64$ possible arrangements. This is more than enough for the coding of the 20 amino acids. These four-base triplet codon combinations are what Gitt and his colleagues mean by the "Quarternary Triplet Code." Transfer RNA (tRNA) helps to decode the message of mRNA and picks up and transports the correct sets of amino acids (anticodons) that complement the codons on the mRNA strand. Codon and anticodon are then slotted into place by ribosomal RNA (rRNA). Ribosomal RNA is the RNA component of ribosomes, the molecules that catalyze protein synthesis. Ribosomes are responsible for joining the correct sequence of amino acids together to make the protein. They run across the mRNA until they encounter a stop codon, at which point it leaves the mRNA molecule and releases the protein for use in the cell. The process of changing information from the language of RNA into the language of amino acids is called translation.

DNA contains complex and specific information for protein synthesis. This occurs with astounding precision, but mistakes happen. What typically happens is the improper pairing of nucleotides, which occurs at a rate of about 1 for every 100,000 nucleotides. Like any factory, the cell contains a quality control department that inspects the final product looking for defects. The cell contains specialized enzymes that recognize imperfections, remove the wrong nucleotides, and replace them with the correct ones. A defective protein may also be stripped down to its component parts for reuse. However, some replication errors evade the detection and repair process. When this happens, we call the result a mutation. Most mutations are neutral, others are deleterious, but a rare few form the raw material of natural selection by which living things adapt to their changing environments.

Protein Folding

The flow of genetic information being translated into a protein entails getting the right amino acid chain and folding it into the correct three-dimensional shape. In the Journal of Theoretical Biology, biochemists Denton, Marshall, and Legge take us on a fascinating journey through the intricate process of protein folding, and argue that the processes determining their structure could not have evolved piecemeal by natural section: "It is more than anything else the complex hierarchic structure of the folds—their being composed of clearly defined substructures and submotifs combined together into what appear seemingly to be irregular complex hierarchic wholes, the sort of order which is so characteristic of that of a machine or artifact—which conveys the irresistible feeling that such forms *could not possibly be natural or lawful*" [my emphasis].[6] They go on to hypothesize that the self-organizing structures folding within the cell are governed by a rich "vocabulary of words" (information). They do not speculate on whom or what is responsible for imparting that information, but we know that lifeless atoms cannot write their own information-conveying software.

Figure 12.2: Protein Folding

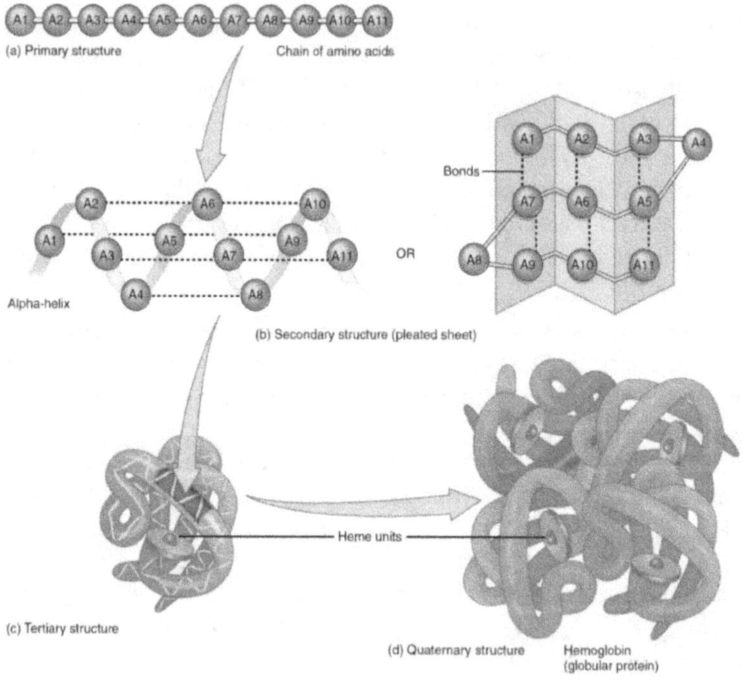

(a) Primary structure — Chain of amino acids

Alpha-helix

(b) Secondary structure (pleated sheet)

(c) Tertiary structure

Heme units

(d) Quaternary structure — Hemoglobin (globular protein)

Every protein requires genes to specify the correct sequence of hundreds of amino acids before the folding can take place, which is an immense amount of information. As shown in Figure 12.2, there are four stages by which a chain of amino acids are folded to become a functioning protein (in this case hemoglobin): primary, secondary, tertiary and quarternary; fancy names for first, second, third, and fourth. The primary structure identified in Figure 12.2 is the sequence of amino acids held together by their peptide bonds. The secondary structure is the protein beginning to fold according to various types of hydrogen bond. Each amino acid interacts with the others, and either twists into a corkscrew-like alpha helix or takes the shape of a folded beta sheet. Certain proteins called chaperones act as catalysts facilitating the correct assembly, but do not constitute a part of the assembled product. No cell can survive without chaperones, but since chaperones are proteins, what facilitated correct folding before chaperones existed? This is one of many chicken-or-egg paradoxes faced by naturalistic accounts of life, since no evolutionist claims that proteins and chaperones arrived on the scene simultaneously enjoying the relationship they have.

During the tertiary stage the protein is folded into a precise 3-D structure specific to its intended function. This requires a number of forces working together to produce interactions of groups of amino acids. Each fold represents an energy value, and the emerging protein is fighting nature's tendency toward disorder. It is therefore necessary for thermodynamic mechanisms to adjust it to its stable state. In the quaternary stage, a number of "local" amino acid chains from the tertiary structures fold together into a global quarternary, structure. It has been estimated that under unguided (random) conditions the number of possible configurations of a 100 amino acid chain is 10^{70}. If this is multiplied by the minimal time required to find one configuration (10^{-11} seconds) at random, this means a folding time of about 10^{52} years.[7] But of course, proteins fold on the order of milliseconds to seconds because of the information contained in the DNA.

This description of protein folding greatly oversimplifies a massively complicated process. An article proclaiming IBM's building of a new computer with huge computing power called Blue Gene provides a clue as to how complicated. Noting that protein folding holds the key to understanding the basics of how life works. "The scientific community considers protein folding one of the most significant 'grand challenges'—a fundamental problem in science or engineering."[8] It also stated that: "Blue Gene's massive computing power will initially be used to model the folding of human proteins, making this fundamental study of biology the company's first computing 'grand challenge.'" A *New York Times* article quotes a senior researcher in IBM's computational biology center as saying that it will take "Blue Gene about a year to simulate on the computer the folding of a single protein. How long does it take the body to fold one? Less than a second. It is absolutely amazing the complexity of the problem and the simplicity with which the body does it every day."[9] If great scientific minds built this remarkable computer capable of more than 1,000 trillion operations per second a year to merely simulate what our internal "commuters" do in less than a second thousands of times each day, how great must the Mind be that built us?

The Wonders of the Human Cell

In Darwin's day, living cells were thought of as simple blobs of protoplasm that keep our bodies glued together. We now know that they are collections of super-efficient factories containing thousands of entities working in tandem to provide for our physical needs. Cells are three-dimensional living structures formed from a one-dimensional string of instructions, and we don't fully understand how this happens. There are over 200 different kinds of cells in our body that perform different tasks; some are brain cells, others make bone, muscle, and hair, and others make red or white blood cells. These living factories divide without losing their cohesiveness, but they must die (about 300 million every minute) in a form of cellular suicide known as apoptosis, or programmed cell death. This is a normal part of cellular development by which enzymes destroy the DNA in the nucleus, with the debris being cleaned up by scavenging vacuum cleaners called macrophages

The information in DNA must be stored, transcribed, and translated, and then the output must be inspected, packaged, and sent to its proper destination. All this busy activity requires an awful lot of hardware and software packaged in this marvel of precision atomic engineering. Anyone who takes a class in cell biology is (or should be) awed and mystified by the living cell's marvel of design fitted into a space many times smaller than the period at the end of this sentence. There are textbooks of 500 pages or more on cell biology, so it is impossible to tell the whole story in a couple of pages. Getting lost in too much detail takes the mind off the big picture anyway. It is the big take-away picture I want to paint, and that is a sense of wonder at the marvelous design and stunning complexity of the cell. Bill Bryson captures his awe of the cell in four sentences: "Every cell in nature is a thing of wonder. Even the simplest are far beyond the limits of human ingenuity. To build the most basic yeast cell, for example, you would have to miniaturize about the same number of components as are found in a Boeing 777 jetliner and fit them into a sphere just 5 microns [0.00019685 of an inch] across; then somehow you would have to persuade that sphere to reproduce." If that's not enough to inspire awe, he goes on to write: "But yeast cells are as nothing compared with human cells, which are not just more varied and complicated, but vastly more fascinating because of their complex interactions."[10]

Each one of our trillions of cells is a marvel of nanotechnology. To accomplish their goals, they must work on a myriad of subordinate molecular goals that somehow collectively "know" what the ultimate goal is; that is, keeping their host body alive and healthy. These non-stop super-factories contain everything needed to supervise, plan, construct, package, and transport proteins, and they continually make copies of themselves over the lifespan of its host organism.

DNA: God's Book of Life   125

Figure 12.3: The Human Cell and its Major Parts

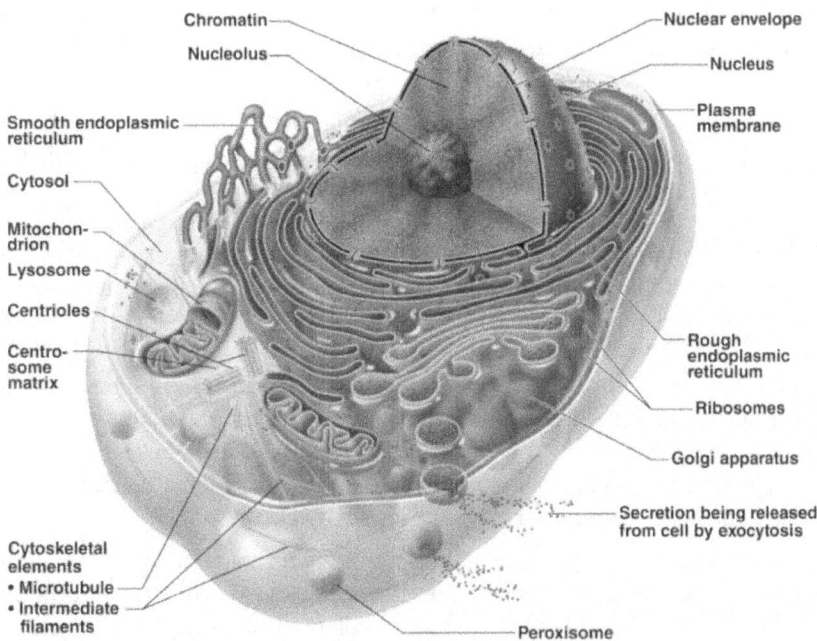

As is the case with any factory, the cell has many structures required to fulfill its purpose (see Figure 12.3). We start with the cytoskeleton; the structural foundation of the cell that determines its shape. It is also an assembly conduit directing the organelles and other substances around the cell. The nucleus is the control center from which the boss sends out instructions (mRNA) about what proteins are needed at the present time, and the ribosomes are the workers that will build them on the assembly line. The Golgi apparatus bundles proteins as they are synthesized, and the lipid molecules that form the cell's membrane serve as security guards monitoring which substances are allowed in and out of the cell, and through which entrance or exit. The factory floor is the cytoplasm, and the vacuoles are a kind of warehouse that stores the nutrients a cell needs and storing waste products awaiting disposal, protecting the cell from contamination. The endoplasmic reticulum is part of the quality control mechanism that inspects the finished product to ensure that only correctly folded proteins are sent to their destination, and the lysosomes are the janitorial staff that break down cell waste and discard it.

All systems require energy to function, and cells get theirs from over 1,000 organelles per cell called mitochondria. Mitochondria use cellular oxygen to convert chemical energy from food into adenosine triphosphate (ATP), and have their own separate genomes and multiply automatically when their cells need more energy. ATP's energy is stored in chemical bonds that can be opened and the energy redeemed. It has been noted that: "The efficiency of the transport of electrons in the last stage of ATP synthesis is 91%, an efficiency of which engineers can only dream."[11]

It is obvious that cells are not a series of simple building blocks that click into place like Lego cubes, but rather they are systems within systems within systems. It would seem logical that all of these parts would have to arrive on the scene as a whole unit to be functional. If a protein evolved first, how did it arrange itself without the ribosomes; where was the ATP to energize the process? What if it needed repair and there was no repair mechanism? If the repair mechanism evolved first, what would have been its purpose if there was nothing to repair? More importantly, where did any of the things get the information needed to do their jobs from? There are many ingenious naturalistic answers to such chicken-or-egg question; some plausible and some not, but every answer is shot down sooner or later with further research. The bottom line is that the cell is irreducibly complex in that it requires all its parts to be in place for it to function; if any of the interacting parts are removed the entire system breaks down.

Is there a super intelligence behind the information contained in DNA? I have a book of matches in front of me that tells me to "Close cover. Keep away from children." No one doubts that the information content consisting of 29 letters arranged in orderly sequence is the product of an intelligent mind. How likely is it that the human genome's 3 billion-plus letters arranged in orderly fashion, and conveying immensely more complex information, is not the product of a mind with infinite intelligence? Francis Collins calls DNA the "language of God," and offers its mind-boggling complexity as a compelling argument for His existence.[12]

When Collins calls DNA a language, it is not meant as a metaphor. It is literally a language of meaning just as any other language. It contains all the principles of human language, as physicist and information scientist Hubert Yockey points out: "Information, transcription, translation, code, redundancy, synonymous, messenger, editing, and proofreading are all appropriate terms in biology. They take their meaning from information theory and are not synonyms, metaphors, or analogies."[13] The amazingly complex and exquisitely orchestrated processes occurring in the cell are ultimately driven by information, but from a materialist point of view, these elegantly designed processes, unmatched by anything even dreamed of by humans, is the result of multiple pure chance events cooperating with the laws of physics and chemistry. Yet biologists are at a loss to reconcile the information-driven and chemically indeterminate nature of DNA with materialism. This is not a criticism since science *must* look for natural explanations regardless of an individual scientist's religious convictions; it cannot stop and conclude that God did it. We know that He did, but He leaves it up to science to discover how. This is the intellectual, and indeed, the *spiritual* excitement of science.

The complexities of cellular biology have led many scientists and philosophers to accept God, including Antony Flew. In 2004, Flew shocked the atheist world by announcing he had come to believe in God. It was as if Pope Francis announced that he had become a Muslim, because Flew was atheism's pope; a man who had written many books and articles peddling it (and socialism) for 50 years. When asked in an interview with Benjamin Wiker if he had "heard a voice" leading him to God, Flew replied that he did not have a sudden "Road to Damascus" conversion, but rather a "two-decade migration," as the evidence for God accumulated. There were two particularly strong factors that led him to God. One was the increasing number of famous scientists who affirm that there has to be a super Intelligence behind the complexity of the universe, and the other was the amazing complexity of DNA and the utter impossibility of a non-intelligent source of the origin of life:

> I believe that the origin of life and reproduction simply cannot be explained from a biological standpoint despite numerous efforts to do so. With every passing year, the more that was discovered about the richness and inherent intelligence of life, the less it seemed likely that a chemical soup could magically generate the genetic code. The difference between life and non-life, it became apparent to me, was ontological and not chemical. The best confirmation of this radical gulf is Richard Dawkins' comical effort to argue in *The God Delusion* that the origin of life can be attributed to a "lucky chance." If that's the best argument you have, then the game is over.[14]

In his 2007 book, *There is a God*, Flew traces his intellectual journey from atheism to by following the data wherever it may lead him. As if to vindicate the olds saying that: "A man who is not liberal in his youth has no heart; if he is still liberal in his adulthood he has no head," Flew tells us that in his youth he was a "hotly energetic left-wing socialist" but abandoned it in his thirties.[15] Abandoning this Godless creed is the first step toward abandoning atheism, and after Flew did, he became a vigorous defender of free-market capitalism, and much later, came to accept God.

Allister McGrath is another world-famous former atheist who fell to the early allure of Marxism, although he wasn't as tardy as Flew in rejecting it. He saw the falseness of it fairly quickly when at Oxford University studying molecular biophysics, in which he received a PhD. It was during this time, he wrote, "I was discovering that Christianity was far more intellectually robust than I had ever imagined. I had some major rethinking to do, and by the end of November [1971], my decision was made: I turned my back on one faith [Marxism] and embraced another."[16] The wondrous ordered complexity he saw in the structure of the cell was part of the reason he became a committed Christian.

Endnotes

1. Clinton, W. 2000, np.
2. Gitt, W., Compton, B. and Fernandez, J., 2011, p. 166.
3. Rands, C., Meader, S., Ponting, C., and Lunter, G., 2014.
4. Clark, D. and Pazdernik, N. 2009, p. 239.
5. Nicholl, D. 2008, p. 187.
6. Denton, M., Marshall, C., and Legge, M. 2002, p.339.
7. Karplus, M. 1997, np.
8. IBM. 1999, np.
9. Lohr, S. 1999, np.
10. Bryson, B. 2003. p. 372.
11. Gitt, W. Compton, B. & Fernandez, J. 2011, p. 297.
12. Collins, F. 2006.
13. Yockey, H. 2005, p. 6.
14. Wiker, B. nd, np.
15. Flew, A. & Varghese, R., 2007, p.33.
16. McGrath, A., 2010, p. 81.

CHAPTER THIRTEEN
Abiogenesis:
The Mother of all Scientific Puzzles

> "Once we see that the probability of life originating at random is so utterly minuscule as to make it absurd, it becomes sensible to think that the favorable properties of physics on which life depends are in every respect deliberate, even to the limit of God."
> Sir Fred Hoyle, world-famous physicist

The Mystery of Life's Beginning

Physics is the most basic of the sciences, dealing as it does with the most fundamental elements of reality. When physicists feel that they have exhausted materialist resources, many cross the permeable boundary between science and religion to ponder deep philosophical questions such as: "What does it all mean?" As we move from physics to chemistry and biology, we enter more materialist territory, because chemists and biologists do not work on problems that lead them to ask questions of ultimate reality. An exception to this are origin of life (OoL) scientists. The origin of life is a more intractable problem for materialists than even the origin of matter. We know a lot about the constituents of matter derived from humble hydrogen atoms forged by the energy of the Big Bang. Physics is certainly an intellectually taxing subject, but a few beautiful equations containing fewer than 10 symbols describe such things as gravity, force, and the unity of matter and energy. Take the equation for gravitational attraction: $F = Gm_1m_2/r^2$. G is a fixed number called the gravitational constant, m_1 and m_2 represent the masses of the two mutually attracted objects, and r^2 is the squared distance between them. This simple little equation describes a force that determines planetary orbits, and builds galaxies, stars, and planets, but we cannot begin to describe the complex operation of a human cell this way.

Materialism proposes that life arose from dead matter, and uses the term abiogenesis for the hypothetical process by which chemical evolution became biological evolution. How did this happen? No one has a clue. The leap from non-living matter to living matter would require a set of random non-living molecules to spontaneously arrange themselves completely undirected in a very specific way before we could call the result alive. Atoms spontaneously assemble from other atoms, but atoms have a lower energy configuration than their constituent parts because when they fuse to form a more complex element, energy is released to make the new element stable. The assembly of organic molecules, on the other hand, requires an increase of energy, not a decrease. Combined organic molecules are inherently unstable rather than stable, and they require a constant source of energy. Energy is supplied to living things by ingesting nutrients and metabolizing them. Living things also reproduce their kind, so there must be a system of replication as well. All living systems must thus possess a minimum of two things for them to be characterized as such: metabolism and reproductive capacity. The huge challenge confronting OoL researchers is not only how inanimate matter could be converted into an animate system, but also which came first, a self-replicating system or a metabolic system. Before life existed, how did these things that are essential to all living systems, and produced only by these living systems, come into being?

The DNA/RNA molecules of life must be enclosed in a protective membrane because chemistry in the environment outside the cell is hostile to them. They must also have a metabolic mechanism to draw energy from their environment, and must be able to reproduce themselves. These requirements are interdependent and must be online simultaneously. Not surprisingly, there is a mountain of chicken-or-egg problems in OoL research, the complexity of which may be gauged by the fact that 150 theories of abiogenesis were published between 1957 and 2000, and more have arrived on the scene since then.[1] We will examine the most popular theories: the "RNA world," the "metabolism first," and the "information first" hypotheses in the next chapter. This chapter highlights just a few of the difficulties faced by OoL researchers.

Nobel Laureate Francis Crick, the co-discoverer of the structure of DNA has stated that: "An honest man, armed with all the knowledge available to us now, could only state that in some sense, the origin of life appears at the moment to be almost a miracle, so many are the conditions which would have had to have been satisfied to get it going."[2] Paul Davies made the same point: "Many investigators feel uneasy about stating in public that the origin of life is a mystery, even though behind closed doors they freely admit that they are baffled."[3] OoL researchers Trevors and Abel wrote an article in *Cell Biology International* aptly titled "Chance and necessity do not explain the origin of life." In their article, they surveyed the many speculations offered to explain the immense gap between prebiotic chemistry on a lifeless Earth and the stunningly complex instructions contained in DNA, and concluded: "Contentions that offer nothing more than long periods of time offer no mechanisms of explanation for the derivation of genetic programming. No new information is provided by such tautologies. The argument simply says it happened. As such it is nothing more than blind belief."[4]

A prime example of such blind belief is Richard Dawkins' statement: "At some point a particularly remarkable molecule was formed by accident. We will call it the Replicator. It had extraordinary property of being able to create copies of itself. Replicators began not merely to exist, but to construct for themselves containers, vehicles for their continued existence."[5] Dawkins then drops the messy problem of how this happy accident occurred like a hot potato and launches into a discussion of DNA. His skip-over (which he hopes goes unnoticed) reminds us of the situation in physics before the 1930s pertaining to the origin of the universe—just accept an eternal universe as a brute fact that needs no explanation and move on.

Life arose either by spontaneous generation or Divine creation, as George Wald, Nobel Laureate in physiology and medicine, admitted. Exposing his blind materialism, he choose to "believe the impossible:" "We cannot accept that [Divine creation] on philosophical grounds; therefore, we choose to believe the impossible: that life arose spontaneously by chance!"[6] Wald later became a deist upon contemplating the remarkable fitness of the universe for life and the mystery of human consciousness. In the *International Journal of Quantum Chemistry,* he wrote: "It has occurred to me lately—I must confess with some shock at first to my scientific sensibilities—that both questions might be brought into some degree of congruence. This is with the assumption that mind, rather than emerging as a late outgrowth in the evolution of life, has existed always, as the matrix, the source and condition of physical reality—that the stuff of which physical reality is composed is mind-stuff."[7] We call the mind from which the stuff of physical reality came, God.

Another Nobel Laureate, biochemist Christian de Duve, wrote that "If you equate the probability of the birth of a bacterial cell to that of the chance assembly of its component atoms, even eternity will not suffice to produce one for you."[8] Even eternity! Despite this, de Duve believed that life is a lucky accident somehow forged by chance and necessity, the proof being simply that we are here! To say that something occurred by chance means that it did not happen by necessity or design. To say that something occurred due to necessity is to claim that it could not have been otherwise. In this context, it means that life had to happen given the laws of physics and chemistry, but numerous first-class scientists maintain that that information is the key to life, and that information is not reducible to the laws of physics or chemistry.

In 1969, Dean Kenyon, one of the foremost biophysicists of our time, published *Biochemical Predestination,* with Gary Steinman, in which it was proposed that abiogenesis was not only possible but inevitable. They argued that the chemical properties of amino acids attracted each other and formed long chains that became the first proteins which eventually would lead to LUCA (the Last Universal Common Ancestor). The book was warmly received by biologists and chemists thrilled with the idea from leading biophysicists that life could emerge from non-life. *Biochemical Predestination* dominated the OoL life landscape for a long time, partly because of Kenyon's impeccable scientific credentials (PhD in biophysics from Stanford and Postdoctoral Fellow in Chemical Biodynamics), and perhaps predominantly because it ruled out intelligent design. But after a career trying unsuccessfully to determine how organic molecules could self-organize, Kenyon came to the conclusion that: "We have not the slightest chance of a chemical evolutionary origin for even the simplest of cells...so, the concept of the intelligent design of life was immensely attractive to me and made a great deal of sense, as it very closely matched the multiple discoveries of molecular biology."[9] Kenyon, a former atheist, became a practicing Christian, dragged to God by his science.

The Oparin-Haldane Hypothesis and the Miller-Urey Experiment

The earliest scientific speculation about the OoL came around the time the Big Bang was emerging as an alternative to the static universe in the early1930s. Alexander Oparin and J.B.S. Haldane separately concluded that life could not have formed from an oxygen-rich atmosphere such as that currently bathing the Earth. Oxygen interferes with reactions that transform simple organic molecules into more complex ones by stealing electrons. They therefore posited that Earth's early atmosphere must have been "reducing;" that is, one in which there is little or no oxygen present, and one that easily produces chemical reactions. Such an atmosphere was considered rich in hydrogen and other compounds that easily donate atoms to other substances, such as methane and ammonia. These reducing gases were considered the major components in the so-called "primordial soup" by which chance and necessity was believed to produce the first molecules for life in the form of chains of amino acids necessary to build proteins.[10]

The reducing atmosphere hypothesis was widely accepted, and in 1952 the famous Miller-Urey experiment was conducted. Harold Urey and his student Stanley Miller created a closed system of flasks in the lab containing the reducing gases assumed to constitute the Earth's early atmosphere—water, methane, ammonia, and hydrogen. A Bunsen burner served as a heat source, and electrodes provided a continuous electric spark, mimicking lightning, provided the catalytic force. After about a week, a tar-like sludge was produced in the flask containing five amino acids. Amino acids are the building blocks of life, but the distance from simple amino acids to proteins, never mind to DNA and a cell to house it, could be measured in light-years. Amino acids do not live, and the fact that they combine to make proteins only in a specific way presents a huge problem for OoL researchers.

The appeal of a reducing atmosphere was that in such an atmosphere little energy is needed to form carbon-rich molecules; life's scaffolding. Carbon-rich molecules were thus assumed to predate life. However, geologists have found no carbonates (chemical compounds derived from carbonic acid or carbon dioxide) in rocks dating back to the early Earth, indicating that most carbon dioxide was still locked in the atmosphere. Moreover, calcium carbonates found in limestone is composed of the skeletal remains of various sea organisms, indicating that life existed before carbonates: "The sedimentological and geochemical evidence thus indicates a biogenic origin of the C [carbon] forming the graphite globules."[11] Furthermore, in a reducing oxygen-free world there would be no ozone layer, thus allowing large amounts of ultraviolet radiation to reach the Earth, "making delicate chemical reactions on the planet's surface very difficult."[12] Oxygen thus presents a paradox to OoL scientists because either its presence or its absence stymies prebiotic molecule formation. On the one hand, the absence of oxygen means no ozone protection, and on the other, its presence interferes with needed reactions.

Another reason that scientists now know that the early Earth did not have a reducing atmosphere is the evidence provided by a mineral called zircon found in igneous rock. Zircon is formed from magma oozing from volcanic activity mixing with surrounding rocks. Zircon does not interact chemically, making it both highly durable and the most reliable method of geological dating. It is also very reliable for determining what gases were present in the ancient atmosphere. Scientists have concluded from examining many zircon samples that the atmosphere on the early Earth had an oxygen level very close to that of the present-day Earth. This was the nail in the coffin of the reducing atmosphere hypothesis: "We can now say with some certainty that many scientists studying the origins of life on Earth simply picked the wrong atmosphere."[13]

From Amino Acids to Proteins: The Chirality Problem

Even if amino acids can be made under strict laboratory conditions, it's a far cry from making them self-assemble into chains to form proteins because: "From a chemical point of view, proteins are by far the most structurally complex and functionally sophisticated molecules known."[14] Amino acids are monomers ("one part") that must bond together into large molecular chains called polymers ("many parts") to form functioning proteins in a process called polymerization. If monomers did polymerize in the hypothesized prebiotic soup, they would quickly break apart. Biological chemists frequently point out that there is no evidence that a primordial soup ever existed, and even if it did: "Polymerisation into RNA requires both energy and high concentrations of ribonucleotides [the building blocks of RNA]. There is no obvious source of energy in a primordial soup. Ionizing UV [ultraviolet] radiation inherently destroys as much as it creates."[15]

Living things must extract energy from the environment, and unguided polymerization runs afoul of the second law of thermodynamics. Because polymerized molecules have already reacted (by forming from monomers), they are at thermodynamic equilibrium. In a state of thermodynamic equilibrium, no further changes can occur because there is no free energy intrinsic to the system that would allow it to do so. Free energy can only be supplied to a living thing by a mechanism that can harvest energy from the environment (food, water, sunlight) to counteract the decaying effects of the second law; only then can it break free of its shackles. The problem is, of course, that a system must already be alive for it to possess such a mechanism.

Thermodynamic equilibrium is illustrated by a steaming hot cup of coffee and a glass of ice water in a room. Left sitting, the coffee will transfer heat to the room and to the ice water, and the room will transfer heat to the ice water until coffee, water, and room will be at the same temperature. They will maintain a constant temperature in the absence of further heat from outside the system. When a physical system reaches its lowest energy state, such as the coffee no longer transferring heat to the room and water, and vice-versa, the system is in equilibrium. By definition, this means that no further change can take place since low energy is associated with stability and not change. The hypothesized molecules in the prebiotic soup would have had "no internal free energy that would allow them to react further. Life is not just about replication; it is also a coupling of chemical reactions."[16]

Getting amino acids to polymerize to produce a functional protein confronts the chirality problem. *Chiral* comes from the Greek for "hand." Two amino acids that are alike in structure and function are also distinct from each other when they are mirror images. One version is labeled D ("dextro") for right-handed, and the other L ("levo") or left-handed.[17] Note from Figure 13.1 that the two structurally identical amino acids (they have the same atoms: carbon, hydrogen, oxygen and nitrogen) are like your hands. Like your gloves, they may appear to be identical, but you cannot fit your right hand into your left glove. Likewise, D and L acids will not bond. The "COOH" and "NH$_2$" stand for "carbolic" and "amino" group, respectively, and the "R" stands for R group. Each of the amino acids has an R group of atoms specific to it and bonded to the central atom that confers on them their particular chemical properties. Just like hands and gloves, chemical reactions driving our cells only work with molecules of the correct handedness.

When amino acids are found in nonliving material, or when synthesized in the laboratory, they come equally in D and L forms. Biologists call equal D and L acids a racemic. There are an equal number of Ds and Ls in nature, but a homochiral set of amino acids is necessary for life. That is, all amino acids must be left-handed and all sugars (ribose) must be right-handed in order to produce DNA and RNA. The molecular locks of life can only be opened by molecular key with the proper handedness; nothing else will fit. Even one right-handed amino acid would destabilize the DNA helix (it is the correct chirality that gives the helix its shape) and it would not be able to form long chains of information.

Given that nature always produces a racemic, what is the probability that even an unrealistically short protein of 100 amino acids (the largest known protein is titin, which has over 34,000 amino acids) would form by chance from all left-handed monomers from a natural racemic of monomers? Because the odds of either an L or a D in nature are 50-50, it is $0.5 \times 0.5 \times 0.5$...and so on 100 times, which turns out to be 1 in 10^{31} tries. Astrobiologists Plaxco and Gross use a longer chain, and inform us that it "is highly improbable that a random chemistry could produce a polymer molecule that contained monomers of only one-handedness. To be precise, the probability of achieving homochirality in a 189-unit polymer from an equal-molar mixture of left- and right-handed monomers is 1 in 2^{189} (1 in 8×10^{56})!" [18] Given this, they write: "The current genetic code seems far more highly optimized than one would expect were it simply an accident."[19]

There is yet another problem in addition to the mind-numbing statistical improbability of achieving homochirality, and that is back again to the second law of thermodynamics. One Biochemist notes: "Consider one of the simplest steps in the origin and evolution of life, the choice of one chiral form over a racemic mixture. Thermodynamics do not permit this initial step. They dictate full racemization of all non-completely racemic mixtures. In this respect, weak nuclear interactions seemingly do not obey the second law."[20] He goes on to show why this is so via from quantum physics, but he is basically saying that a naturalistic origin of a homochiral chain of amino acids is impossible because it violates the second law.

Even if chance managed to achieve homochirality, we still have the problem of chemical reactivity. In the present context, chemical reactivity refers to how fast amino acids react with others. If you have a number of L-amino acids in the lab and allow them to interact with the expectation that they will eventually form a functioning protein chain, we find that the most reactive acid will link up first and the least reactive will line up last. Given the hundreds, or even the thousands, of L-amino acids that have to line up in a precise way to get a functional protein, it is no surprise that this never happens by unguided processes. Yet it happens every minute of every day guided by the information content of DNA.

Figure 13.1: Illustrating Chirality

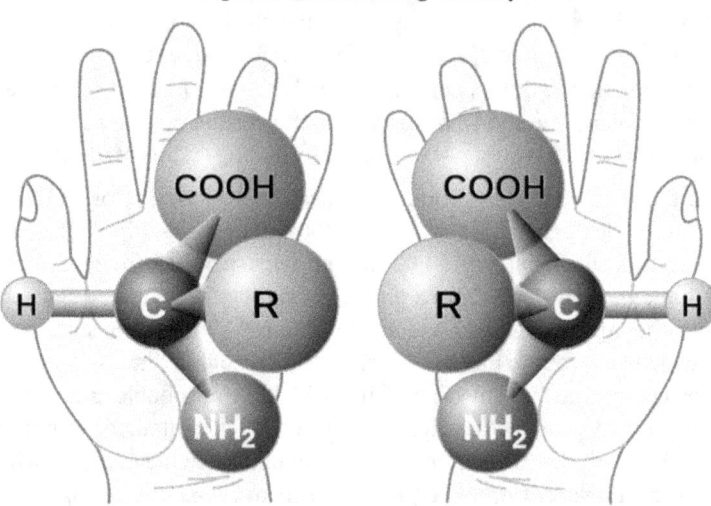

Organic chemist Charles McCombs informs us that: "the polymer chain found in natural proteins and DNA has a precise sequence that does not correlate with the individual components' reaction rates."[21] In other words, amino acid chains are not ordered by their rate of reaction. Unguided, amino acids will attach to either end of the chain randomly according to their reaction rates, but DNA assures that they line-up in a precise sequence regardless of their reaction rates. That just does not happen without information. If only the laws of physics and chemistry determined the sequence we would not be around to think about it since "the precise sequence by random chemical reactions is unthinkably unlikely."[22] The problems attending efforts to formulate a materialist account of life are so overwhelming because, as molecular chemist Steven Benner informs us: "An enormous amount of empirical data have established, as a rule, that organic systems, given energy and left to themselves, devolve to give useless complex mixtures."[23] Unguided organic reactions in a pool of chemicals form a gooey tar, a problem appropriately known as the asphalt problem. Benner lists a number of other seemingly unresolvable paradoxes that "suggest that it is impossible for any non-living chemical system to escape devolution to enter the Darwinian world of the 'living.'"[24]

The Multiverse and Panspermia

To address the absurd improbability of life emerging from non-life, some OoL scientists have taken a page from the M-theory handbook for a quick-and-easy way to get rid of fine-tuning—just posit a multiverse. Evolutionary biologist

Eugene Koonin calculates the enormous improbability for the simultaneous emergence of translation (the process by which the ribosome uses RNA as a template to make protein) and replication: "the probability that a coupled translation-replication emerges by chance in a single O-region [observable region of the universe] is P< 10^{-1018}. Obviously, this version of the breakthrough stage can be considered only in the context of a universe with an infinite (or, in the very least, extremely vast) number of O-regions."[25] Let's not overlook the fact that this is the probability of getting just replication and translation; you still have to get these functions enclosed in a cell with all its complex interdependent parts and get their functioning started. Furthermore, you still have to show how translation-replication can emerge by natural means on this planet, but never mind the messy chemistry; just concentrate on the really big number of "O-regions" and we can get back to blind chance—problem solved!

In a lecture at the Royal Institute in London, Sir Fred Hoyle agreed that the complexity of life does not lend itself to chance: "So if one proceeds directly and straightforwardly in this matter, without being deflected by a fear of incurring the wrath of scientific opinion, one arrives at the conclusion that biomaterials with their amazing measure or order *must be the outcome of intelligent design.* No other possibility I have been able to think of in pondering this issue over quite a long time seems to me to have anything like as high a possibility of being true." [my emphasis][26] Hoyle looked to the heavens as a way out of the conundrum; not to God, but to the idea of panspermia ("seeds everywhere"). Panspermia is the notion that the cosmos is teeming with life, and that it hitched to Earth on a comet, asteroid, or meteor, or else intelligent aliens sent out protected spores to fill the cosmos. In their book *Evolution from Space*, Hoyle and his colleague Chandra Wickramasinghe calculated the probability of getting the 20 standard amino acids to line up correctly, of obtaining a suitable sugar backbone for DNA and RNA, and of functioning enzymes. They combined these immense probabilities to arrive at a probability vastly smaller than any other encountered in this book: "there are about two thousand enzymes, and the chance of obtaining them all in a random trial is only one part in $(10^{20})^{2000} = 10^{40,000}$, an outrageously small probability...this simple calculation wipes the idea entirely out of court." [27]

Even if we were to accept the notion of panspermia, it does not solve the OoL puzzle; it merely moves its origin elsewhere in the vastness of space where the same $10^{40,000}$ problem remains. This number vastly exceeds Dembski's probability boundary, so even if we conceive of a multiverse with an almost infinite number of universes with perhaps multiple trillions having some potential for life, $10^{40,000}$ still wipes blind chance entirely out of court. Even Hawking's 10^{500} universes could not accommodate such a huge improbability. Ever the enigmatic thinker, Hoyle was aware that moving the origin of life elsewhere does not solve the problem of how it arose, but he noted that it required "intelligent control." In Gert Korthof's review of Hoyle's *The Intelligent Universe*, he quotes Hoyle's words: "Even after widening the stage for the origin of life from our tiny Earth to the Universe at large, we must still return to the same problem that opened this book—the vast unlikelihood that life, *even on a cosmic scale*, arose from non-living matter. It is apparent that the origin of life is overwhelmingly a matter of arrangement by intelligent control. *Unintelligent natural selection is only too likely to produce an unintelligent result*" [my emphasis].[28]

Hoyle left unanswered the nature of this intelligent controller. This is an indication of the enigmatic nature of this brilliant scientist because he has many statements in his books and articles in which we may envision him struggling with himself not to mention God while using metaphors that strongly suggest that he had God in mind (note his words in the epigraph). Recall from Chapter Five that Hoyle was led to conclude that some "super-intellect had monkeyed with the physics, as well as the chemistry and biology" when confronted with the exquisite fine-tuning of the carbon-making process in stellar nucleosynthesis. His colleague, Chandra Wickramasinghe, however, did invoke God in a way Christians understand him: "From my earliest training as a scientist, I was very strongly brainwashed to believe science cannot be consistent with any kind of deliberate creation. That notion has been painfully shed. At the moment I can't find any rational argument to knock down the view that argues for conversion to God . . . Now we realize the only logical answer to life is creation—and not accidental random shuffling."[29] The more scientists allow themselves to break free of their faith in blind materialism to ponder deep metaphysical questions, the more likely they are to come to the same conclusion as Wickramasinghe.

Endnotes

1. Świeżyński, A. 2016.
2. In Lim, R., 2017, p. 58.
3. Davies, P., 2003, p. xxiv.
4. Trevors, J. and Abel, D., 2004, p. 736
5. Dawkins, R. 1989, p, 15.
6. Wald, G., 1954, p.48.
7. Wald, G., 1984, p.1.
8. De Duve, cited in Andrews 2017, p. 248.
9. Kenyon, D., 2002, p. 35.
10. Meyer, S., 2003.
11. Rosing, M. 1999, p. 676.

12. Ward, P. and Brownlee, D., 2000, p. 62.
13. Rensselaer Polytechnic Institute, 2011.
14. Alberts, B. et al, 2015, p. 109.
15. Lane, N., Allen, J., and Martin, W. 2010, p. 272.
16. Ibid., p. 272.
17. Pross, A., 2012, p. 27.
18. Plaxco, K., and Gross, M. 2006, p. 114.
19. Ibid., p. 129.
20. Garay, A., 1993, p.168.
21. McCombs, C. 2004, p. iii.
22. Ibid., iii.
23. Benner, S., 2014, p. 341.
24. Ibid., p. 342.
25. Koonin, E., 2007, p.19.
26. In Johnson, D., 2009, p. 89.
27. Hoyle, F., & Wickramasinghe, C., 1981, pp.19-21.
28. Korthof, G., 2006, np.
29. In Seckbach, J., and Gordon, R., 2009. pp. 343-344.

CHAPTER FOURTEEN

Molecules, Membranes, and Information

> "In the presentation of a scientific problem, the other player is the good Lord. He has not only set the problem but also has devised the rules of the game – but they are not completely known, half of them are left for you to discover or to deduce."
> Erwin Schrodinger, Nobel Laureate physicist

The RNA World Hypothesis

As Schrodinger says in the epigraph of this chapter, the Lord has left mysteries for us to "discover or deduce," so despite the immense improbability of abiogenesis, science must never give up its pursuit despite apparently insurmountable difficulties. We keep trying because God does not make His creation undecipherable; He wants us to discover it, and the proof of this is how far we have come in doing so. There are two major hypotheses of abiogenesis: the RNA-world and the metabolism-first hypotheses. There is also a third alternative forcefully emerging that is conducive to theism, we might call the information first hypothesis. The RNA-world hypothesis is addressed first because it is currently the most popular model.

RNA molecules are crucial for life because they transcribe, translate, and build proteins according to instructions from DNA. Some components of RNA have been synthesized in the lab by intelligent chemists who know how to control reaction conditions, and how to select the right component in the right order for each reaction. Because chemists can do this, it is assumed that, given enough time, nature can do the same thing. Unlike chemists, nature is not intelligent and cannot think ahead, but it would require the same intelligent control, order, and selectivity, to foresee the end product. Crucially, the bits and pieces destined to end up an RNA molecule must be kept away from water because water is "inherently toxic to polymers (e.g. RNA) necessary for life."[1] Astrobiologists Neveu, Kim, and Benner inform us that: "Even the monomers of RNA have problems. In water, deamination reactions convert cytosine to uracil, adenine to hypoxanthine, and guanine to xanthine, in each case destroying information carried by the nucleobase."[2] So much for Darwin's "warm little pond" or ocean thermal vents.

DNA chemist Robert Shapiro likens the assumption that dumb nature can duplicate what clever chemists have accomplished in the lab to a golfer who had just finished 18 holes of golf. Before going home, he places the ball on the first tee assuming that it will eventually make its way back to the 18th hole driven by natural forces such as wind, rain, earthquakes, and tornadoes. It is so overwhelmingly improbable (let's be honest and say impossible) that the ball will make it to the 18th hole by natural forces that no scientist would believe that it could. If our scientist did find it resting snuggly in the 18th hole, he would naturally assume that an intelligent agent had put it in there.[3] Yet, in the horrendously complicated business of forging the first self-replicator, it is taken on faith that it must be within the realm of probability.

DNA, RNA, and proteins work as a unit. DNA stores information, RNAs reads it, and proteins do the necessary enzymatic work of catalyzing reactions. DNA requires enzymes to replicate, but enzymes are proteins that can only be synthesized by DNA; neither can exist without the other. To get this system, left-handed amino acids have to line up just right at exactly the same time and place as the right-handed nucleic acids arranged themselves. Then these molecules must hook up to form a functioning, inseparable, irreducible whole. The probability of this occurring without intelligent guidance is almost beyond calculation. Indeed, it is acknowledged by OoL researchers that the DNA-RNA-protein system is far too complex to have arrived spontaneously as a system, so they want to determine which came first. RNA-world proponents go back to the idea of a primordial soup in which free-floating nucleotides (they are not found this way in nature), all of which must be right-handed, somehow came together in just the right order to form an RNA molecule. Having assumed that it did, many biologists thought this would solve the chicken-or-egg problem since RNA can store genetic information, self-replicate, and perform the enzymatic activity of proteins; an impressive repertoire of functions.

The RNA-world has many supporters, but many detractors also. Biochemist Harold Bernhardt calls the RNA-world hypothesis "the worst theory of the early evolution of life (except for all the others)."[4] He notes that RNA is too complex to have arisen by chance, is inherently unstable, and that the best ribozyme replicase (a molecule that catalyzes its own replication) created in the lab is about 190 nucleotides in length. There is no way to know whether such a ribozyme replicase existed naturally in early environments, and 190 nucleotides is "far too long a sequence to have arisen through any conceivable process of random assembly." Bernhardt notes that it requires between 10^{14} and 10^{16} randomized RNA molecules "as a starting point for the isolation of ribozymic and/or binding activity in *in vitro* selection experiments, completely divorced from the probable prebiotic situation."[5] Biochemist Charles Kurland is impressed by the creation of synthetic ribozymes, but asks: "Why are there no examples of naturally occurring protein-free ribozymes to link the postulated RNA world to the modern cellular world?"[6] He also notes that synthetic ribozymes run at *one-millionth* of the rate of natural catalyzation.

A reviewer of Bernhardt's article wrote that such experiments and their "relationship to the prebiotic world is anything but worthy of 'unanimous support'. There are several serious problems associated with it, and I view it as little more than a popular fantasy."[7] Others have also cast aspersions on the RNA-world hypothesis: it "has been reduced by ritual abuse to something like a creationist mantra," and that the RNA world is "an expression of the infatuation of molecular biologists with base pairing in nucleic acids played out in a one-dimensional space with no reference to time or energy."[8] Time and energy are major stumbling blocks for the RNA-world hypothesis. Finally, after surveying the many difficulties with the RNA-world hypothesis, Jesse McNichol concluded in *Biochemistry and Molecular Biology Education* that: "Because of these seemingly insurmountable difficulties, the idea of a 'perfect accident' insinuates itself into to logic of RNA world proponents."[9]

Biologists Robertson and Joyce dismiss the claim that the RNA hypothesis solves the chicken-or-egg problem: "To say that the RNA World hypothesis 'solves the paradox of the chicken-and-the-egg' is correct if one means that RNA can function both as a genetic molecule and as a catalyst that promotes its own replication."[10] The gene/enzyme double-duty idea presents a problem because the two roles require contradictory properties. An enzyme, being a protein, must be reactive and fold or it is useless, while an RNA molecule carrying information must do neither because its information would become gibberish. The efficiency and fidelity of replication must also be sufficient to produce viable copies at a rate exceeding the rate of decomposition of the parent molecule, which is problematic given the inherent instability of RNA. No natural ribozyme that can catalyze such a reaction has been found, and lab-made ribozymes carry the reaction out much too slowly to keep up with the degradation of the parent molecule. This indicates the obvious: it needs a source of outside energy for catalysis.

Ignoring the destructive effect of water on RNA, it has been proposed that hydrothermal vents (black smokers) deep in the ocean serve as this energy source. Besides these destructive effects, others have noted that, "several issues relating to black smokers as sites of life's origin are problematic, among them their extreme temperature (more likely to break down organics than form them), their low pH, their short lifetimes and their lack of compartmentalization, with its dismal consequence of irretrievable dilution into the ocean."[11] Another biochemist writes: "From a chemistry perspective, the most reasonable scenario to spontaneously form stable membranes would be at temperatures below 60°C, with near neutral pH...This is in stark contrast [to hydrothermal vents, which are boiling (>100°C)]."[12] Neutral pH means acid-base balance. A fluid that is more acidic relative to its base content is harmful to organisms.

The premise of OoL research is that life is just a matter of getting the physics and chemistry right, after which biology will take over. While life must be consistent with the laws of physics and chemistry, it cannot be derived from them. Life runs on the information content of DNA, but information means nothing without an interpreter. Just as information on a DVD disk needs a DVD player to convert the tracks into images and sounds, the information contained in the genes must have the cellular machinery to transcribe messages into proteins. One without the other would be totally useless. Andrew McIntosh, a thermodynamics expert, writes of the irreducible complexity "involved in creating the DNA/mRNA/ribosome/amino acid/ protein/ DNA-polymerase connections. All of these functioning parts are needed to make the basic forms of living cells to work...It is against the known principles of thermodynamics in physics and chemistry for this to happen spontaneously."[13]

Even if we grant the existence of self-replicating RNA surrounded by the right amino acids, a protein cannot be made unaided by the necessary cellular components since RNA contains only raw information. A blueprint cannot make anything without "workers" who understand it and can assemble what it represents. These workers (enzymes, chaperones, etc.) and a protein's "consumers" (receptor proteins) are an interdependent protein system. Even simple bacteria contain hundreds of different proteins, and humans possess millions. Work done by The Human Proteome Project (HPP) illustrates the problem. The HPP is an international project with a different group of scientists responsible for each of the 22 somatic chromosomes, one each for the X and Y sex chromosomes, and one for mitochondrial DNA. Researchers examined the proteome (the complete set of proteins expressed by an organism) of 276 coding genes on chromosome 18 and found 100,000,000 blood plasma proteins and 10,000 copies per liver cell.[14] Billions of amino acids have to line up in the proper sequence just to make the proteins for blood plasma and liver cells; think of the trillions that have to line up "just right" for all our body parts. These many problems have forced some OoI scientists to abandon the RNA-world hypothesis and turn to the metabolism-first hypothesis.

Metabolism-First Hypothesis

Metabolism is the mechanism by which all living things circumvent the second law of thermodynamics by harvesting outside energy from the environment. It refers to all chemical processes that occur in living cells that enable organisms to grow and thrive. Metabolism converts food into energy to fuel cellular processes. It is an extraordinary mechanism that functions with purposeful intent. Biochemist David Abel notes that: "Metabolism is the most highly integrated, holistic, conglomerate of organized formal functions known to science." He then asks" "How did life get so organized and goal-oriented out of an inanimate prebiotic environment that could care less about function or useful work? Chance and necessity cannot pursue function, let alone such an extraordinary degree of cooperative work."[15]

For metabolism to work there must be a lipid membrane boundary between the inner cell and the outside world. We have already briefly discussed the staggering complexity of the whole cell, but now we concentrate on the complexity of its lipid/protein overcoat. It is not enough to possess a protective membrane like the simple plastic bag you brought the goldfish home in. The membrane must be sufficiently complex to allow vital elements to enter the cell, and cell products to exit. This is called "compartmentalization." Simple short-lived lipid membranes can be made in the lab, but even the smallest environmental change (the presence of organic solvents, increased salt concentration, temperature change, etc.) destroys them.[16] The importance of compartmentalization implies that the cell would have to come before metabolism, for what is the point of metabolism unless you have an organism with compartmentalized cells for it to sustain? The membrane is far from a simple sac holding together the contents of the cell. It is a double-layered lipid/protein membrane of great complexity. As Figure 14.1 shows that the membrane acts as a castle with multiple drawbridges that selectively allow the entry of resources required by the castle's residents and the exit of things needed outside the castle walls.

The metabolism-first model is animated by the problems plaguing the RNA-world model. That is, RNA is too inherently unstable and too complex to have arisen spontaneously, and that the catalytic repertoire of RNA (its enzyme activity) is too limited. To try to get around these problems, the metabolism-first hypothesis proposes that the spontaneous formation of simple molecules, such as the compound formed from carbon dioxide and water called acetate, triggered life. Such a primitive cell is assumed to have contained proteins (ignoring the difficulties of making them) that possessed a crude non-genomic replication capacity, and subsequent evolution processes led to the accumulation of organic molecules that could serve as catalysts for more complex molecules of RNA and DNA.[17] However, OoL organic chemist Addy Prost says that the metabolism-first model confronts that pesky second law of thermodynamics again: "How would metabolic cycles form spontaneously from simple molecular entities, and, more importantly, how would they maintain themselves over time? We run yet again into that thermodynamic brick wall."[18]

The simplest of living cells is so complex that OoL researcher Lynn Margulis informs us that "[The smallest bacterium] is so much more like people than Stanley Miller's mixtures of chemicals, because it already has these system properties. So to go from a bacterium to people is less of a step than to go from a mixture of amino acids to that bacterium."[19] The point is that a bacterium already has the replication and metabolic capacity to evolve into something else (even though it hasn't, since bacteria have remained bacteria since their first appearance), whereas dead unguided amino acids have no way of making the necessary genetic material. To put it plainly, life must already exist to produce all the things which are necessary for life to exist: life comes only from life. Evolutionary biologist Eugene Koonin has concluded that it would require a "minimal gene set consisting of ~250 genes" for the simplest of known bacteria to have life.[20]

Chemist, and lifelong OoL researcher Leslie Orgel, characterized the metabolism-first scenario as a kind of "if pigs could fly" chemistry based on little more than the deficiencies of the RNA-world hypothesis. He writes that "The most serious challenge to proponents of metabolic cycle theories—the problems presented by the lack of specificity of most nonenzymatic catalysts—has, in general, not been appreciated. If it has, it has been ignored. Theories of the origin of life based on metabolic cycles cannot be justified by the inadequacy of competing theories: they must stand on their own." [21]

Research by Vasas, Szathmáry, and Santos has shown that metabolic systems such as those proposed by metabolism-first proponents are unable to retain information (no genome) about their composition to allow them to evolve toward a metabolic pathway. In other words, they do not contain hereditary information by which they could pass on details of their composition to progeny. Commenting on both the RNA-world and metabolism-first scenarios, they maintain that: "Both schools acknowledge that a critical requirement for primitive evolvable systems (in the Darwinian sense) is to solve the problems of information storage and reliable information transmission."[22]

Figure 14.1: Part of a Human Cell Membrane

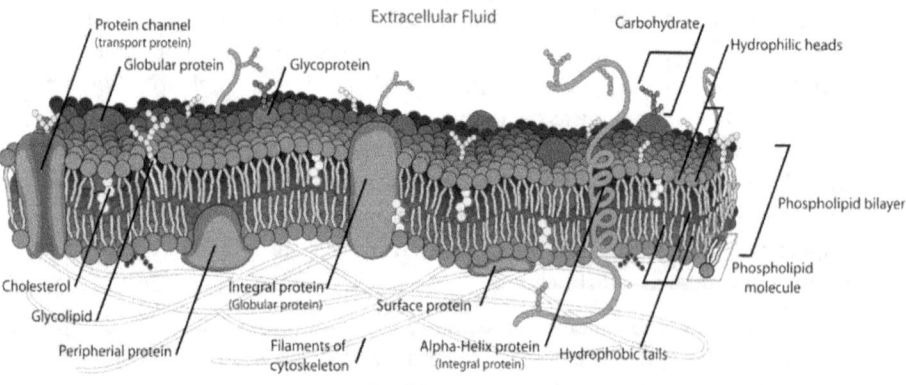

All life requires an energetic force, and thus metabolism-first researchers, like their RNA- world counterparts, suggest that life first arose in the hot vents deep in the ocean spewing out heat and various kinds of inorganic molecules. However, in addition to the problems mentioned earlier regarding the effect of water on RNA, according to a 2017 paper in the journal *Advances in Biological Chemistry*, this scenario also runs afoul of the second law of thermodynamics. The authors state that "all molecules near the heat source, the hot water, will be equivalently heated up. They will then move away to cooler parts, and whatever reactions that occurred in the hot parts will just cease and cold, unreactive (dead) products will float off and dilute."[23]

Information: The Recipe for Life

Problems such as these are why some OoL scientists now argue for top-down causation in the form of information: "the key distinction between the origin of life and other 'emergent' transitions is the onset of distributed information control, enabling context-dependent causation, where an abstract and non-physical systemic entity (algorithmic information) effectively becomes a causal agent capable of manipulating its material substrate."[24] Information holds the key to the mystery of the nature and origin of life, argue astrobiologist Sara Imari Walker and astrophysicist Paul Davies: "Although it is notoriously hard to identify precisely what makes life so distinctive and remarkable there is general agreement that its informational aspect is one key property, and perhaps the key property. The manner in which information flows through and between cells and sub-cellular structures is quite unlike anything else observed in nature."[25] Walker and Davies redefine the OoL problem by shifting from a chemical "hardware" point of view to a "software" information point of view.

Walker and Davies view life as emerging in a phase transition from a bottom-up reductionist chemical process to one of top-down information flow and management: "The origin of life may thus be identified when information gains top-down causal efficacy over the matter that instantiates it."[26] Information is non-material, so how can it affect the material? Timo Henny provides a clue: "Davies claims that life's defining characteristics are better understood in terms of information. This is not as absurd as it may seem. Energy is abstract, yet we have little trouble accepting it as a causal factor."[27] Although energy is defined as the ability of one physical system to do work on another; we *infer* its existence by its effects, but we do not know what it really *is*. Information is like this; it transfers knowledge of what to do from one system to another, and thus we can define it and affirm its existence and usefulness without being able to say categorically what it is. Information is always created by intelligence, and only a mind of infinite wisdom could create the information necessary for life.

In another much-cited later paper, Walker and Davies appear to agree with this. They developed a model in which they talk of the fine-tuning of information and note that if the pathway from chemistry to life is the result of "fixed dynamical laws, then (our analysis suggests) those laws must be selected with extraordinary care and precision, which is tantamount to intelligent design: it states that 'life' is 'written into' the laws of physics *ab initio* ["from the beginning"]. There is no evidence at all that the actual known laws of physics possess this almost miraculous property."[28]

All molecules in an organism are in an information-rich cooperative relationship with all other molecules by sending and receiving information on which they must act or else break down. Information transfer can only occur when both sender and receiver are intelligent enough to know what the information entails. Take the marvelous machinery of the Krebs cycle. The Krebs cycle is the cellular respiration system by which glucose is broken down in the presence of oxygen to produce cellular energy. Every movement we make and breath we take induces a series of complicated chemical reactions involving electrons changing a series of enzymatic molecules into others. If you have seen a schematic image of the Krebs cycle you will appreciate its marvelous complexity. It is impossible to imagine how this system could have been cobbled together higgledy-piggledy by molecular tinkering, for from where did animals get the energy to move and breathe before it came online? No human bioengineer given an eternity of time could produce such a fine-tuned system.[29]

Each stage in the Krebs cycle involves messages that command "do this!" While the carriers of information are material (transformative chemical reactions in the Krebs cycle or the paper and ink of a book), the information (its meaning) itself is not. Gitt, Compton, and Fernandez provide us with a definition of what they call "universal information:" "Universal Information (UI) is a symbolically encoded, abstractly represented message conveying the expected action(s) and the intended purpose(s). In this context, 'message' is meant to include instructions for carrying out a specific task or eliciting a specific response."[30] They apply this definition of information to the DNA/RNA/protein synthesizing system in hierarchical fashion. They begin with *cosyntics*, which is a set of recognizable abstract grammatical symbols that have been put together as words to construct a syntactically correct sentence. The DNA code is a set of abstract symbols (the base letters: AGCT) with syntactic rules, but sets of letters do not by themselves necessarily convey meaning. Meaning is determined by the natural language of a system. For instance, physicist Hubert Yockey points out that the phrase: "'O singe fort' has no meaning in English, although each is an English word, yet in German it means 'O sing on,' and in French it means 'O strong monkey.'"[31]

To mean something in the language of DNA, the sequence of nucleotides must have specificity for a meaningful message. This leads to the next stage in the information hierarchy: *semantics*, which means that the sequence of letters must form a sentence containing meaning. The RNA codons are instructions for specific amino acids and their specific sequence. The next stage is *pragmatics*, which means that the message requires a reply on the part of the receiver. When ribosomes read the instructions from mRNA, they respond by transporting specific amino acids to the site where specified protein is constructed. The final point of the information hierarchy is *apobetics*, which is the result or purpose of the information. According to Gitt, Compton, and Fernandez, the DNA/RNA protein synthesizing system achieves the highest of purposes: "A living, functioning organism is achieved at the highest level of purpose."[32]

Commenting on the information first hypothesis, John Lennox states: "This proposal, that information be regarded as a fundamental quantity, has profound implications for our understanding of the universe. But it is not new, it has been around for centuries. 'In the beginning was the Word'...all were made by Him."[33] Werner Gitt reaches the same conclusion when he observes that the question of "'How did life originate?" is inextricably linked to the question: "Where did the information contained in all those base sequences in the genetic code come from? Anybody who wants to make meaningful statements about the origin of life would be forced to explain how the information originated. All materialist views are fundamentally unable to answer this crucial question."[34] The amount of information in DNA is so truly astounding that an article in *Science* called it "the ultimate information storage device." The article further states that: "Our genetic code packs billions of gigabytes [one gigabyte equals nine billion bytes, which are units of data consisting of eight binary digits, or bits] into a single gram. A mere milligram of the molecule could encode the complete text of every book in the Library of Congress and have plenty of room to spare."[35]

In a sophisticated book-length argument, Gitt argued the impossibility of information arising from dead matter. This was the conclusion of multiple OoL scientists at the International Conference of the Origins of Life held in Germany. Gitt put this in his Theorem 28: "There is no known law of nature, no known process, and no known sequence of events which can cause information to originate by itself in matter."[36] He later identifies the author of this sophisticated information content as God.[37] Gitt is too much of a scientist to offer this as a God-of-the-gaps argument. He is simply saying that whatever mechanistic process science might uncover, the vital information ruling them is abstract and non-mechanistic, and has been provided by an intelligent agent. Ernst Chain, Nobel Laureate in medicine and physiology, notes that behind the intellectualized OoL speculation is the desire to explain God away: "I have said for years that speculations about the origin of life lead to no useful purpose as even the simplest living system is far too complex to be understood in terms of the extremely primitive chemistry scientists have used in their attempts to explain the unexplainable that happened billions of years ago. God cannot be explained away by such naïve thoughts." [38]

What can we conclude about OoL research thus far? There was considerable optimism after the 1953 Miller-Urey experiment that it would be relatively easy to kick-start life, but optimism has slowly faded to pessimism. Even Urey admitted that while he believes in abiogenesis, he does not do so by dint of evidence, but by faith. He remarks: "All of us who study the origin of life find that the more we look into it, the more we feel that it is too complex to have evolved anywhere. But we believe *as an article of faith* that life evolved from dead matter on this planet. It is just that its complexity is so great, that it is hard for us to imagine that it did" [my emphasis].[39] We have gained an immense amount of chemical and biological knowledge in the process of OoL research, and that is a big plus, but OoL theories continue to checkmate one another. It is fair to say that the millions of hours that have been spent in experimentation and calculation since the Miller-Urey experiment have resulted in a clearer understanding of the immensity of the problem rather than its solution. OoL theories are confronted with numerous chicken-or-egg problems, and no sooner than one research team thinks they have solved some part of the problem another team comes along and checkmates them. Others have resorted to "multiverse-of-the- gaps" arguments to explain the impossibility of life's naturalistic origin on Earth. This does not mean that scientists must stop trying to discover the rules of science by which Hoyle's "super-intellect" created all. Science must look for natural explanations regardless of an individual scientist's religious convictions; it cannot stop and conclude that God did it. We know that He did by "devising the rules of the game," as Erwin Schrodinger said in the epigraph to this chapter, but he left it up to science "to discover or to deduce" how He did it. Perhaps by pursuing their theories more of them will come to appreciate God's handiwork and join with the many physicists who have been dragged by their science to God.

Endnotes

1. Benner, S. 2014, p. 342.
2. Neveu, M., Kim, H., and Benner, S. 2013, p. 394.
3. Shapiro, R. 2007, np.
4. Bernhardt, H. 2012, p. 1.
5. Ibid., p. 7.

6. Kurland, C., 2010, p. 866.
7. Bernhardt, H. 2012, p. 7.
8. Kurland, C., 2010, p. 870.
9. McNichol, J., 2008, p. 257.
10. Robertson, M., and Joyce, G., 2012. p. 7
11. Lane, N., Allen, J., and Martin W., 2010, p.273.
12. Glasco, D. 2016, p. 219.
13. McIntosh, A. 2009, p. 370.
14. Ponomarenko, E., Poverennaya, E., Ilgisonis, E. et al, 2016.
15. Abel, D. 2011, p. 123.
16. Benner, S. 2014.
17. Gupta, N., Agogino, A., & Tumer, K. 2006.
18. Pross, A. 2012, p. 107.
19. In Horgan, J. 2012.
20. Koonin, E. 2000, p. 99.
21. Orgel, L., 2008, p. 0012.
22. Vasas, V., Szathmáry, E., and Santos, M., 2010, p. 1470.
23. Busby, C., and Howard, C., 2017, p. 172.
24. Walker, S., and Davies, P. 2013, p. 7.
25. Ibid., p. 1.
26. Ibid., pp. 5-6.
27. Hannay, T. 2019, p. 428.
28. Walker, S., and Davies, P. 2016, p. 8.
29. de Castro Fonseca, M., et al, 2016.
30. Gitt, W., Compton, B. and Fernandez, J., 2011, p. 70.
31. Yockey, H. 2005, p. 6.
32. Gitt, W., Compton, B. and Fernandez, J., 2011, p.169.
33. Lennox, J. 2009, p.177.
34. Gitt, W., 2006, p.99.
35. Bohannon, J. 2012.
36. Gitt, W. 2006, p.106.
37. Ibid., p. 99.
38. In Clark, R. 1985, pp. 147-148.
39. Persaud, C. 2007, p. 84.

CHAPTER FIFTEEN
Cracks in Neo-Darwinism: Micro is not Macro

> "There is a Divine Providence over and above the materialistic happenings of biological evolution."
> John Eccles,
> Nobel Laureate in physiology and medicine

Darwin's Doubters

Evolutionary scientist Scott Gilbert tells us that "The modern synthesis [of Darwinism and genetics] is remarkably good at modeling the survival of the fittest, but not at modeling the arrival of the fittest."[1] This alerts us to the fact that we are moving from the confused science of abiotic "*arrival* of the fittest" to the "remarkably good" and supposed settled science of "the *survival* of the fittest." Some members of a species may be "fitter" than others, but that tells us nothing about how they arrived on the scene to take part in the evolutionary competition. Many take it for granted that the theory of evolution is a settled scientific law of the kind we see in physics, but this is far from true. Paleoanthropologist Bernard Wood states that the theory is so widely accepted because there are images everywhere, from cereal boxes to textbooks, showing a straight-line progression from fish, to retile, to ape, and on to a human striding gracefully into the future: "Our progress from ape to human looks so smooth, so tidy. It's such a beguiling image that even the experts are loath to let it go. But it is an illusion."[2]

Fred Hoyle also had serious problems with "remarkably good" Darwinism. In his book, *Mathematics of Evolution*, he writes in his own inimitable way that Darwinists have replaced God, who produced rabbits in ways too mysterious understand, to believe "that rabbits had been created by sludge, by methods too complex for us to calculate and by methods likely enough involving improbable happenings. Improbable happenings replace miracles and sludge replaced God."[3] Hoyle explains in mathematical terms why so many Darwinian claims are outside the realm of probability. He does not deny that small-scale changes within a species occur (microevolution); his argument is with macroevolution. He argues that species adapt only within narrow limits. That is, they produce variation only within their kind; rabbits cannot become rhinos, even at their prodigious reproduction rate. Hoyle concludes: "The mistaken extrapolation from evolution in the small to evolution in the large that followed the Darwinian theory of 1859 led society into a bog which has only grown deeper in the passing years."[4]

Another dissenter, Nobel Laureate physicist Robert Laughlin, writes about evolution in ways that suggest "evolution-of-the-gaps" arguments abound in biology. He notes that the theory of macroevolution is ideological because it cannot be tested and because it tends to prevent thinking rather than stimulating it. He calls it an "anti-theory" used both to paint over embarrassing findings and to legitimize questionable ones: "Your protein defies the laws of mass action? Evolution did it! Your complicated mess of chemical reactions turns into a chicken? Evolution! The human brain works on logical principles no computer can emulate? Evolution is the cause!"[5] Then we have Richard Lewontin, who describes Darwin's theory of natural selection as "hopelessly metaphysical," and asks "For what good is a theory that is guaranteed by its internal logical structure to agree with all conceivable observations, irrespective of the real structure of the world?"[6]

How do we reconcile these arguments with Theodosius Dobzhansky's oft-quoted statement that "Nothing in biology makes sense except in the light of evolution."?[7] Dobzhansky was one of the true giants of 20th-century biology, and both a Christian and an evolutionist. He believed that science does not preclude evolution having an author or an ultimate goal: an Alpha and Omega. Hoyle, Laughlin, and Lewontin, were right, but so was Dobzhansky, because they had different versions of evolution in mind. Lewontin's claim does not mean that he doubts microevolution because, as he says, one cannot imagine any observation that would disprove natural selection as a cause of changes in organisms within their kind. He only views the theory of natural selection as metaphysical rather than scientific because "Natural selection explains nothing because it explains everything [about microevolution]."[8] Dobzhansky's view was also limited to the kind of factual evolution scientists are able to observe and study, but Darwinism claims more than it can demonstrate when venturing beyond this level.

If you believe that Darwinism is disputed only by religious fundamentalists, you will have to account for the more than 1,000 doctoral-level scientists, mostly biologists, who signed a statement expressing their skepticism. The *Scientific Dissent from Darwinism* statement reads: "We are skeptical of claims for the ability of random mutation and natural selection to account for the complexity of life. Careful examination of the evidence for Darwinian theory should be encouraged." Among the comments of the signatories is that of mathematician Colin Reeves, who wrote that Darwinism may have been plausible in the 19th century, but "what we have learned since the days of Darwin throws doubt on natural selection's ability to create complex biological systems—and we still have little more than hand waving as an argument in its favor."[9] Of course, a lot more scientists would sign a statement affirming their belief in Darwinism, but headcounts do not settle scientific issues. I mention the *Dissent* only to show that Darwinism in not in the same league as theories such as the laws of thermodynamics, the atomic theory of matter, or the germ theory of disease.

I have written articles and book chapters on evolutionary theory myself without giving serious thought to its problems. Not being involved in basic evolutionary research, I relied on mainstream evolutionists for my information. This is how we all accept science with which we are not intimately acquainted. We defer to specialists, but go to other specialists for second opinions if we have doubts. Because Darwin doubters are seen as antiscientific nincompoops, biologists with doubts may be reluctant to voice them. Given this, when I began to have doubts about macroevolution, I went back to the source and reread relevant parts of Darwin's *The Origin of Species*.

Charles Darwin: Atheism, First Cause, and Teleology

What about Darwin himself, the man Richard Dawkins credits with making it intellectually respectable to be an atheist? Although Darwin was embittered by the loss of his nine-year-old daughter, Annie, he never called himself an atheist. He did call himself a theist, however, as in the following passage, written some 33 years after the publication of *The Origin of Species*:

> Another source of conviction in the existence of God, connected with the reason and not with the feelings, impresses me as having much more weight. This follows from the extreme difficulty or rather impossibility of conceiving this immense and wonderful universe, including man with his capacity of looking far backwards and far into futurity, as the result of blind chance or necessity. When thus reflecting I feel compelled to look to a First Cause having an intelligent mind in some degree analogous to that of man; and I deserve to be called a Theist.[10]

In an 1879 letter to John Fordyce, Darwin preempted Dawkins' claim that belief in evolution logically leads to atheism: "It seems to me absurd to doubt that a man may be an ardent Theist & an evolutionist." Darwin's friend, Asa Gray, a Harvard botanist, like Dobzhansky, was both an evolutionist in his belief that species exhibit "descent with modification," and a theist who affirmed that God is the ultimate Creator. Darwin reiterated in his letter to Fordyce that he was no atheist, although he fluctuated in his level of belief: "In my most extreme fluctuations I have never been an atheist in the sense of denying the existence of a God."[11] He never warmed back up to Christianity, however, and denied the divine revelation of the Bible, and so his beliefs may be characterized as deistic, despite his assertion that he "deserve(s) to be called a Theist."

According to biologist Stephen Freeland: "He [Darwin] did not reject the idea that the laws of nature (including natural selection) stemmed from an Ultimate Cause, nor did he deny that natural selection could lead predictably to sentient or humans; he simply denied that the pool of variation on which natural selection worked was directly manipulated by a higher hand"[12] Darwin wrote in *The Origin of Species* that: "To my mind it accords better with what we know of the laws impressed on matter by the Creator, that the production and extinction of the past and present inhabitants of the world should have been due to secondary causes, like those determining the birth and death of the individual."[13]

Darwin's "secondary causes" recalls St. Augustine's words in *De Genesi ad Literam* (V.4:11; my emphasis): "It is, therefore, *causally* that Scripture has said that earth brought forth the crops and trees, in the sense that *it received the power of bringing them forth*. In the earth from the beginning, in what I might call the roots of time, God created what was to be in times to come." Augustine maintained that the natural properties of the earth that make crops and trees possible are secondary to the primary cause, immanent in the laws of nature from the beginning of the universe. There is little difference between Augustine and Darwin on this. Both talk about primary laws "impressed on matter by the Creator," and both recognize that secondary causes can change matter "in times to come."

Dawkins views natural selection as a process devoid of purpose: "Natural selection, the blind, unconscious, automatic process...which we now know is the explanation for the existence and apparently purposeful form of life, has no purpose in mind. It has no mind and no mind's eye. It does not plan for the future. It has no vision, no foresight, no sight at all."[14] Dawkins is at odds here with Darwin, who did see a purpose and a goal in evolution. In the penultimate page of *The Origin of Species*, Darwin wrote: "Hence we may look with some confidence to a secure future of equal inappreciable length. And as natural selection solely by and for the good of each being, all corporeal and mental endowments will tend to progress towards perfection."[15] Although I have never seen him mention it, Dawkins surely knows that Darwin was a teleologist; a believer in an end, purpose, or goal. In a letter written to T. H. Farrer, Darwin wrote: "If we consider the whole universe, the mind refuses to look at it as the outcome of chance—that is, without design or purpose."[16] Furthermore, throughout his notes, articles, and books Darwin used the terms "Final Cause" consistently to refer to God as the ultimate explanation of everything.

Evolution by Natural Selection

Microevolution—what Hoyle described as common-sense evolution—is doubted by no one. It is an established fact of critical importance in many fields such as medicine, where mutations in bacteria result in their resistance to drugs. Macroevolution, on the other hand, is doubted, and even categorically denied, by a number of first-rate scientists. As Hoyle noted, the problem emerges when we extrapolate the fact of small-scale evolution to explain large-scale evolution. Microevolution is defined as "variation within prescribed limits of complexity, quantitative variation of already existing organisms," and macroevolution as "large-scale innovation, the coming into existence of new organs, structures, of qualitatively new genetic material."[17] Darwinists aver that the accumulation of small *quantitative* changes of microevolution *within a species*, eventually results in the large *qualitative* changes of macroevolution and a totally new species. Through this process of small accumulations, a small shrew-like creature is said to have become the ancestor of all mammals, including us. This was supposed to have happened quickly (in evolutionary terms) after the extinction of the dinosaurs some 65 to 66 million years ago, making it easier for mammals to survive and reproduce in large numbers.[18]

Darwin's insight was that populations of organisms grow until they strain the ability of the environment to support all members. The production of excess offspring results in a struggle for existence in which only the fittest survive. In evolutionary biology, "fitness" refers to the potential (a probability; not a guarantee) of a particular genotype for leaving more offspring in subsequent generations relative to other genotypes. Darwin noted that there is a considerable degree of variation among organisms within species with respect to phenotypes (differences in disease resistance, aggressiveness, color, size, speed, cunning, etc.), some of which confer advantages in terms of survival and reproductive success. The precedents of natural selection are: (1) phenotypic trait variation in a breeding population with (2) consistent fitness differences between phenotypes, and (3) heritability of phenotypic trait(s). Variants of a trait sometimes give their possessors an edge in the struggle for resources and mates in prevailing environmental conditions. The result of this process is a change in trait frequency distributions across generations. The edge, whatever it may be, meant that those organisms possessing it would be more likely than those not possessing it to survive and reproduce, thus passing on the genetic edge to future generations.

The arrival of a new advantageous trait is the result of genetic mutations—changes in the DNA sequence of a gene caused by errors in DNA replication the cell's repair mechanism failed to catch. Most mutations are neutral, but many others are deleterious in that they reduce fitness by increasing susceptibility to a variety of disorders. On rare occasions, a beneficial mutation arises that increases an organism's fitness, and if it is sufficiently advantageous it will go to fixation.

Fixation occurs when an advantageous mutant allele completely replaces all other alleles contributing to a given trait after a certain number of generations. Skin color and lactose tolerance are examples of fixated alleles among Northern European populations. People who moved out of Africa into Europe about 40,000 years ago brought their dark skin with them, which is advantageous in sunny latitudes, but not in northern latitudes. Dark-skinned people in northern latitudes received insufficient sunlight for them to synthesize vitamin D, so natural selection favored the evolution of two genetic solutions; pale skin that absorbs sunlight more efficiently, and lactose tolerance which enabled them to obtain vitamin D in milk products.[19] Darwin called this type of process natural selection, because it is nature (the environment existing at the time) that "selects" the favorable variants and preserves them in later generations. This example is, of course, only of natural selection "within kind."

Icons of Evolution Debunked: Junk Genes

A long-standing icon of evolution is the existence of so-called "junk DNA" in the human genome. Junk DNA was so named because about 98 percent of the human genome does not code for proteins, and was taken as support for evolution as a random, purposeless, process. Non-coding regions of the DNA were seen as genetic fossils that once had a function but now don't. This argued against an intelligent creator since no designer worth his salt would leave such garbage floating around doing nothing in his creation. The decade-long project called the Encyclopedia of DNA Elements (ENCODE), a consortium of 442 researchers from 32 different institutions around the world, ended such talk for most biologists. After five years of lab work and the equivalent of 300 years of computer time, ENCODE found that 80 percent of the human genome serves some function in regulating when, how, and where a gene is activated, with the promise of new technology finding function for the remainder. Ewan Birney, the lead researcher of ENCODE, notes that: "By carefully piecing together a simply staggering variety of data, we've shown that the human genome is simply alive with switches, turning our genes on and off and controlling when and where proteins are produced."[20]

ENCODE has identified about 10,000 stretches of DNA containing non-coding genes that make a variety of RNA molecules that regulate the actions of protein-coding genes. *Scientific American*'s Stephen Hall remarks: "The ENCODE project has revealed a landscape that is absolutely teeming with important genetic elements—a landscape that used to be dismissed as 'junk DNA.'"[21] This was not good news for evolutionists, and a donnybrook ensued between supporters and detractors of ENCODE. In a 2013 lecture, Dan Graur, a virulent critic of ENCODE because of its implications for Darwinism, said that if our genomes are devoid of junk DNA: "then a long, undirected evolutionary process, cannot explain the human genome. If, on the other hand, organisms are designed, then all DNA, or as much as possible, is expected to exhibit function. If ENCODE is right, then Evolution is wrong."[21] Molecular biologist Jonathan Wells comments on Graur's attitude: "In other words: 'If ENCODE is right, then evolution is wrong.' But for Graur, evolution *can't* be wrong. His solution to the problem? Kill ENCODE"[22]

Geneticist Nessa Carey pours scorn on the practice of calling DNA "junk" merely because it is not responsible for protein-coding. She asks us to imagine visiting a car factory in which only two people (the 2% protein-coding genes) were involved in building cars and another 98 percent (the non-coding DNA) sitting around idly doing nothing. No one would think that only these two people were needed to run the factory, and she asks why we think that it is not ridiculous to think this way about the genome. Cars are the endpoint of the factory as proteins are the endpoints of the genome; neither could be produced without the "junk." Carey says that while two people can build a car, just as two-percent of the genome can build proteins, they cannot sustain the whole process alone: "and certainly can't turn it into a powerful and financially successful brand. Similarly, there's no point having 98 people mopping the floors and staffing the showrooms if there's nothing to sell. The whole organization only works when all the components are in place. And so it is with our genomes."[24] The genome is irreducibly complex and requires all its components for the system to function.

One kind of DNA "junk" evolutionists identified as being antithetical to supernatural creation of the genome are transposons, or "jumping genes." Transposons make up about a quarter of the human genome and are short sections of DNA that "jump" around and insert themselves into new DNA sites. Jumping genes used to be considered worse than junk. They were considered parasites because they apparently jump around randomly playing havoc, but new research finds that they are actually critical regulators of the first stages of embryonic development when cells are dividing vigorously. In fact, a key transposon was found to be an enhancer of a gene called SOX9, which is critical for male sex development from an inherent female form (true of all mammals) such that deleting it results in a chromosomal XY male developing a female form with ovaries instead of testes.[25]

Transposons can have deleterious effects if not controlled, however. Biologists have found that a protein called Serrate, which functions in the processing of RNA, is involved in the regulation of transposons. Plant biologist Zeyang Ma marvels at the fine-tuning of what he calls the "beautiful and elegant natural design" of the balancing act between Serrate and transposons. Noting that transposons have positive functions but can be harmful if their expression is too high, he writes: "Transposon genes must be tightly controlled by balanced forces to allow low but essential expression. I'm still very much impressed by the beautiful and elegant natural design that the plant uses a single protein to fine-tune the gene expression level."[26]

Evolutionists may be shocked and disappointed with the unexpected revelation that "junk genes" have a function, but a recent massive study noted the functional importance of the non-coding regions of the genome and concluded that without this "junk" the genome would be useless.[27] This was noted in 1998—before the Human Genome Project was completed and before ENCODE—by intelligent design proponent William Dembski, who wrote: "On an evolutionary view we expect a lot of useless DNA. If, on the other hand, organisms are designed, we expect DNA, as much as possible, to exhibit function."[28] Dembski made this prediction on the basis of simple design logic; that is, any information processing system such as the genome cannot function if it contains useless parts. Imagine if your computer consisted of a bunch of useless parts just jumping around randomly. Dembski's is another successful anthropic prediction.

Epigenetics

The human genome is often referred to as a "blueprint," which connotes a simple one-to-one mapping from the design to the finished product. It conjures up the image of a predetermined form waiting only for a developmental process to make it apparent, like a negative in a darkroom waiting to become a photograph. But people are not machines that are assembled according to instructions on a blueprint; rather, they are organic beings that develop gradually over time, since a fixed genome would have no way of anticipating the future demands of organisms. The genome has the ability to generate emergent properties and complex features without having to go back to the DNA sequencing drawing board. The developmental circumstances a person encounters affect the functioning of the genome by altering the process of genetic transcription and by wiring and rewiring its brain in ways only God Almighty could predict. Thus, there is a new focus on gene regulation in response to environmental challenges that goes beyond the regulation inherent in DNA sequences. The science of epigenetics (the prefix "epi" means "above" or "beyond") explores these regulatory processes.

Epigenetics is defined as "any process that alters gene activity without changing the DNA sequence."[29] It can be viewed as providing the software by which organisms respond genetically to their environments without having to change the DNA hardware. Epigenetic modification of DNA affects the ability of the DNA code to be read and translated into proteins by making the code either more accessible or less accessible.[30] DNA is like an author of a book who provides information to a publisher; epigenetics is like a book editor that may edit and modify the words for greater clarity when warranted. Epigenetics plays a huge part in the biology of all organisms. It, and epistasis (gene interactions whereby the effect of one gene is dependent on the presence of one or more others than "modify" it), are the reasons that while we may have fewer genes than worms or tomatoes, we are immeasurably more complex. In embryonic development, epigenetics is a mechanism by which our trillions of cells containing identical DNA express different parts to become any of hundreds of different cell types.

The difference between genetics and epigenetics is like the difference between writing a book and reading it. The text is stored information that is the same in all the copies, but readers may interpret the story a bit differently based on their temperament, personality, intelligence, and past experiences. Likewise, epigenetics allows for different interpretations of the DNA text according to the internal and external conditions under which it is read. We may also think of the genome as an orchestra, with genetic polymorphisms capable of producing a variety of music and epigenetics conducting the performance: "The epigenetic 'conductor' controls when and what 'instruments' (genes) are to be activated and when they are to be silenced, and when they are activated, the gusto with which they may be played and what other instruments will accompany, augment, and modify the 'music' they make."[31]

Figure 15.1: Methylation and Acetylation (Histone Modification)

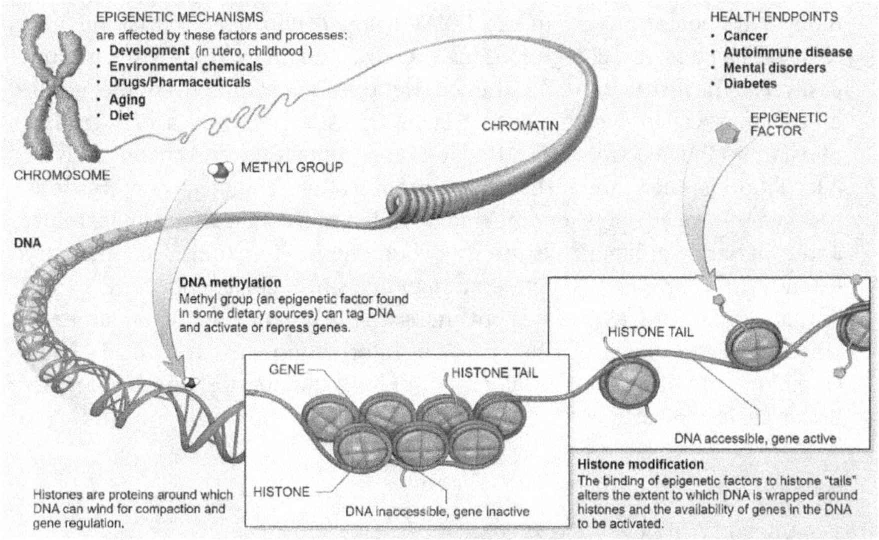

EPIGENETIC MECHANISMS
are affected by these factors and processes:
• Development (in utero, childhood)
• Environmental chemicals
• Drugs/Pharmaceuticals
• Aging
• Diet

HEALTH ENDPOINTS
• Cancer
• Autoimmune disease
• Mental disorders
• Diabetes

CHROMOSOME
METHYL GROUP
CHROMATIN
EPIGENETIC FACTOR
DNA

DNA methylation
Methyl group (an epigenetic factor found in some dietary sources) can tag DNA and activate or repress genes.

GENE
HISTONE TAIL
HISTONE TAIL
DNA accessible, gene active

Histones are proteins around which DNA can wind for compaction and gene regulation.
HISTONE
DNA inaccessible, gene inactive

Histone modification
The binding of epigenetic factors to histone "tails" alters the extent to which DNA is wrapped around histones and the availability of genes in the DNA to be activated.

Epigenetic regulation of genetic activity is accomplished by two main processes: DNA methylation and histone acetylation. Methylation may be permanent or semi-permanent (it is retained during cell replication and may be passed on to future generations) while acetylation is transient. DNA methylation occurs when an enzyme called DNA methyltransferase attaches a methyl group of atoms to a cytosine base. The initial process of reading the DNA code is carried out in the cell nucleus by the enzyme RNA polymerase (RNAP). When a signal is received to manufacture a protein, RNAP runs along the DNA strand "reading" the recipe for that protein, but if RNAP runs into the methyl group it can go no further, thus the code is not read—no transcription order, no protein. This is illustrated in the main portion of Figure 15.1.

An example of methylation is the switch from thymine in DNA to uracil in RNA. Thymine is actually methylated uracil. Methylated uracil allows for the effective recognition and repair of potentially harmful mutations (cytosine can mutate to uracil), and which increases the stability and efficiency of DNA replication. It protects the DNA by shielding it from enzymes that can break it down, and defends it from invading bacteria and viruses. RNA does not need much protection because it is shorter-lived that DNA and produced in greater quantities, making it less likely to be harmed by any "bogus" uracil mutated from cytosine. RNA thus uses the energetically cheaper "legitimate" uracil (de-methylated thymine) to do its work.

The inserts in Figure 15.1 describe acetylation, or histone modification. Acetylation involves enzymes called histone acetyltransferaces (HATs) transferring an acetyl group of atoms that bind to the amino acid lysine tail of the histone (the protein cores around which the DNA is tightly wrapped). This reduces lysine's attraction to the DNA's charged phosphate backbone. The reduced electrostatic charge loosens the DNA, enabling the RNAP to more easily read the instructions.[32] At the completion of transcription, another group of enzymes called histone deacetylases (HDACs) removes acetyl groups, reinstating the electrostatic attraction and repressing chromatin activity. Acetylation is, therefore, a short-term modification. The take-away lesson is that genes have an epigenetic "memory" in the sense that what you do, think, and experience in life affects the way your genome functions. This memory extends to your parents and grandparents since epigenetic markers are heritable. So once again we are converging on the notion that the fundamental currency of the universe is information. Information written by God's hand assembles and directs the development of both the physical universe and its biological inhabitants.

Endnotes

1. Cited in Whitfield, 2008, p. 282.
2. Wood, B. 2002, p. 44
3. Hoyle, F. 1999, p. 3.
4. Ibid., p. 139.

5. Laughlin, T., 2005, pp. 168-169.
6. Lewontin, R., 1972, p.18.
7. Dobzhansky, T., 1973, p. 125.
8. Lewontin, R., 1972, p.181.
9. Colin Reeves' comment made with in the *Dissent from Darwinism Statement*, np.
10. Darwin, C., 1892, p. 61.
11. Darwin, C., 1879, np.
12. Freeland, S., 2008, p. 290.
13. Darwin, C., 1982, p. 458.
14. Dawkins, R., 2006, p. 9.
15. Darwin, C. 1903, p.395.
16. Darwin, C., 1982, p.459.
17. Lennox, J., 2010, pp. 101-102.
18. O'Leary, M., et al, 2013.
19. Beleza, S., et al, 2012.
20. National Institute of Heath. 2012.
21. Hall, S. 2012.
22. Graur, D. 2013.
23. Wells, J. 2017, p.130.
24. Carey, N. 2015, p. 3.
25. Gonen, N. et al. 2018.
26. In Physics.Org. 2018.
27. Bartha, I. et al, 2018.
28. In Meyer, S. 2009, p. 407.
29. Weinhold, B. 2006, p. 163.
30. Gottleib, G. 2007.
31. Walsh, A. 2009, p. 51.
32. Renthal W. and Nestler, E. 2009.

CHAPTER SIXTEEN

The Problems of Information, Devolution, and Time

> "One of the reasons I started taking this anti-evolutionary view, was that it struck me that I had been working on this stuff for twenty years and there was not one thing I knew about it."
> Colin Patterson: Senior Paleontologist,
> British Museum of Natural History.

Necessity and Information

Chance is ruled out as an explanation of macroevolution, but what about necessity, the other half of the interplay between chance and necessity in which Darwinists place their faith? Natural selection, the process that generates order in the genome by preserving advantageous alleles and eliminating harmful ones, is considered the necessity half. Without this winnowing process, mutation would yield only disorganization and extinction because of the many harmful mutations. However, natural selection preserves; it does not innovate. It is not a force like electromagnetism in the sense that it acts on something to produce an effect. It does not induce genetic variation; it *reacts* to it by preserving any favorable mutations that happen to arise. Natural selection is thus a *consequence* of the Darwinian "struggle for survival," not its cause.

Darwinists want more from necessity than mere reaction to random events, however. The term "necessity" implies that something is predestined and could not be otherwise, such as the fact that apples always fall downward from the tree. Evolutionists claim that life was predestined by complicated physics and chemistry resulting in the organic emerging from the inorganic. However, evolutionary biologist George Williams notes that the problem with evoking physicochemical necessity is that: "Evolutionary biologists have failed to realize that they work with two more or less incommensurable domains: that of information and that of matter... The gene is a package of information, not an object. The pattern of base pairs in a DNA molecule specifies the gene. But the DNA molecule is the medium, it's not the message."[1] For instance, the three-letter code for the amino acid arginine is AGC, but AGC is not arginine; it is the information required to make it. There are no laws of physics or chemistry that say those three letters must code for that particular amino acid. Similarly, physicist Vincent Bauchau says: "The sequence on a string of DNA is not determined by the laws that govern the physical and chemical properties of DNA. If it was so, the string could not contain any information. For DNA to work as the carrier of genetic information, it was necessary that this molecule acquire the capability to change its sequence arbitrarily...there is nothing from chemistry or physics that can be used to derive the function of DNA. This function is irreducible."[2]

Physicist Hubert Yockey informs us why many principles of biology are not reducible to the laws of physics and chemistry. Even to construct the simplest organism requires more genetic information "than the information content of these laws. The existence of a genome and the genetic code divides living organisms from nonliving matter. There is nothing in the physico-chemical world that remotely resembles reactions being determined by a sequence and codes between sequences."[3] Thus, we see two remarkable things about the DNA code: (1) It has the ability to change itself in response to environmental conditions, and (2) The information content in the genome is far greater than the information content of physicochemical laws. What is *not* remarkable about this ingenious code is to say that it had an intelligent designer. The DNA code is *information,* and information is non-material. It depends on matter for storage and movement, but it is not defined by the biochemistry of the molecules used for these purposes. Rather, it is DNA information that defines the operations of matter and energy.

Mathematician Amir Aczel reminds us that DNA is a molecule way beyond what we normally think of as a chemical model because it is far too complex. He asks: "Was it perhaps the power, thinking, and will of a supreme being that created this self-replicating basis for life...How do we define something that can stretch, split along its middle, and clone itself—as a chemical molecule or as a living thing?" [4] The DNA code is not a simple unidirectional code like a computer's binary code; it is unfathomably more complex. Rather, DNA is bidirectional and relays different messages when read in opposite directions. Multiple sets of instructions are embedded in the DNA; a phenomenon known as overlapping genes.

Overlapping genes basically means that the nucleotides at the end portion of one gene sequence overlap with a second sequence, with the letters of the first sequence having a different meaning than they have in the second. A sequence of nucleotides can thus code for more than one gene product by using different reading frames. Stephen Meyer likens this to "Russian Dolls; dolls within dolls...because there are genes within genes," and makes an analogy with spy codes. A spy message read by an enemy might seem to be about mundane matters down on the farm, whereas a friendly reader in possession of the cipher key would read it as intended by the sender. Within the cell, RNA, proteins, and enzymes work together to access, identify, and transcribe the correct "spy" message. Thus: "The presence of these genes imbedded within genes (messages within messages) further enhances the information-storage density of the genome and underscores how the genome is organized to enhance its capacity to store information."[5]

A team of genetic researchers noted that the "Maintenance of dual-coding regions is evolutionarily costly," and that dual overlapping protein-coding is "virtually impossible by chance."[6] Placing the adverb "virtually" before adjective "impossible" allows them space to avoid the charge of advocating intelligent design, and to think that maybe, just maybe, the codes somehow wrote themselves. You only have to think of the brainpower involved in writing a spy's encrypted message to dismiss this absurd notion. The spy must use standard language so that both friend and foe can understand it, and both the manifest and hidden messages must convey real meaning. If the manifest message did not, it would arouse enemy suspicion. If the hidden message did not also convey real meaning, it would be useless to a friendly reader. Of course, there is no intention to deceive in the genome; the spy analogy is just a way to reveal the ingenuity of the code that allows its molecular readers to understand multiple messages both backwards and forwards.

The ability of DNA to convey information relies crucially on its freedom from chemical determinism. If it were not free of this restraint its information conveying capacity would be destroyed. Because the information in DNA is not reducible to the laws of physics or chemistry, its information content cannot originate from them any more than the information in a newspaper originates in the chemistry of ink: "Instead, the genetic code functions as a higher-level constraint distinct from the laws of physics and chemistry, much like a grammatical convention in a human language."[7] Using an analogy of magnetic letters on a metal surface, Stephen Meyer points out that they can be combined and recombined to form any sequence of words. The law of electromagnetic attraction determines that magnetic letters will stick to a metal surface, but the law does not determine their arrangement into meaningful sequences; an intelligent agent must do that. Likewise, there are no laws of physics or chemistry that dictate the arrangement of DNA bases into meaningful biological sentences. The same chemical bond occurs between any base and the DNA backbone, and each can attach to any site on it with equal facility. There are thus two features of DNA that show its self-organizing properties are not explained by the laws of physics and chemistry: "(1) There are *no* bonds between bases along the information-bearing axis of the molecule and (2) there are no *differential* affinities between the backbone and the specific bases that could account for variations in sequence."[8]

John Haught also emphasizes DNA's freedom from physio-chemical determinism: "The specific sequence of the 'letters' in the DNA of any particular organism consists of an informational arrangement that cannot be reduced without remainder to chemistry. This is necessarily the case, for if DNA were the product of chemical determinism alone there would be only one kind of DNA molecule, when in fact an indefinite number of arrangements of the 'letters' in DNA molecules is chemically possible."[9] It is always the case that natural laws will, by definition, predictably and reliably produce the same results, so the arrangement of DNA bases would be the same each time. That they do not illustrates the specified complexity of the information content of DNA and its freedom from physio-chemical necessity.

Macroevolution and the Problem of Time

At a 2016 Royal Society Meeting in London, biologist Gerd Muller accused Darwinists of excluding the "big questions." He notes that microevolution theory performs very well, and if evolutionary explanations would be confined to this level, he explains, there would be no controversy. He chides evolutionists for habitually taking the success of small-scale evolution as the "explanation of *all* evolutionary phenomena" and points out that "a wealth of evolutionary phenomena remains excluded. For instance, the theory largely avoids the question of how the complex organizations of organismal structure, physiology, development or behavior—whose variation it describes—actually arise in evolution." [10] Nowhere in the Darwinian literature do we find a discussion of how many stepwise accumulations of advantageous mutations are necessary for the complex differences in body and mind that exist between a shrew and a Shakespeare. Although 65 to 66 million years ago is a long time, it is hardly enough time to produce the necessary game-changing mutations, as a number of articles in molecular biology journals attest.

The Problem of Waiting Time

Enzymes are important biomolecules that catalyze biochemical reactions in living things. Two biochemists recently showed that without a particular enzyme, the catalytic reaction essential to creating the building blocks of DNA and RNA would take 78 million years. That's not the time for the enzyme to evolve, but just the time it would take to react. They remark on another enzyme essential for the biosynthesis of hemoglobin and chlorophyll: "Now we've found a reaction that—again, in the absence of an enzyme—is almost 30 times slower than that. Its half-life—the time it takes for half the substance to be consumed—is 2.3 billion years, about half the age of the Earth. Enzymes can make that reaction happen in milliseconds." Despite these astounding findings, the authors retain their Darwinian pedigree by remarking that: "It makes you wonder how natural selection operated in such a way as to produce a protein that got off the ground as a primitive catalyst for such an extraordinarily slow reaction."[11] Indeed it does.

How long would it take for one lowly enzyme, not just to react, but to evolve from one form to another? Molecular biologists Gauger and Axe designed an experiment to determine how long it would take to make such a conversion with a minimum of seven nucleotide substitutions via random mutations. They estimated that it would take 10^{30} generations to get a paralogous (different but descending from the same ancestor) enzyme with a new fold. This is a timescale way beyond life on Earth. At the time their article was written (2016), the universe was 4.23 x 10^{17} seconds old, or 423 followed by 15 zeros; 10^{30} is 1 followed by 30 zeros. Thus, two enzymes cannot be reconfigured through a gradual process of mutation and selection if given all the seconds that have ticked by since the Big Bang. Gauger and Axe cite others who have found similar results, and argue that a Darwinian explanation is inadequate to explain the results of such studies, noting that: "Its deficiencies become evident when the focus moves from similarities to dissimilarities, and in particular to functionally important dissimilarities—to innovations. The extent to which Darwinian evolution can explain enzymatic innovation seems, on careful inspection, to be very limited."[12]

Similarly, in the journal *Theoretical Biology and Medical Modelling*, John Sanford and his colleagues developed mathematical models to determine "the waiting time" necessary to form a specified string of nucleotides in a hominid population of 10,000 individuals by mutation and selection under ideal conditions. They note that a new gene contributing to modern humans evolving from whatever primitive creatures they were hypothesized to have been, would require millions of advantageous mutations. Many lines of evidence support the conclusion that to get nucleotide strings of moderate length by trial and error would take more time than the age of the universe. They conclude: "In small populations, the waiting time problem appears to be profound, and deserves very careful examination. To the extent that waiting time is a serious problem for classic neo-Darwinian theory, it is only reasonable that we begin to examine alternative models regarding how biological information arises."[13]

OoL theorist Brig Klyce is also intrigued by the impossibly long time it would take for life to kick-start itself from random strings of amino acids banging against one another to make useful proteins. He calculated the probability of finding actual useful proteins out of all possible random—and useless— proteins to be about 1 in 10^{500}. He continues: "Imagine that every cubic quarter-inch of ocean in the world contains ten billion precellular ribosomes. Imagine that each ribosome produces proteins at ten trials per minute (about the speed that a working ribosome in a bacterial cell manufactures proteins). Even then, it would take about 10^{450} years to probably make one useful protein."[14] This is billions of times longer than the age of the universe.

Mutations and Devolution

Organisms devolve as well as evolve. John Lennox observes that experiments of selective breeding of many thousands of generations of fruit flies (they live a maximum of 50 days) produce nothing but weird fruit flies with features that are maladaptive rather than adaptive. Moreover, they quickly achieve genetic homeostasis; that is, their gene pool runs out of variation capacity. He also notes that studies of 30,000 generations of E. coli produce only harmful *de*volutionary results, losing many of the building blocks of RNA, rather than beneficial results. Biochemist Michael Behe notes that "The lesson of E. coli is that it's easier for evolution to break things up than to make things."[15]

Relatively simple organisms such as bacteria have an easier shot at gaining favorable mutations than humans because they reproduce exponentially more rapidly and have exponentially larger populations. Bacteria evolve adaptations that make them resistant to antibiotics, but these are adaptations within kind, and no new body parts have evolved to make them other than what they were when they first arrived on the living landscape about 3.5 billion years ago. Populations of mammals did become much larger after the dinosaurs were no longer around to feast on them, so mutations rates would have gotten much greater. However, evolution requires adaptive mutations, not maladaptive ones, and the latter are many times more common. Thus, while elevated mutation rates help advantageous traits to spread throughout a population faster, it also hurts by increasing maladaptive mutations. In the journal *BIO-Complexity*, three microbiologists note that of the hundreds of different amino acids that distinguish between enzymes with different functions, if more than the tiniest fraction is important for making the enzymes different, "then it may be effectively impossible for undirected mutations to stumble upon the right combinations for functional conversions…The problem for evolutionary explanations is that the very special circumstances needed to achieve even weak conversions in the lab translate into highly unrealistic evolutionary scenarios."[16]

And this is just the difficulty of the mutation and natural selection of lowly enzymes or unrealistically small strings of nucleotides. How about the literally millions of such mutations—with intermediary mutations more likely to be maladaptive than adaptive—required to go from Hoyle's "sludge" to the genius of Hoyle himself? This becomes exponentially unlikely when we realize that to produce a new phenotypic trait such as an arm or an eye via Darwinian scenarios requires genetic innovation to control metabolic pathways, and such innovation requires countless *coordinated* sequences of enzymatic steps, not innovation in one isolated enzyme.

Macroevolution, Speciation, and the Cambrian Explosion

At the heart of most scientific objections to macroevolution is speciation; the evolution of new and distinct species. Species are groups of interbreeding animals that cannot reproduce with animals not of their kind. It is easy to demonstrate local adaptions within a species, but to demonstrate speciation is quite a different matter, although evolutionists beg to differ. A book published by the National Academy of Sciences (NAS) stated: "A particularly compelling example of speciation involves the 13 species of finches studied by Darwin on the Galápagos Islands, now known as Darwin's finches."[17] Notwithstanding the fact that Darwin's finches remained finches and can interbreed, none of these so-called "species" are distinct. Like humans, they simply vary in minor physical ways within their kind (they are ecomorphs). In a 2015 *Biological Reviews* article, two evolutionists wrote: "We suggest that morphological clusters represent locally adapted ecomorphs, which...have been confused with, species...Thus the pattern of morphological, behavioural and genetic variation supports recognition of a single species of *Geospiza* [finches], which we suggest should be recognized as Darwin's ground finch."[18] Thus, the NAS was pushing a fiction on us that they knew to be false, which led law professor Phillip Johnson to write, "When our leading scientists have to resort to the sort of distortion that would land a stock promoter in jail, you know they are in trouble."[19]

If speciation is legitimate, we should find innumerable transitional fossils, but we do not. As with the case of Collin Patterson quoted in the epigraph of this chapter, N. Heribert Nilsson expresses his disappointment that his life's work for the search of transitional fossil forms bore no fruit: "My attempts to demonstrate evolution by an experiment carried on for more than 40 years have completely failed... The fossil material is now so complete that it has been possible to construct new classes, and the lack of transitional series cannot be explained as being due to scarcity of material. The deficiencies are real, they will never be filled."[20] Eminent paleontologist Stephen J. Gould admitted that it is a "trade secret" of paleontology that fossils that could possibly be considered transitional forms are exceedingly rare, and adds: "The evolutionary trees that adorn our textbooks have data only at the tips and nodes of their branches...in any local area, a species does not arise gradually by the gradual transformation of its ancestors; it appears all at once and 'fully formed.'"[21]

Gould wrote those lines to argue against Darwinian gradualism and to promote his own theory of punctuated equilibrium. The gist of the theory is that mating populations are at evolutionary equilibrium for many generations, and this stasis is occasionally punctuated by rapid bursts of change. Gould's theory is more consistent with the fossil record than Darwin's gradualism, but he did not specify a mechanism by which it worked akin to Darwin's natural selection. However, recent work has proposed that Hsp90 (heat shock protein 90) encourages long-term species stasis by preventing the expression of mutations, but when or if Hsp90 is compromised, the accumulated mutations are released. Of course, only advantageous mutations are of use, and we know that deleterious mutations are far more common, as one study with fruit flies demonstrated. Many hundreds of generations were bred under normal conditions, and then researchers induced a drastic change in their climate. After a few generations of trying to adapt to the new conditions, a menagerie of monsters appeared: "When the genetic variations usually suppressed by Hsp90 began to express themselves, major changes developed in the insects' body plans. Some insects began to sprout weird limbs from different wings, some thick-veined wings, others deformed eyes or legs."[22] This sounds a lot more like devolution than evolution.

Punctuated equilibrium was also an effort to explain the conundrum of the Cambrian Explosion. It is called an explosion because after waiting for animals to arrive on the planet for three billion years, they seemed to arrive all at once 500 to 550 million years ago, with no ancestral fossils to be found in the geological record. The Cambrian conundrum was recognized by Darwin in *The Origin of Species*. Darwin noted the problem posed to his theory by the sudden appearance of numerous animal forms with no ancestors to be found in the fossil record. He wrote: "If numerous species, belonging to the same genera or families, have really started into life all at once, the fact would be fatal to the theory of descent with slow modification."[23]

The problem Darwin noted in 1859 is still with us. Biologists Peterson, Dietrich, and McPeek note that the Cambrian explosion changed the Earth's biota in profound and fundamental ways: "going from an essentially static system billions of years in existence to the one we find today, a dynamic and awesomely complex system whose origin seems to defy explanation." They also note that the rock strata prior to the Cambrian explosion show no signs of previous life forms other than single-cell organisms and that: "numerous animal phyla with very distinct body plans arrive on the scene in a geological blink of the eye, with little or no warning of what is to come in rocks that predate this interval of time."[24]

Not only did these body plans arrive abruptly, their basic form has remained the same. As marine paleontologist Jeffrey Levinton tells us: "Evolutionary biology's deepest paradox concerns this strange discontinuity. Why haven't new animal body plans continued to crawl out of the evolutionary cauldron during the past hundreds of millions of years? Why are the ancient body plans so stable?"[25] Levinton sees this as a big problem for macroevolution, writing; "Evolution at the species level continues unabated, but variation in the surviving body plans does not seem to occur."[26] Peterson, Dietrich, and McPeek agree, and note what this means for macroevolution: "Thus, elucidating the materialistic basis of the Cambrian explosion has become more elusive, not less, the more we know about the event itself, and cannot be explained away by coupling extinction of intermediates with long stretches of geologic time, despite the contrary claims of some modern neo-Darwinists."[27]

The Tree of Life

The "tree of life" is a metaphor Darwin used to describe the relationships between organisms, both living and extinct, and to show that all life descended from a common source. As Darwin described it:

> The affinities of all the beings of the same class have sometimes been represented by a great tree. I believe this simile largely speaks the truth...At each period of growth all the growing twigs have tried to branch out on all sides, and to overtop and kill the surrounding twigs and branches, in the same manner as species and groups of species have tried to overmaster other species in the great battle for life."[28]

There is no denying that this was a brilliant hypothesis, but it appears to have failed. Darwin admitted that the available data did not support his theory, but he was optimistic that future data would because he believed that the number of intermediate varieties must be "truly enormous." Over 100 pages later, however, he asks: "Why then is not every geological formation and every stratum full of such intermediate links? Geology assuredly does not reveal any such finely graduated organic chain; and this, perhaps, is the most obvious and gravest objection which can be urged against my theory. The explanation lies, as I believe, in the extreme imperfection of the geological record."[29]

However, 160 years later, after thousands of paleontologists and anthropologists have spent hundreds of thousands of hours digging and delving at hundreds of geological sites around the world, we still lack true intermediate varieties, although some claim that an odd tooth or bone fragment represents the evolutionary Holy Grail. *Homo sapiens* did not arrive full-blown at the Cambrian, but our appearance was very abrupt according to the dean of evolutionary biology, Ernst Mayr: "The earliest fossils of *Homo*, *Homo rudolfensis* and *Homo erectus*, are separated from *Australopithecus* by a large, unbridged gap. How can we explain this seeming saltation [sudden appearance of new genetic characters]? Not having any fossils that can serve as missing links, we have to fall back on the time-honored method of historical science, the construction of a historical narrative."[30] This "historical narrative" for the transition from *Australopithecus* to *Homo* is to study handy non-transitional fossils (and perhaps claim them as transitional) and to simply assume that the "large, unbridged gap" was, in fact, bridged. In the absence of fossil evidence, claims about the transition to *Homo* are inferences made by studying the non-transitional fossils we do have, and then assuming that a transition simply must have occurred.

Having failed to demonstrate the tree of life by the fossil record, scientists swung from the fossil tree to the molecular biology tree. The branch of molecular biology involved in this endeavor is molecular systematics. Molecular systematics scientists use genomic molecules from a number of species and examine genetic sequences to see how closely they may be related. The more closely related two species are, the more closely their genomic material is assumed to match. The ultimate goal is to construct a tree of common ancestry of all species down to the roots. It is a grand idea, but despite huge amounts of data subjected to powerful statistical techniques, conflicting versions of the tree are common. One suite of genes reveals one tree, while another suite reveals a different tree; trees using DNA produce different results from trees using RNA, and trees using microRNA suggest something else again. Some trees place humans in the same lineage as elephants; other in a lineage leading to worms, and other trees reveal that half of our genes have one evolutionary history, and the other half a different evolutionary history.[31] We also see data fudging when things don't go as planned. Molecular biologists Rokas and Carrol report that one study "omitted 35% of single genes from their data matrix, because those genes produced phylogenies at odds with conventional wisdom."[32]

In short, Darwin's tree of life presents many difficulties for evolution. Many have concluded that it has all been a fool's errand. Evolutionary biologist Eric Bapteste notes that: "For a long time the holy grail was to build a tree of life"...A few years ago it looked as though the grail was within reach. But today the project lies in tatters, torn to pieces by an onslaught of negative evidence."[33] A trio of morphologists (the branch of biology that deals with the form and structure of organisms) wrote: "As morphologists with high hopes of molecular systematics, we end this survey with our hopes dampened. Congruence between molecular phylogenies is as elusive as it is in morphology and as it is between molecules and morphology."[34] None of this means that biologists have given up on testing Darwinian ideas, but the more they learn more evidence we accumulate for our inference to the best explanation.

Endnotes

1. Williams, G. 1992, p. 11.
2. Bauchau, V., 2006, p. 36.
3. Yockey, H., 2005, p. 2.
4. Aczel, A. 1998, p. 88.
5. Meyer, S. 2009, p. 463.

6. Chung, W., Wadhawan, S. et al. 2007.
7. Meyer, S. 2009, p. 240.
8. Ibid., p. 244.
9. Haught, J. 2008b, p. 76.
10. Müller, G, 2017, p. 3.
11. Wolfenden, R. 2008, np.
12. Gauger, A., and Axe, D., 2011, p.13.
13. Sanford, J., Brewer, W., et al, 2015, p. 27.
14. Klyce, B. Nd; np.
15. In Lennox, J., 2010, p. 110.
16. Reeves, M., Gauger, A., & Axe, D., 2014. pp. 10-11.
17. Ayala, F., et al, 1999, p. 10.
18. McKay, B. and Zink, R., 2015, p. 689.
19. Johnson, P., 1999, np.
20. In Nitardy, C., 2012, p. 60.
21. Gould, S, 1977, p. 14.
22. ABCScience, 1998.
23. Darwin, C., 1982, p. 309.
24. Peterson, K., Dietrich, M., and McPeek, M., 2009. p. 736.
25. Levinton, J. 1992, p. 84.
26. Ibid., p. 91.
27. Peterson, K., Dietrich, M., and McPeek, M., 2009. p. 736.
28. Darwin, C. 1982, p.171.
29. Ibid., p. 292.
30. Mayr, E. 2004, p. 198.
31. Maxmen, A. 2011.
32. Rokas, A., & Carroll, S. 2006, p. 1902.
33. In Ebifegha, M. 2009, p. 36.
34. Patterson, C., Williams, D., and Humphries, C. 1993, p. 179.

CHAPTER SEVENTEEN

God of the Gaps, Intelligent Design, and Theistic Evolution

> "God is not an alternative to science as an explanation, he is not to be understood merely as a God of the gaps, He is the ground of all explanation: it is his existence which gives rise to the very possibility of explanation, scientific or otherwise."
> John Lennox: Oxford mathematician and philosopher

God of the Gaps: The Atheist's Empty Ploy

"God of the gaps" is a pejorative phrase used against those who purportedly introduce God into a scientific argument. A God-of-the-gaps argument boils down to saying that a gap in scientific knowledge is proof of God's hand in the matter. This argument is illustrated in a well-known cartoon in which two physicists are standing looking very puzzled at a blackboard full of equations with a tag on the end reading "Then a miracle occurs!" People have always created gods to explain natural phenomena they did not understand. The ancients created Thor and his hammer to explain thunder, Poseidon to explain droughts or floods, and Zeus, who threw lightning bolts upon the Earth to show his anger. Atheists extrapolate these absurd ancient *created* gods to modern natural theology practiced by eminent scientists and philosophers seeking evidence of an *uncreated* God from what we *do* know, and not from our ignorance.

19th-century evangelist and biologist Henry Drummond was annoyed by efforts both to place God into gaps in scientific knowledge and to divorce Him from science: "There are reverent minds who ceaselessly scan the fields of nature and the books of science in search of gaps—gaps which they fill up with God. As if God lived in gaps!"[2] Theologian Dietrich Bonhoeffer issued a similar warning against the error of using God as a stop-gap for holes in our knowledge, for when the frontiers of science are pushed back then: "God is being pushed back with them, and is therefore continually in retreat. We are to find God in what we know, not in what we don't know; God wants us to realize his presence, not in unsolved problems but in those that are solved."[3]

As John Lennox points out in the epigraph, God is the ground of all explanations, but he is not Newton's cosmic fiddler. Discovering the laws of nature is the task God has set for science. "God of the gaps" is a lame phrase of abuse because the gap is not science's door slowly closing with each advance and squeezing God out, it is a gap growing ever wider as science advances to let Him in. How can anyone view our ever-expanding knowledge of the grandeur of the universe as undermining He who created it? How can we not see God beyond the widening expanse that stunningly reveals a universe so complex that only a mind of a supreme intelligence could bring into being? This is Einstein's cosmic curtain opening ever wider as science advances: "The more I study science, the more I believe in God."

As scientists drink near to the bottom of Heisenberg's glass, many recognize God. None have inserted God into unresolved gaps, but they see advances in science pointing toward God rather than away. They have observed the incredible fine-tuning of the laws of nature, the vast information content of DNA, and the intricate nanotechnology of the living cell and have ventured beyond science in their efforts to understand it all. Except for multiverse proponents, scientists have ruled out chance for these things as far beyond the available resources of probability. More and more are sounding like Nobel Laureate physicist Joseph J. Thomson: "As we conquer peak after peak we see in front of us regions full of interest and beauty, but we do not see our goal, we do not see the horizon; in the distance tower still higher peaks, which will yield to those who ascend them still wider prospects, and deepen the feeling, the truth of which is emphasized by every advance in science, that 'Great are the Works of the Lord'."[4]

Is Intelligent Design Guilty of God-of-the-Gaps Reasoning?

Intelligent Design (ID) is a scientific endeavor that claims the existence of an intelligent cause behind the universe and the origin of life are testable hypotheses. Proponents of ID believe in a designer God, but such a belief is not necessary, as Michael Behe notes: "The conclusion that something was designed can be made quite independently of knowledge of the designer. As a matter of procedure, the design must first be apprehended before there can be any further question about the designer."[5] We have noted that prominent atheist scientists have also argued for the intelligent design of the origin life given the immense improbability of a naturalistic origin, but locate this intelligence in areas other than the creative work of a personal God. Instead, they invoke panspermia, intelligent aliens, or perhaps a sentient universe manifested in its forces and laws; a doctrine known as pantheism ("All is god;" "Nature is my god.").

No group of scientists has been accused more of God-of-the-gaps reasoning than ID scientists. Physicist Nathan Aviezer, who is a Christian, accuses ID of such reasoning. Aviezer says that he never inserts God into his scientific work, and that "The religious person never invokes the supernatural as the explanation of and physical phenomenon ...the framework in which God interacts with the physical world is within the laws of nature."[6] This is precisely the operating assumption of ID scientists whose whole enterprise is based on using the tools of science to look for evidence of design in nature. ID scientists always remain within the framework Aviezer identifies, and you will never see "Then a miracle occurs" in their equations.

What do ID scientists mean by "design?" Anything designed includes three things: contingency, complexity, and specification, and excludes necessity and chance. William Dembski, who has a PhDs in mathematics and philosophy, has written many books and articles on cause and effect and how to discover hallmarks of ID. He notes that there are three mechanisms that bring about an effect: natural law, chance, and design. To attribute ID to anything it is necessary to rule out both necessity (natural law) and chance for its existence. Dembski notes that the simplest explanations are those that appeal to natural laws because they do not allow for contingency since things must happen the way they do. Explanations involving chance are more complicated because they are both contingent and involve probability. The most complicated explanations appeal to design, because: "they admit contingency but not one characterized by probability."[7]

The existence of something that does not appeal to necessity or chance means that its existence is contingent on conscious choice. Take the existence of space shuttles and skyscrapers. Both are contingent in that they didn't have to exist according to any law of nature. We rule out chance next because we know that the specified complexity of space shuttles or skyscrapers are the products of ID. Knowing this, of course, tells us nothing about the specifics of their assembly. Claiming that they are the product of ID is not an explanation of how or why these things were built; it is merely an inference based on what we know about reality. The obvious reality is that all things that are not just random clumps of "stuff" are generally designed by intelligent agents for a purpose.

Atheists would agree to this with reference to space shuttles and skyscrapers, but deny that the universe with all its stupendous complexity is also intelligently designed for a purpose. The ID argument is not a "science-stopper" any more than noting that space shuttles or skyscrapers are intelligently designed is a science stopper. We still have to investigate the details of how these things came to be, and in common with all scientists, this is what ID scientists do. Scientists who construct their claims on a system of beliefs to which they are deeply attached and refuse to countenance alternative are the real science stoppers. If we think we already know with certainty how things are and fail to question our assumptions, we miss the opportunity for true scientific discovery.

Aviezer's primary target in his article is microbiologist Michael Behe's claim about the irreducibly complex design of the bacterial flagellum, a whip-like appendage by which bacteria navigate through their environment. Behe has been studying flagella for decades using standard scientific methods, so it is a mystery why Aviezer invokes God-of-the-gaps reasoning to describe Behe's work. Atheist philosopher of science Bradley Monton agrees, and maintains that Behe's irreducible complexity argument is not God-of-the-gaps reasoning because Behe is not arguing that the flagellum could not possibly arise naturalistically, but is rather: "giving positive reasons that the sequence of events that would have to happen for irreducibly complex systems like the bacterial flagellum to arise via an undesigned process is an improbable sequence, and hence the design hypothesis should be taken seriously."[8]

Other secular scientists have defended ID after giving it serious thought. A notable example is a task force of nine distinguished scholars from various universities formed by the president of Baylor University to assess the scientific legitimacy of William Dembski's work. This extraordinary step was taken in response to faculty animosity that arose when Dembski became the director of the newly formed Michael Polanyi Center. Much to the chagrin of Baylor's censorious faculty, this committee of scholars concluded that: "research on the logical structure of mathematical arguments for intelligent design has a legitimate claim to a place in the current discussions of the relations of religion and science."[9]

ID is also wrongly accused of God-of-the-gaps thinking because it supposedly makes no falsifiable predictions. This is strongly denied by ID scientists. In his magisterial 611-page book *Signature in the Cell*, Stephen Meyer writes that ID theory "merely claims to detect the action of some intelligent cause (with power at least the equivalent to those we know from experience) and affirms this because we know from experience that only conscious, intelligent agents produce large amounts of specified information."[10] Meyer lists a number of hypotheses derived from ID theory, including the possibility of effectively demonstrating that "large amounts of functionally specified information do arise from purely chemical and physical antecedents." [11] This would falsify one of ID's hypotheses, but we have seen that there is an emerging consensus that complex specified information, such as that contained in DNA, cannot arise from the laws of physics and chemistry. A design inference in ID theory is thus not triggered by any phenomenon that we cannot currently explain, but rather from what we know about cause and effect. It is triggered when an event defies probability and when it conforms to a meaningful complex specified functional pattern. Recall that Dembski correctly predicted from ID theory that function would be found for what geneticists used to call "junk DNA" a decade before it was found they were wrong.

Is Intelligent Design Anti-Evolution?

Another criticism of ID is that it is anti-evolution. Dembski denies this: "Intelligent design does not claim that living things came together suddenly in their present form through the efforts of a supernatural creator."[12] ID recognizes that microevolution is the only reasonable explanation for the life forms we see around us, and that no one denies it. In fact, ID uses the same principle that Darwin adopted to explain historical events. That is, when trying to explain past events, scientists identify causes that produce an effect today and extrapolate those causes to past events on the reasonable assumption that if they work now, they must have worked then. Darwin observed that natural selection produces small changes within species in a relatively short period of time and extrapolated from this to conclude that it could produce large-scale changes over longer periods of time. He thus concluded that natural selection was "causally adequate" to produce not only variation within species but entirely new species as well.[13] The value of the ID enterprise is that it challenges Darwinism to account in naturalist terms for the innumerable genetic changes required for macroevolution to occur. Non-ID science seems to content itself by saying, as Richard Dawkins did in *The God Delusion*, that it was "a lucky chance." Surely this is an evolution-of-the-gaps argument.

ID scientists investigate biological patterns showing signs of intelligence, direction, and purpose. It takes the mind-boggling improbabilities of life forming from non-life as powerful signs of intelligence, and places special emphasis on the irreducible and specified complexity of DNA, the cell, and other biological features. Geneticist Joseph Kuhn notes that ID investigates: "Irreducibly complex systems involving thousands of interrelated specifically coded enzymes"...At an absolute minimum, the inconceivable self-formation of DNA and the inability to explain the incredible information contained in DNA represent fatal defects in the concept of mutation and natural selection to account for the origin of life and the origin of DNA."[14]

Investigating ID is not a science-stopper; it is a science-mover because it challenges others to show that something allegedly designed could be the result of necessity or chance. Atheist philosopher, Thomas Nagel, argues this point in his *Mind and Cosmos: Why the Materialist Neo-Darwinian Conception of Nature Is Almost Certainly False*. Nagel argues that ID should be taken seriously and deserves our gratitude for challenging a scientific worldview that he considers entirely ideological.[15] Nagel was hammered mercilessly as a "traitor to science" for his defense of ID because to challenge Darwinism borders on sacrilege. Yet, in an article on Nagel's book and its critics, Michael Choroser wrote: "The odd thing is, however, that for all of this academic high dudgeon, there actually are scientists—respected ones, Nobel Prize-winning ones—who are saying exactly what Nagel said, and have been saying it for decades." [16]

It is unfortunate that many scientists think that ID proponents believe the act of creation froze everything in place as it exists today with no evolutionary changes "within kind." It is macroevolution with which ID has issues. In an interview with *UCBerkley News*, Nobel Laureate physicist Charles Townes views the belief that ID is anti-evolution is "totally illogical," and asserts that ID is eminently scientific. He notes that: "This is a very special universe: it's remarkable that it came out just this way...I do believe that God has a continuing influence—certainly his laws guide how the universe was built. Now, that design could include evolution perfectly well. Evolution is here, and intelligent design is here, and they're both consistent."[17]

ID's powerful arguments have attracted many former Darwinists, including Gunter Bechly, a German paleontological biologist. As the leading evolutionist in Germany, Bechly was invited in 2009 to organize a museum exhibit in Stuttgart to celebrate the bicentennial of Darwin's birth. Among the exhibits on display, Bechly featured an old-fashioned weight scale showing a number of anti-Darwinian books in one pan and Darwin's *Origin of Species* in the other. Naturally, Darwin's book left the combined weight of the other books dangling in the air. This powerful visual symbol was designed to show that all contrary evidence is impotent against the weight of Darwin's theory.

However, Bechly decided to read those dangling books and began to have gnawing doubts about his commitment to Darwinism. The upshot was that he rejected Darwinism and became a Christian. Bechly proclaims that he is a theist who strongly rejects atheism and ontological materialism/naturalism, and that: "I have not become a theist *in spite* of being a scientist but *because* of it. My 'conversion' was based on a critical evaluation of empirical data and philosophical arguments, following the evidence wherever it leads. I am skeptical of the Neodarwinian theory of macroevolution and support Intelligent Design theory for purely scientific reasons."[18] Bechly is a scientist who follows the data where they lead instead of blindly adhering to ideology. Perhaps other critics of ID should approach ID books with similar open-mindedness.

No ID in Class: It May Make Students Think about God

With so many secular scientists, including Nobel Laureates, pointing out flaws in Darwinian macroevolution, why are schools forbidden to teach those flaws?

Phillip Johnson provides a wry answer given by Chinese paleontologist Jun Yaun Chan: "In China we can criticize Darwin but not the government. In America you can criticize the government but not Darwin."[19] Johnson was commenting about the brouhaha that followed the Kansas Board of Education's decision to omit macroevolution from the curriculum and include ID instead. A number of scientific and secularist organizations filed suit against the board, and after a series of hearings a federal judge ruled against the board in *Kitzmiller v. Dover Area School District* (2005). In a 139-page ruling, Judge John Jones ruled that ID is not science, and that it teaching it would violate the establishment clause of the First Amendment of the United State Constitution.

It is sheer hubris for a non-scientist to state that scientific articles questioning macroevolution published by both ID proponents and non-proponents in peer-reviewed journals is not science. Moreover, the establishment clause of the First Amendment Jones finds to be violated forbids only the United States Congress from establishing a national religion, as its wording plainly states: "Congress shall make no law respecting an establishment of religion." Jones' ruling is tantamount to equating the Kansas Board's decision with Congress doing just that. Nevertheless, Jones' decision was made in accordance with the Supreme Court's belief that any reference to God should be purged from the public square. I have documented how the Supreme Court has managed to use the establishment clause to eviscerate the free exercise clause forbidding Congress from inhibiting the free exercise of religion in my book *The Gavel and Sickle*.

On the other hand, in *Elk Grove Unified School District et al v. Newdow et al.* (2004), the Supreme Court ruled that religion also includes nontheistic belief systems. To be consistent with this ruling, a school that teaches Darwinism must also permit teaching ID if it is to maintain religious neutrality. Because *Elk Grove* was decided a year before *Kitzmiller*, Jones ignored legal precedent, although calling non-theistic belief system religious is a stretch, but it is a stretch made by the highest court in the land. It was also made by French physician, zoologist, and microbiologist Pierre-Paul Grassé, who asserts that Darwinism is indeed a non-theistic religion: "Directed by all-powerful selection, chance becomes a sort of providence, which, under the cover of atheism, is not named but which is secretly worshipped."[20]

Evidently, the courts consider it acceptable for the schools to inculcate a naturalist worldview at the expense of the intellectually exciting give and take of contending views. Immunologist Scott Todd tells us why ID is excluded for the academy, *despite* evidence to the contrary: "it should be made clear in the classroom that science, including evolution, has not disproved God's existence...*Even if all the data point to an intelligent designer, such a hypothesis is excluded from science because it is not naturalistic*" [my emphasis].[21] In effect, this means that scientific evidence against macroevolution cannot be construed as evidence for ID, because ID is ruled out by fiat. Accordingly, even if macroevolution were to be conclusively ruled out by naturalistic science, it would have to seek out some other non-designed, non-purposive, explanation of the existence of life because God is strictly off limits.

Theistic Evolution

ID is not the only scientific alternative to a non-theistic view of life. Its major theistic rival is theistic evolution (TE), which is the belief that God created all living things using the process of evolution in ways that conform to orthodox accounts. TE accepts both micro and macroevolution, but unlike secular scientific accounts, it denies that the process is undirected and purposeless. Many scientists associated with TE belong to the BioLogos Foundation established by the geneticist and physician Francis Collins. The eminent agnostic evolutionist, Stephen Jay Gould, noting the many scientists who are religious, wrote that that: "Either half of my colleagues are enormously stupid, or else the science of Darwinism is fully compatible with religious belief – and equally compatible with atheism."[22] Unlike ID, TE is accepted by all mainstream Protestant denominations and by the Catholic Church, so it must have strong persuasive punch.

To crudely analogize the difference between ID and TE: ID's image of God is that of an omniscient engineer methodically planning ahead, and leaving little to chance, while the TE image is that of a supremely gifted artist, slowly creating a painting of great beauty with some "happy accidents" along the way. TE proponents are unimpressed with the idea of irreducible complexity, and chide ID proponents for limiting God's reach. Kenneth Miller, a cell biologist and devout Christian, accuses ID of having a narrow view of the capabilities of nature and of God, saying that ID hobbles God's: "genius by demanding that the material of His creation ought not to be capable of generating complexity. They demean the breadth of His vision by ridiculing the notion that the materials of His world could have evolved into beings with intelligence and self-awareness." [23]

TE thus accepts a chain of naturalistic events leading to complex life, but affirms God's guiding hand in the process. God's chosen method of bringing life into existence was to endow nature with the creative power to organize itself. Although I have concentrated on ID because of its challenges to Darwinian orthodoxy, I do not dogmatically assert that it is superior to TE in providing a God-centered account of life. There are brilliant scientists in both camps, and I cannot presume to be their judge on the matter. The reason I devote so little space to TE is simply that it accepts mainstream evolutionary accounts and adds God as both the Agent who designed and brought the universe into being, and who stipulated the purpose for which it was created.

TE is not just an "Add God and stir" approach to science, although some TE scientists are content to leave it at that. Others want to provide an adequate naturalistic explanation for God's guiding hand in evolution, and some have appealed to quantum mechanics. Quantum physicist Robert Russell makes the TE case in his theory of NIODA (non-interventionist objective divine action) in which he sees continuous creation arising indirectly from: "God's direct action of sustaining in existence quantum systems and their properties during both their time evolution and their irreversible interactions."[24] Another quantum physicist, Amit Goswani, also sees God operating at the quantum level: "The idea of a God as an agent of downward causation has emerged in quantum physics."[25] These scientists introduce the notion of "special providence," by which God acts through the indeterminism of quantum mechanics to guide the natural world without violating the natural order. Special providence is noninterventionist divine action that complements general providence: God's continuous upholding of natural order of the Universe.

Oxford philosopher Ignacio Silva argues that this view amounts to a reducing God to a "cause-among-causes" in that He is constrained to work with quantum phenomena: "God is bound by nature to act within the laws of nature."[26] Silva argues that this demotes God to a necessary but not sufficient cause because He does not act autonomously. Another way of looking at this is physicist and Anglican priest John Polkinghorne's appeal to *kenosis*. *Kenosis* is the idea that God willingly surrenders some of His authority and grants nature the power to make itself by natural forces. He writes: "The play of life is not the performance of a pre-determined script, but a self-improvisatory performance by the actors themselves....God shares the unfolding course of creation with creatures, who have their divinely allowed, but not divinely dictated, roles to play in is fruitful becoming."[27] Polkinghorne's God is one who is constantly working creatively through the unfolding of the inherent potentialities in nature. This was Darwin's view; all things unfold through "the laws impressed on matter by the Creator."[28]

For many TE proponents, evolution is God's way of maintaining epistemic distance between Himself and humanity. TE scientists maintain that while we can see God's hand in evolution, we have to look long and hard. Philosopher and theologian John Hick argues that this epistemic distance is necessary for humans to voluntarily come to God. Hick writes: "the reality and presence of God must not be borne in upon men in the coercive way in which their natural environment forces itself upon their attention. The world must be to man, at least to some extent, *etsi deus non daretur*, 'as if there were no God.' God must be a hidden deity, veiled by his creation."[29] I believe this to be the best argument for TE's acceptance of mainstream evolutionary science—God cannot be so obvious that we have no choice but to accept him.

Although ID and TE scientists are often at odds, there is a common ground in that there may be no real contradiction in seeing evolution both as a God-guided semi-autonomous naturalistic mechanism, as TE scientists affirm, and the many marvelous designs that ID scientist reveal. Leading TE proponent Francis Collins acknowledges that evolution could be directed, noting that: "evolution could appear to us to be driven by chance [TE], but from God's perspective the outcome would be entirely specified [ID]. Thus, God could be completely and intimately involved in the creation of all species, while from our perspective ... this would appear a random and undirected process"[30] Both ID and TE camps see the Creator's plan and purpose unfolding over time; whether by direct intervention in which specified irreducible complexity was present at the beginning or by the unfolding of His laws over time.

Does it really matter which? Whatever science eventually discovers, there is nothing it could show that would cast doubt on God as the author of the information coded the universe and in the book of life. We may never know how He did it, and perhaps despite its seemingly insuperable problems, macroevolution did proceed from sludge to Shakespeare after all. Albertus Magnus, 13th-century scientist, philosopher, and theologian, gets the last word on this: "In studying nature we have not to inquire how God the Creator may, as He freely wills, use His creatures to work miracles and thereby show forth His power; we have rather to inquire what Nature with its immanent causes can naturally bring to pass."[31]

Endnotes

1. Kant, I. 1999.
2. Drummond, H., 1894, p. 333.
3. Bonhoeffer, D., 1972, p.311.
4. In Gaither, C. and Cavazos-Gaither, A., 2001, p. 128.
5. Behe, M. 1996, p. 197.
6. Aviezer, N. 2010, p 8.
7. Dembski, W. 1998, p. 100.
8. Monton, B. 2009, p. 115.
9. In Olasky, M. and Perry, J. 2005, p. 197.
10. Meyer, S., 2009, pp. 428-429.
11. Ibid., p. 429.
12. Dembski, W. 2006, p. 314.
13. Meyer, S. 2009, p. 160.
14. Kuhn, J. 2012, p 46.
15. Nagel, T. 2012.

16. Choroser, M. 2013.
17. Powell, B. 2005.
18. Bechly, G. nd.
19. Johnson, P. 1999.
20. Grassé, P. 1977, p. 107.
21. Todd, S. 1999, p. 423.
22. Gould, S. 1992, p. 119
23. Miller, K. 1991, p. 26.
24. Russell, R. 2008, p. 590.
25. Goswami, A. 2014, p. 22.
26. Silva, I. 2015, p. 107.
27. Polkinghorn, J. 2001, p. 94.
28. Darwin, C. 1982, p. 458.
29. Hick, J. 1977, p. 281.
30. Collins, F. 2006, p. 205.
31. In Walsh, J. 2013, p. 338.

Chapter Eighteen
The Human Body: The Temple of God I

> "There is a fundamental mystery in my personal existence, transcending the biological account of the development of my body and my brain. That belief is in keeping with the religious concept of the soul and with its special creation by God."
> John Eccles:
> Nobel Laureate in physiology and medicine

I Am Fearfully and Wonderfully Made

St. Augustine once remarked that: "Men go abroad to wonder at the height of mountains, at the huge waves of the sea, at the long courses of the rivers, at the vast compass of the oceans, at the circular motion of the stars, and pass by themselves without wondering."[1] We do indeed often express admiration and fascination with many natural phenomena such as magnificent sunsets and beautiful waterfalls, and man-made objects such as super-computers and space shuttles, without ever realizing that we are more worthy of wonder than anything else we wonder about. We are amazing engineering creations in every way, from the elegant sophisticated operations of our brains, our immune systems, and muscles to the stunning complexity with which all parts are wired together. Psalm 139:13-15 notes how God laid the foundation of our being: "For you created my inmost being, you knit me together in my mother's womb. I praise you because I am fearfully and wonderfully made; your works are wonderful, I know that full well. My frame was not hidden from you when I was made in the secret place." You are a miracle of divine orchestration; God's masterpiece; the reason for which He created the universe. The more scientists learn about the human body, from genome to brain, the more awestruck they become.

Yes, we live in the most advanced structure in the universe; one that has the ability to repair itself, and to make others of its kind. It contains a chemical plant far more complex than any built by man, and its interconnected biomechanical and biochemical engineering converts the food we eat into energy and living tissue that repairs itself, and does thousands of other things automatically every second of every day. It has a conscious computer called a brain that is immeasurably more sophisticated than any man-made computer, and a library we call a genome, which contains much more information than any physical library on Earth. If we take the conscious information processes such as thought, speech, and voluntary movements, combined with unconscious ones (automatic, chemically controlled processes), it: "involves the processing of 10^{24} bits [of information] daily. This astronomically high figure is higher by a factor of 1,000,000 than the total human knowledge of 10^{18} bits stored in all the world's libraries."[2]

Imagine the mind-numbing complexity of something that processes 10^{24} bits of information every day. Even one-trillionth that number exceeds anything engineers ever dreamt of building. There are multiple functional systems within systems in the body, so we are forced to limit discussion to the most fascinating ones that pose the biggest problems for materialistic explanations of their design. Science does an excellent job of explaining how the body functions, but not how the basic body plan arose, and why it exists at all. The design of any one of our body systems is more than enough to infer a master biochemical engineer behind them "monkeying" with the physics, chemistry, and biology. We begin with the deep biological mystery (and miracle) of how the human form is possible at all. We are the products of sexual reproduction, but how sexual reproduction came on the scene when asexual reproduction is much more efficient poses a deep conundrum for materialists.

From Zygote to Baby

Just as the universe began with a Big Bang ignited by our Heavenly Father and spread out in all its magnificence, so did you. Your personal big bang was ignited by your earthly father when one of his approximately 40 million to 2.5 billion sperm cells won the frantic race to merge with your mother's egg. The instant that single sperm penetrated the egg, a chemical on the egg blocked all other sperm entering. At this point of life, the merger of sperm and egg is called a zygote, which has the genetic information (half from dad and half from mom) to enable you to spread out from that microscopic zygote to become the magnificent and complex creature that you are. I am not trying to flatter you; you really are far more complicated than any machine ever made. As the only creatures made in God's image, we are all awesome because we are spiritual beings endowed with a soul. We are the only objects in the universe that are also subjects who write poetry and symphonies, invent methods to plumb the depths of the cosmos or the tiniest molecule, build cathedrals and spaceships, and seek to know our Maker.

The single-cell zygote is the first step in becoming an adult human being with 35 to 40 trillion cells. In the first step of human growth, the zygote begins to develop into a hollow ball of cells called the blastocyst. The blastocyst travels down the mother's fallopian tube, and in 3 to 5 days-time implants itself on the walls of her uterus. Some cells of the blastocyst develop into the embryo and others into the placenta, from which the developing human will get its nourishment from its mother. The embryo's cells then continue to divide into distinguishable human features.

How do these rapidly dividing cells know what to become, since they are all stem cells contain the same DNA and thus can potentially become any body part? We all know that the XX or XY sex chromosomes determine if the embryo is to become a girl or a boy. A single gene on the Y chromosome called the SRY (sex-determining region of the Y) gene determines your sex; if you have it, you will be male, if not, you will be female.[3] The process of sex differentiation requires a cascade of molecular processes in what was initially an embryonic hermaphrodite. The SRY gene only determines the internal and external sex organs; the rest of the human body plan males and females have in common must develop according to the body plans carried on the autosomal (non-sex) chromosomes.

The rapidly multiplying cells must migrate to their proper place to become bone for the skull, make the brain, the skin, eyes, ears, heart, lungs, and the other body parts. Undifferentiated stem cells become differentiated by exposure to different signals from both inside and outside the cells—chemicals, extracellular proteins, hormones, and neighboring cells—that turn genes on or off. Receptors on cell surfaces read these signals and respond appropriately. Genes that are turned on make proteins specific to the body part they make, and once that happens, genes coding for other parts are forever shut down in those cells.

The master regulators of the process of building a body to the correct specification are a group of 39 genes organized in clusters called homeobox, or Hox genes. Hox genes are expressed according to their position within the gene cluster, and are themselves activated by proteins encoded by earlier genes. Once activated, they begin laying out the body's architecture in orderly fashion along the embryo's head-to-tail axis. The body plans of all animals are governed by Hox genes.[4] Darwinists use the ubiquity of Hox genes to argue for the common ancestry of all life forms, but theists argue that we should expect this commonality given their common origin from the same Creator. The complexity of it all is truly amazing, for how does the developing child manage to balance all the finely-tuned kinetic, chemical, cellular, and genetic signals involved? Nobel Laureate in physiology and medicine, Sir Charles Sherrington, who saw God manifested in everything in the natural world, expressed his astonishment at this miracle of organization thus: "The whole astonishing process achieving the making of a new individual is thus an organized adventure in specialization on the part of countless co-operating units. It does more than complete the new individual; it provides for the future production of further individuals from that one."[5]

Contrary to Sherrington's amazement at God's creations, abortion activists echo the atheist refrain that humans are nothing special. If humans are nothing special, the human fetus, is only "potentially" human, and until it becomes one it is a mere assemblage of tissue to be sucked from the womb and discarded with hospital waste. The Association of American Physicians and Surgeons disagrees, noting that: "It is undisputed as a matter of science that a new, distinct human organism comes into existence during the process of fertilization—at the moment of sperm-egg fusion—and before implantation of the already-developing embryo into the uterine wall." In other words, the zygote contains all the genetic information that is present in the embryo, the fetus, the baby, and the adult man or woman. As such, the zygote is as deserving of its life as any person at any seamless stage of development because life is a journey, and we are never complete, but always becoming. Just because we are further along that journey than our sons and daughters, does that make us more human than they are, or they more human than their children? How about the human being in a young woman's womb; is it any less human than she just because it has not yet taken its first steps?

It is indisputable from a scientific point of view that the zygote/embryo/fetus is not developing *into* a human being; it is developing *as* a human being. It has its own special suite of genes which, unless it is one of an identical pair of twins, is unique in the world. Abortionists make the claim that the developing child in a woman's womb is part of her body, and that no one but she and her physician should decide whether it lives or dies. But while the zygote/embryo/ fetus is temporarily a part of the woman's body, it is not a tumor, an appendix, gall bladder, or boil, all of which are entirely hers for she and her physician to talk about. While granting that the new human being is for a short time residing inside its mother, and thus is a part of her, it is also *apart* from her as a distinct human being with a genome (except in the instance named above) that is unique in the world.

The Bible states in a number of places that a human life is always a human being from conception. For Instance, Jeremiah 1:5 says: "Before I formed you in the womb I knew you, before you were born I set you apart; I appointed you as a prophet to the nations," and Psalm 139:16 says: "Your eyes saw my unformed substance; in your book were written, every one of them, the days that were formed for me, when as yet there was none of them." According to passages such as these, God's eternal mind knows all His creatures *before* they are physically conceived, and that the human body was formed from "unformed substance" according to the model fashioned in the mind of God.

Nuclear physicist Gerald Schroeder is just as awed as Charles Sherrington by the complexity of "becoming" guided by the chemistry of human body, which he views as just as extraordinarily fine-tuned as the universe. He writes about the millions of fibers in our optic nerves, the billions of in our retinae, the millions of nerve endings we have for smell, hearing, taste, touch, and many other wonderful and complex features that he believes defies the randomness of mindless natural selection because: "all this three-dimensional structure arises somehow from the linear, one-dimensional information contained along the DNA helix."[6]

Sex: Evolution's "Queen of Problems"

The zygote marks the beginning of each human life, but why is there is such a thing as a zygote? The existence of zygotes is an incredibly improbable thing because they are cells containing mixtures of two genomes resulting from the sex act of males and females. On a geological time-scale, sexual reproduction is relatively new on the reproductive scene. Asexual reproduction worked just fine for many millions of years before sex arrived. Life appeared on planet Earth just as soon as it could about 3.8 billion years ago in the form of single-celled organisms called prokaryotes (bacteria and their close cousins, archaea). While bacteria and archaea are incredibly complex in their own right, they are far less complex than organisms classified as eukaryotes. Eukaryotes are organisms with membrane-bound nuclei and cell organelles (compartments within a cell such as mitochondria and endoplasmic reticulum), that prokaryotes lack. Prokaryotes are haploids; that is, they possess only a single set of chromosomes, but humans are eukaryotes; that is, diploids possessing two sets of chromosomes. Each parent donates one half of their chromosomes that combine when sperm meets egg to make 46 in their offspring.

Many forms of life propagate themselves asexually, some both sexually and asexually, but the higher forms of life exclusively reproduce sexually. We know how it is done, but how sex arrived on the stage has been called evolution's "queen of problems." Except for the origin of life and the emergence of human consciousness, no other problem has sown such confusion among Darwinists. The conundrum is: why substitute something as time-consuming and energy-depleting as sex when the more efficient asexual reproductive method had been in place for about two billion years prior to its arrival? Then there is the perplexing issue of how the mechanisms involved in sexual reproduction are maintained given the advantages of simple cell division in asexual reproduction. Natural selection is supposed to preserve mechanisms that work, and not to eliminate them and start all over with something more complex.

Another puzzling issue is that sexual reproduction exists in most species of plants and animals, yet flora and fauna are lines of evolution as divergent and independent as they could be. The same happy accident is assumed to have occurred in both life forms, which is an awful lot to swallow. Theories of why sexual reproduction is with us abound, but they are more like art dressed up in scientific clothing than science itself: "'We have tons of theories, and some are completely crazy,' says Alexey Kondrashov, an evolutionary geneticist at Cornell University."[7] And Meirmans and Strand wrote in 2010 that: "One scientist counted more than 20 different theories in the early 1990s, and the number of theories has certainly not decreased since then."[8]

Because nature puts such a high premium on genetic fidelity, asexual reproduction should be the way to go since it transmits the entire parental DNA intact, whereas sexual reproduction involves genetic shuffling of separate gametes from male and female genomes. Sexual reproduction must be doing something very important, however, since it has been maintained for millions of years. The major evolutionary theory about the advantages of sexual reproduction over asexual reproduction is the Red Queen hypothesis. "Red Queen" is a metaphor taken from Lewis Carroll's book *Through the Looking Glass* in which the Red Queen, who has been running with Alice for a long time without getting anywhere, stops and pants: "it takes all the running you can do to keep in the same place."

In evolutionary biology, the metaphor is used to denote the coevolution of species, and to emphasize that species must continually change ("keep running") to keep up with predators and/or prey. If a species stops running (stops evolving) it loses the competition with species that do not. The hypothesis thus describes an evolutionary arms race between predator and prey or between host and parasite. For example, gazelles must evolve genetic mutations to help them run faster than cheetahs or they would become extinct. If every gazelle ran faster than every cheetah, however, cheetahs would become extinct, so they too would have to evolve genetic mutations for faster running. Host organisms must also obtain the mutations that allow them to resist parasites, which require the abilities conferred by the shuffling and recombination involved in sexual reproduction.

This is fine as far as it goes, but it describes a process that takes place *after* sexual reproduction is already in the picture, so it hardly satisfies as an explanation of how we got it in the first place. Evolutionary geneticists Root Gorelick and Henry Heng wrote a multidisciplinary review on the topic that concluded that while the recombination of genes in sexual reproduction produces variety *within species*, it does so at a very slow rate, and that the primary function of sex is to keep the species genome intact. That is, at the species genomic level the function of sex is to *reduce* diversity. Their primary point is that sexual reproduction has two distinct functions, the most important of which is to *maintain* the existence of a species by assuring genomic integrity. The second function is to *increase* genetic diversity within a species' conserved genome. In other words, at the species genome level evolution reduces genetic variation in the species while increasing it at the individual gene level. They note that if sex was just for increasing genetic diversity it would not have evolved because asexual systems show higher levels of diversity: "In contrast, the increased diversity at the gene level by meiosis [explained below] is secondary, as the combination of genes contributes to new features of existing systems rather than altering the system in any fundamental way."[9]

This sounds like bad news for macroevolution because Gorelick and Heng are saying that sexual reproduction applies a brake to it while permitting microevolution. Microevolution allows organisms to adapt to their environments in Red Queen fashion while still remaining "within kind." As Gorelick and Heng put it: "Resynthesis of the function of sex based on the genome theory has drastically changed the way we study sex and evolution, with sexual reproduction as the key that distinguishes between drastic genome alteration mediated macroevolution and gene mutation mediated microevolution."[10] Recall that in Chapter 16 Jeffrey Levinton noted more or less the same thing about body plans; that is, while evolution occurs within the species producing new adaptations, the basic species genome is highly conserved.

As with the Red Queen hypothesis, Gorelick and Heng's theory does a good job of explaining the advantages of sexual reproduction once it is here, but neither theory explains *how* it got here. Christians have a perfectly good explanation. Genesis 1:27-28 tells us: So "God created man in His own image; in the image of God He created him; male and female He created them. Then God blessed them, and God said to them, 'Be fruitful and multiply; fill the earth and subdue it; have dominion over the fish of the sea, over the birds of the air, and over every living thing that moves on the earth.'" If Darwinists want to explain the incredible jump from asexual to sexual reproduction, they will have to explain how meiosis evolved from mitosis, a problem that greatly bothered W.D. Hamilton, who was widely recognized as one of the greatest biologists of the 20th century. Hamilton wrote that: "if there is one event in the whole evolutionary sequence at which my own mind lets my awe still overcome my instinct to analyse, and where I might concede that there may be a difficulty in seeing a Darwinian gradualism hold sway throughout almost all, it is this event—the initiation of meiosis."[11]

Mitosis and Meiosis

The cells in our bodies are constantly being replaced at different rates, depending on the tissue, and some tissues such as cells in the cerebral cortex are thought to never be replaced. Cells are replaced by cell division in a series of stages known collectively as the cell cycle. There are two forms of cell division: mitosis and meiosis. Mitosis produces two identical diploid cells with the same genetic information as the parent cell. Prokaryotes and certain primitive eukaryotes use mitosis to reproduce, but in humans, mitosis is only used for growth and replacement of bodily cells; i.e., non-gamete, cells. Meiosis is a much more complicated process; one that keeps many a biologist busy examining its minute details all their careers. In Gorelick and Heng's terms, meiosis is the mechanism by which within-species variation is achieved by shuffling the genes of individuals around while maintaining the species genome intact. It is a beautifully orchestrated process with its own internal wisdom that only a Designer could have implanted.

From Figure 18.1 we note that the first stage of meiosis is regular mitosis, producing two diploid cells with identical genetic information. In the next phase, that cell divides again, and then again, to produce four haploid cells containing half (23 chromosomes) the original amount of genetic information of the parent cell. During this process there is a lot of genetic reshuffling, so these cells are not perfect duplications as in mitosis. Meiosis produces only gametes—female eggs and male sperm.

Since meiosis produces gametes, there are major differences in human male and female meiosis. In males, meiosis is suppressed until they reach puberty and occurs in tubules of the testicles which produce the male gametes—sperm. In females it occurs at birth in cells called oogonia, which are stem cells that migrate into the ovary to eventually (only minuscule few) become the female eggs. Meiosis possesses the miraculous ability to half the chromosomal count to make sexual reproduction by the union of male and female gametes possible. The union of 23 chromosomes for the male and female gametes gives the new organism the full complement of 46 chromosomes. This reshuffling is the basis of the genetic diversity we see all around us.

So, how did we get from the efficient one-step process of mitosis to the complications of meiosis? While Wilkins and Holliday keep their Darwinian pedigree intact by writing: "meiosis almost certainly evolved from mitosis," they admit that it seems impossible because: "It has not one but four novel steps: the pairing of homologous chromosomes, the occurrence of extensive recombination between non-sister chromatids [one of two threadlike halves of a replicated chromosome] pairing, the suppression of sister-chromatid separation during the first meiotic division, and the absence of chromosome replication during the second meiotic division."[12] They admit that such a highly unlikely and complex transition from mitosis to meiosis present a huge challenge to Darwinian explanations of the origin of sex. They seem to have specified irreducible complexity lurking somewhere in their subconscious minds, but dare not let it reach the surface lest they be accused of supporting intelligent design: "While the simultaneous creation of these new features in one step seems impossible, their step-by-step acquisition via selection of separate mutations seems highly problematic, given that the entire sequence is required for reliable production of haploid chromosome sets." [13]

Figure 18.1: Mitosis and Meiosis

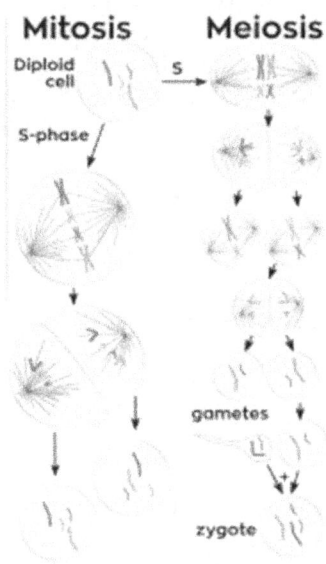

Evolutionist Mark Ridley thinks that sex is evolutionarily absurd. Just think about the meiotic shuffling to obtain a random assortment of genes, and you will agree with him when he writes: "You only have to think of sex to see how absurd it is. The 'sexual' method of reading a book would be to buy two copies [male and female versions], rip the pages out, and make a new copy by combining half the pages from one and half from the other [sexual recombination—half from each parent—to form a new set of chromosomes], tossing a coin at each page to decide which original to take the page from and which to throw away [the random shuffling of the genes during meiosis to get 50 percent]."[14]

Thus, for sexual reproduction to occur according to a materialist view a whole slew of ridiculously implausible things must occur. First, males and females had to evolve; either simultaneously or one sex evolved in order to evolve the other, while still retaining the first. To get males and females, separate X and Y chromosomes had to somehow arise from some unknown ancestral asexual chromosome. The sexes had to evolve the necessary equipment (a penis, a vagina and all the complicated internal sex organs) and the necessary materials (sperm and eggs) to fashion an offspring. Then meiosis would have to evolve from mitosis to ensure that only 50 percent of each parent's genetic material is passed on. All this would have had to occur simultaneously because a male isn't much use without a female (or vice-versa); neither is a penis without a vagina, sperm without eggs, nor the process of meiosis without any of the above. Additionally, the female had to evolve a mechanism to prevent her immune system from destroying the male sperm, which it would otherwise recognize as an antigen (an invader to be destroyed by the immune system). None of these requisites can just hang around for eons of time for the rest to evolve via Darwinian "trial and error." We are indeed "Fearfully and wonderfully made."

Endnotes

1. In Everard, D. 2012, p.152.
2. Gitt, W. 1996, p. 187.
3. de Vries, G. and Södersten, P. 2009.
4. Rux, D. and Wellik, D. 2017.
5. Sherrington, C. 1951, p. 188.
6. Schroeder, G. 1997, p. 187.
7. In Wuethrich, B. 1998, p. 1980
8. Meirmans, S. and Strand, R. 2010, p. s3.
9. Gorelick, R., & Heng, H. 2011, p. 1089.
10. Ibid., p. 1095.
11. Hamilton, W. 1999, p. 419.
12. Wilkins, A. and Holliday, R. 2009, p. 3.
13. Ibid., p. 3.
14. Ridley, M. 2001, p. 209.

CHAPTER NINETEEN

The Human Body: The Temple of God II

> "My strong belief is that God created human beings and therefore he knows about every aspect of the human body. So if I want to fix it, I just need to stay in harmony with Him."
> Ben Carson:
> neurosurgeon and ex-US Presidential candidate

Optimized Perfection

Biologist Jerry Coyne tells us that: "If anything is true about nature, it is that plants and animals seem intricately and almost perfectly designed for living their lives."[1] Note that he says "almost." This alerts us to the fact that certain parts of our bodies are sub-optimal. Atheists pounce on this to ask "What can we expect from bodies cobbled together by trial and error?" Their argument is that an omniscient God would have made the human body perfect. But there is no such thing as a perfect design of *anything;* there is only compromise among multiple competing criteria. Engineer Henry Petroski points out that: "All design involves conflicting objectives and hence compromise, and the best designs will always be those that come up with the best compromise."[2]

Designs of subsystems are constrained by the goal of optimizing the functioning of the whole. Take the last vehicle you bought; what did you want of it? You wanted it to be affordable, safe and reliable, to get good gas mileage, to be agile in performance, and to look good. These are conflicting objectives. If you optimize the first factor, you may buy a used Volkswagen Bug, but forget everything else except good gas mileage. To optimize safety, it would have to be built like a tank, so forget affordability and gas mileage. Automakers would like to optimize all parts, but they must compromise so that the whole is optimized. Your body is like that; some parts may not be of optimal design, but the whole is exquisitely designed. As physicist Chet Raymo remarked: "For all of the improvements an engineer might suggest for the human body, the body is still a thing that no engineer could hope to equal. Fabulously resilient. Capable of stunning feats of endurance. Exquisitely attuned to the environment."[3] We begin our second look at the human body by exploring its protection—the immune system.

The Innate Immune System

When the minuscule zygote has developed into a beautiful baby and is ready to be born, it is confronted with a host of bacteria, viruses, toxins, and other parasites that want to invade its little body. These invaders are collectively called antigens; short for *anti*body-*gen*erating. To defend against these invaders, the child possesses a complex of cells, proteins, and organs with an elaborate communication system that works around the clock called the immune system. The immune system fights off invaders by creating antibodies specific to each antigen.

The child's first line of defense is mom. While the child was gestating in the womb, it began developing its innate immune system—the first line of defense. The innate system is primed by antibodies circulating in the mother's system, and is further primed during the birthing process when bacteria from mom's birth canal are passed on, helping to build the colony of gut bacteria that further contribute to the child's immunity.[4] Most antigens to which the baby is exposed can be handled by the innate immune system, which has receptors on the skin and around vulnerable areas such as the nose and mouth that detect and respond to a limited set of antigens. However, the innate system lacks the "memory" capabilities of a mature system. After birth, the child slowly begins to develop its own system; the adaptive immune system. The adaptive system can marshal a large number of receptors that generate a vast repertoire of defenses by producing specific antibodies that "remember" antigens so antibodies can fight them more efficiently if they invade again.

If asked to choose a single image to symbolize love, it would be the bliss on a young mother's face as she breastfeeds the contented infant snuggled in her arms. This is as close to *agape*—the unconditional love of God for his children; pure selfless love—that most humans will ever know. The look of pure delight on her face is the infant's primordial experience of love: one person taking pleasure in the existence of another. The infant experiences warm tactile sensations snuggled against its mother's skin and closing its lips around her nipple to drink the milk of human kindness. This helps to prime the baby's adaptive immune system, and to protect it from psychological maladies by sending electrochemical impulses along its neural pathways to create our noblest emotions. Breastfeeding is important for mother/infant bonding, which is important for regulating emotional areas of the brain called the limbic system. Ashley Montague notes: "Physiologically, the nursing of her babe at her breast produces in the mother an intensification of her motherliness, the pleasurable care of her child. Psychologically, this intensification serves to further consolidate the symbiotic bond between herself and her child. In this bonding between mother and child, the first few minutes after birth are crucial."[5]

Breastmilk has many immunological benefits for the child. The newborn's inability to mount a fully effective immune response to all antigens in its environment could kill it. Fortunately, newborns receive a 100% safe vaccine in the form of colostrum; mom's "high octane" milk expressed from about day four to about day eight of breastfeeding. After the short period of ingesting this super-food, the infant gets 10 to 14 days of transitional milk, which is a blend of colostrum and thinner breast milk. Transitional milk will eventually be replaced by regular breastmilk which contains many chemical compounds important for brain development. This natural formula contains elements that support the baby's immune system such as proteins, fats, and sugars that cannot be duplicated by formula makers. Breastmilk also transfers many of the mother's developed antibodies to the nursing infant via the milk, and since mothers and their babies are typically exposed to the same antigens, infants gain further protection.[6]

Many of the physiological and psychological benefits of breastfeeding is the result of a hormone called oxytocin, the so-called cuddle chemical. Placing the newborn immediately at its mothers' breast induces suckling, which provokes uterine contractions that help to reduce bleeding and to expel the placenta. Psychologically, breastfeeding combines the panoply of sight, sound, smell, touch, with the tangible evidence in the mother's arms that affirms her womanhood, and stimulates the release of oxytocin, thus intensifying the feelings that released it in a felicitous feedback loop. Oxytocin also reduces mothers' sensitivity to environmental stressors, thus allowing for greater sensitivity to the infant. Oxytocin does this by triggering the release of a brain chemical called gamma-amino butyric acid; an "anti-anxiety" molecule. As determined by various physiological measure, lactating mothers show significantly fewer stress responses to their infants' crying and fussing than non-nursing mothers. The sense of emotional warmness generated by oxytocin motivates mothers' desire to pick up their infants, strengthening the mother/infant bond. The nucleus accumbens, a "pleasure center" in the brain, is a target for oxytocin, so lactating mothers are rewarded with a deep sense of calm pleasure.[7]

The Adaptive Immune System

As Figure 19.1 shows, the adaptive immune system is a society of separate but interrelated sub-systems that, like the gallant state of Israel, are constantly on a war footing. To identify and repel invading antigens, it has an army of spies, commandos, generals, and brigades of fierce soldiers to protect the body from harm. Invasion can occur at many different sites, so these protectors are dispersed throughout the body. Some of the organs of the immune system, such as the appendix and tonsils, were once considered vestigial organs and as evidence for macroevolution; remnants of evolutionary history that were once useful but no longer are. Just as scientists are more and more finding that what used to be called "junk genes" are functional, they are finding functionality in "junk organs."

A key feature of the adaptive immune system is the ability to recognize self from non-self. Like distinctive uniforms of armies, each cell in your body carries a distinctive molecule marker that identifies it as "self" as opposed to cells from another person. An exception to this is male sperm. The female's chemistry identifies sperm as a "defector" and accepts it as friendly. The other exception is the embryo. Because the embryo is a new individual with its cells tagged as such, they are foreign markers not recognized by its mother's cells as self. To prevent the mother's immune response from attacking and destroying the embryo, her immune response must be suppressed, but not completely, or else her health, and the baby's, would be in jeopardy. Maternal immune suppression is thus localized at the implantation site of the uterus.[8] The requirements of embryo and mother survival requires a fine-tuned balance that must come online simultaneously to be effective.

The key players in the immune system are white blood cells, or lymphocytes. Lymphocytes make up less than one percent of blood content and are made in the bone marrow. Once manufactured, the new recruits travel to their home bases—the lymph organs—where they await orders to sally forth to fight. As well as repelling invaders, white blood cells identify and destroy dead self-cells. Lymphocytes come in two varieties: T-lymphocytes (T-cells) and B-lymphocytes (B-cells). T-cells recognize antigens by their "uniforms" (the surface shape of their molecules). In some circumstances, T-cells kill an invader directly, in others it sends signals to headquarters via chemical messengers called cytokines so the immune system as a whole can decide the most effective weapon(s) to use, including killer T-cells, B-cells, and macrophages. Macrophages ("big eaters") are giant circulating white blood cells that under the microscope look like a nightmare from a horror movie. These slithery blobs are the shock troops of the immune system that reach out and gobble up invader cells and self-cells damaged by invaders and release their protein fragments that are then attacked by killer T-cells.[9]

When B-cells encounter these fragments, they manufacture antibodies. Each antibody matches a specific antigen in the same way that a key fits into a specific lock, and in doing so marks it for destruction. Once millions of lock-specific antibodies are made, the immune system remembers the invading locks if they encounter them in the future and can more quickly and efficiently destroy them. This activity is guided by another type of T-cell called a helper T-cell. The helper T-cell is like the field commander coordinating the activity of all other cells and supervising the making of different types of cytokines. When invaders have been surrounded, pounded, and shredded, there must be a mechanism to tell the good guys to cease fire—to down-regulate the system—lest it continue pouring troops into the fray that may inadvertently attack their own cells. Suppressor T-cells cells perform this task when they chemically perceive lots of enemy wounded and dead around and release their own chemical signals to cease attacking.[10]

In the early 1980s, scientists made the intriguing discovery that lymphocytes contain receptors for certain types of messaging brain chemicals, and that macrophages synthesize and release another kind of brain chemical. This led to the idea that the brain and the immune system comprise a single integrated defense system that influence one another. The scientific community was reluctant to accept this because the two systems are separated by the blood-brain barrier (a semipermeable barrier which prevents potentially harmful materials carried by the blood from entering the brain).

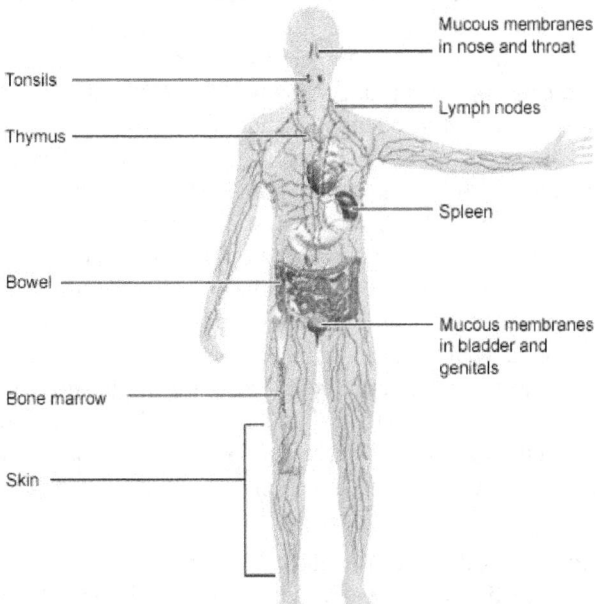

Figure 19.1: Organs of the Immune System

Because the two systems use a common language to communicate, research persisted, and a 2017 review of the literature concluded: "It is now fully recognized that the nervous system and the immune system do not function independently from one another, but that the central nervous system utilizes cellular and molecular elements of immune communication for its own purposes, while the immune response is regulated in part by neuroendocrine mediators produced by immunocytes."[11]

This is another functionally integrated and exquisitely engineered system that cannot be explained by Darwinian mechanisms. A Google Scholar search typing in "immune system" turned up 2,550,000 results. These articles describe in minute detail everything known about it, but microbiologist Michael Behe notes that: "We can look high or we can look low, in books or in journals, but the result is the same. The scientific literature has no answers to the question of the origin of the immune system."[12] One immunologist marvels at the complexity of the system and remarks: "Presumably, the evolution of our immune system is what has allowed us to prevail."[13] The adverb "presumably" is one of many such words that serve to dismiss evolutionary alternatives by fiat; it evolved, period, case closed, now shut up!

The Cardiovascular System: Pipeline of Life

Because cells depend on blood for survival, the cardiovascular system is the first functioning system we develop. Blood and blood vessels begin forming about 14 days after conception, and the developing heart starts pumping within 21 days. This amazing pump has four chambers, two at the bottom that pump blood carrying oxygen out of the heart, and two at the top that receive the returning blood. Returned blood is sent to the lungs to collect more oxygen and to drop off carbon dioxide, which is then exhaled. With each heartbeat, the heart sends blood around the body's pipelines (arteries that carry the blood away from the heart and veins that carry it back) which, if laid end-to-end, "would stretch out from between 60,000 to 100,000 miles. The system is so efficient that every cell in the body is serviced in the process of circulation in an amazing 20 seconds."[14] Physician Sherwin Nuland notes that "the cardiac cycle is no more than yet another of those wonders of coordination and timing upon whose flawlessness and predictability all human life depends."[15] He also tells us that this cycle repeats itself at least 2.5 billion times in an average lifetime and pumps a million barrels of blood, and this momentous task relies "on the performance of trillions of chemical reactions during every instant of its function."[16]

According to an article in *Bloomberg Businessweek*, the most efficient and well-mapped pipeline system carrying the "life-blood" of a community is that of Flanders, the 5,221 square- mile portion of northern Belgium. This huge area has about 400,000 miles of water, sewer, gas, telecommunications, electrical infrastructure, and various other subterranean arteries requiring the services of 300 utility companies and thousands of workers, craftsmen, and engineers to service.[17] When we think of the 60 to 100 thousand miles of blood vessels (depending on body size) is packed into a single human being to provide for its necessities, 400,000 miles packed into 5,221 square miles does not seem such an achievement by comparison. Building and maintaining a complex underground system of pipes, valves, and cables bringing vital services to the community requires intelligence and planning. No one would argue otherwise, but when asked about the far more complicated, self-organized and self-sustaining pipelines in the human body, we are expected to believe it is the result of mindless evolution.

Blood itself is remarkable stuff, consisting of different types of cells in a protein-rich plasma fluid. The cells transported throughout the body by the plasma are red blood cells, white blood cells, and platelets. Red cells account for about 45 percent of the blood's volume. If we think of the white blood cells as the body's military, we may think of the red blood cells as its engineers, physicians, craftsmen, and pickup and delivery guys. These are the workers that perform the mundane tasks of carrying life-giving supplies to cells throughout the body, collecting their wastes, and repairing problems that may arise.

Unlike white blood cells that take on all kinds of shapes, red blood cells are shaped like doughnuts without a central hole. They are extremely flexible and are able to bend enough to squeeze through the smallest capillary in single file. Like their white brothers, red blood cells are produced in the bone marrow at the rate of about two million per second. They last only about four months because they expel their nuclei before leaving the marrow, and thus cannot clone themselves. As cells age and die, they are scavenged by white blood cells and broken up. Their iron content is transported back to the bone marrow by a type of protein called transferrins and used again to produce new red blood cells; cell waste is then excreted from the body.[18]

Finally, we have the platelets, which are not actually cells but small fragments of large cells called megakaryocytes. Platelets comprise the bulk of the blood and help the blood clotting process. Blood clotting involves an enormously complex series of biochemical process to which biochemist Michael Behe devotes a chapter of 15 pages in his book *Darwin's Black Box*. Behe notes: "Blood clot formation seems so familiar to us that most people don't give it much thought. Biochemical investigation, however, has shown that blood clotting is a very complex, intricately woven system consisting of a score of interdependent protein parts. The absence of, or significant defects in, any one of a number of the components causes the system to fail: blood does not clot at the proper time or at the proper place."[19]

Clotting involves platelets gathering at the site of an injury, sticking to the lining of the injured area, and forming a platform for coagulation. A protein called fibrin forms a crisscross scaffolding that prevents further bleeding and a platform for new tissue to form. Clotting must be site-specific so that it does not spread to healthy tissue. This would cause heart attacks or strokes, so clotting must have a tightly controlled regulatory system. The importance of tight regulation of the clotting cascade is revealed in studies of so-called "knockout mice" in which researchers inactivate ("knockout") genes of either the go or stop phase of the cascade. Knockout the gene for fibrinogen, the precursor of the "go" protein fibrin, and mice hemorrhage; knockout the gene for another protein called plasminogen, the precursor of plasmin (the "stop" signal) that degrades clots, and clots form throughout the circulatory system, resulting in death.[20] These regulatory genes had to evolve in unison, but how would a mindless process know that the blood would need both an accelerator and a brake?

Neurobiologist Brad Harrub notes that the cardiovascular system is a perfectly balanced, irreducibly complex, feedback system, and that attempts to explain it in Darwinian terms are confronted with many chicken-or-egg dilemmas. For instance, the heart and kidneys require a constant supply of oxygenated blood, but a specialized hormone called erythropoietin produced in the kidneys is required for the proper function of red blood cells. This hormone protects blood cells from destruction, and stimulates bone marrow stem cells to increase their production. The kidneys require red blood cells to deliver oxygenated blood, but they also produce the protein vital to red blood cells. So, which evolved first—kidneys, red blood cells, or erythropoietin? Any part of this perfectly balanced cycle is useless without the others. Harrub notes that: "A close examination of this complex network reveals architectural planning and design that can only be comprehended in light of an intelligent Designer. Our extensive knowledge of the human circulatory system is tremendous evidence of the existence of Almighty God."[21] Additionally, blood is oxygenated in the lungs, and without the lungs the blood would be useless to the body's cells since they need to have oxygen and nutrients brought to them and their waste removed. The heart, lungs, blood, and blood vessels work together to service the cells; without any part of the irreducibly complex whole we cannot exist.

The Eyes: Windows to the World

There are numerous other amazing bodily systems we cannot address for lack of space, but we cannot leave without briefly addressing the eye. The eye is such a complex structure that Darwin himself saw it as an impediment to his theory: "To suppose that the eye with all its inimitable contrivances for adjusting the focus to different distances, for admitting different amounts of light, and for the correction of spherical and chromatic aberration, could have been formed by natural selection, seems, I freely confess, absurd in the highest degree."[22] Eyes are our windows to the wonders of the universe. Photons of light bathe the universe, and we behold its wonders via a vast array of photoreceptors in the retina linked via the optic nerve that collect them. The optic nerves send the photons all the way to the occipital lobe at the rear of the brain that receives, organizes, and interprets the patterns they generate. There are about 120 million photoreceptors called rods in the retina and some six or seven million cones.

Computer engineer John Stevens compares the ability of our eyes to a computer. While today's computing power is very impressive, it cannot begin to match the performance of the human retina. Stevens explains: "Actually, to simulate 10 milliseconds (ms) of the complete processing of even a single nerve cell from the retina would require the solution of about 500 simultaneous nonlinear differential equations 100 times, and would take at least several minutes of processing time on a Cray supercomputer. Keeping in mind that there are 10 million or more such cells interacting with each other in complex ways, it would take a minimum of 100 years of Cray time to simulate what takes place in your eye many times every second."[23]

Although the party-line is that our eyes *must* have evolved from some primitive light-sensitive patch of brain cells, one must wonder what series of random mutations could reasonably explain the simultaneous origin of the optical system linked to nerves that conduct signals to the back of the brain where photon impulses are converted to patterns our minds can comprehend. Biologist Alan Gillen explained it best when he wrote: "No human camera, artificial device, nor computer-enhanced light-sensitive device can match the contrivance of the human eye. Only a master engineer with superior intelligence could manufacture a series of interdependent light sensitive parts and reactions"[24]

Yet, evolutionists are determined to find "flaws" in the eye's design. Guided by his belief that eyes evolved by natural selection, Richard Dawkins mocks their design: "Any engineer would naturally assume that the photocells would point towards the light, with their wires leading backwards towards the brain. He would laugh at any suggestion that the photocells might point away from the light, with their wires departing on the side nearest the light."[25] But the laugh is on him. Research demonstrates that what appears to be wrong wiring is exactly what is required for visual acuity. According to one research team: "we have propagation methods allowing us to show that light guiding within the retinal volume is an effective and biologically convenient way to improve the resolution of the eye and reduce chromatic aberration. We also found that the retinal nuclear layers, until now considered a source of distortion, actually improve the decoupling of nearby photoreceptors and thus enhance vision acuity."[26]

The body is indeed a wonder of design, with all the organs of the body mutually dependent and harmonized so well that things go right most of the time for our allotted three score years and ten. But you must care for that body and not abuse it, as 1 Corinthians 3:16 tells us: "Know ye not that ye are the temple of God, and that the Spirit of God dwelleth in you?" Anyone who ponders the ultimate meaning for our physical existence will find it impossible to believe that the human body with a brain that can think could be cobbled together by a mindless process of natural selection. Cosmologist Alan Sandage came to that conclusion, noting that all parts of a living organism are much too interconnected and complex to have arrived piecemeal: "How does each part know? How is each part specified at conception? The more one learns of biochemistry the more unbelievable it becomes unless there is some type of organizing principle—an architect for believers—a mystery to be solved by science."[27]

Endnotes

1. Coyne, J. 2009, p. 1.
2. Petroski, *H.* 1996, p. 30.
3. Raymo, C. 2002.
4. Parkin, J. and Cohen, B. 2001.
5. Montagu, A., 1978, p. 63.
6. Petherick, A. 2010.
7. Hiller, J. 2004.
8. Dantzer, R. 2017.
9. Marin, I. and Kipnis, J. 2013.
10. Sompayrac, L. 2015.
11. Dantzer, R. 2017, p. 497.
12. Behe, M. 1996, p. 138.
13. Kipnis, J. 2018, p. 397.
14. Nuland, S. 1997.
15. Ibid., p. 213.
16. Ibid., p. 214.
17. Milner, G. 2017, np.
18. Basu, D. and Kulkarni, R. 2014.
19. Behe, M. 1996, p.78.
20. Behe, M. 2001, pp. 687-688.
21. Harrub, B. 2005. p. 82.
22. Darwin, C. 1982, p. 217.
23. Stevens, J. 1985, p. 287.
24. Gillen, A. 2001, p. 99.
25. Dawkins, R., 1986, p. 93.
26. Labin, A. and Ribak, E. 2010, p. 4.
27. Sandage, A. 1985, p. 54.

Chapter Twenty

The Brain: The Little Universe Within

> "We come to exist through a divine act...Each of us is a unique, conscious being, a divine creation...It is the only view consistent with all the evidence."
> John Eccles:
> Nobel Laureate in physiology and medicine.

The Enchanted Loom

The whole human body is a marvel of design, but sitting atop its shoulders is its most awesome feature: the brain. The brain is God's magnum opus; the most complicated and fascinating entity in the universe. Sir Roger Penrose has said that "If you look at the entire physical cosmos, our brains are a tiny, tiny, part of it. But they're the most perfectly organized part. Compared to the complexity of a brain, a galaxy is just an inert lump."[1] We may marvel at the stunning complexity of the human genome, but even it pales in comparison to this walnut-shaped, grapefruit-sized, three-pound mass of tofu-like tissue sitting in our skulls. Genes build brains, but the building is more complex than the builder. The genome provides the information for assembling the brain, and is constantly working within it, but the genome does not possess a conscious mind to think about what it is doing. We may never completely understand the human brain because as has been said many times in neuroscience circles: "If the human brain were so simple that we could understand it, we would be so simple that we couldn't."

Nobel Laureate neuroscientist Roger Sperry rhapsodized about the brain's complexity and mystery when he wrote that: "In the human head there are forces within forces within forces, as in no other cubic half-foot of the universe we know."[2] Although the brain constitutes only two percent of body mass, about 50 to 60 percent of our genes are involved in building it, and it consumes a voracious 20 percent of the body's energy resources.[3] We could exhaust a whole dictionary of sparkling metaphors to sing the praises of this enchanted loom, because within its chemical soup and electrical sparks lie our thoughts, memories, desires, emotions, intelligence, and creativity. Everything we do engages the brain as its electrochemical circuitry captures our genetic dispositions and environmental experiences and blends them into an efficient, self-aware, free-willed human being. The human cell is a breathtakingly complex factory, but the brain is an even more amazing workshop because it is the temporary home of the conscious mind with which humanity reaches out on its journey of discovery.

The human genome contains about 21,000 genes in its three billion-plus base pairs, but it is estimated that the brain contains a sprawling network of 100 billion neurons that receive and process sensory input. The foundation of the brain begins three weeks after conception when a sheet of embryonic cells called the neural plate, folds into the neural tube. At this point we can discern the four "bulges" that will eventually become the central nervous system; the forebrain, midbrain, hindbrain, and spinal cord. The cells in the neural tube differentiate into their assigned parts throughout the first trimester. The organization of these cells requires a series of protein signals that direct waves of migrating cells to their allotted place. When they reach their destination, a protein signal called Reelin signals them to stop: "Each new wave of migrating neurons bypasses the previous wave of neurons such that each new wave of migrating cells assumes the most superficial position within the developing cortex. As each new wave of neurons reaches the top of the cortical plate, it moves into the zone of Reelin signaling and receives the cue to stop."[4] Once neurons are safely embedded in their targeted area, they develop means of communicating with other neurons. To be part of the information processing network, neurons develop dendrites that receive information and axons that pass the information on to other neurons.

There are many other kinds of cells besides communicating neurons in our brains: "The unique character of our brain seems to lie in the existence of many (perhaps as many as forty) different types of neurons, some perhaps specifically human."[5] The most abundant of these are glial cells, which form in the second trimester on the walls of the embryonic neural tube. Glial cells are non-communicating cells that provide important functions for neurons, such as providing physical support, nourishment, maintaining their stability, and forming an insulating sheath of myelin around them. At about 24 weeks, the brain begins to wrinkle, and by about 26 weeks the grooves deepen and it begins to look like a newborn baby's brain.[6]

Collectively, the different brain cell types somehow produce the single most important characteristic that distinguishes humans from the rest of the animal kingdom—a conscious, intelligent mind. Intelligence is a gift from God housed in our brains that has enabled humanity to gain dominion over the Earth, and has allowed our species alone to know and love our Creator. How did we come by this wonderful gift? No one can answer this from a materialist perspective because as physician and geneticist Joseph Kuhn remarks: "The estimation for DNA random mutations that would lead to intelligence in humans is beyond calculation."[7]

Figure 20.1 shows major areas of the brain I briefly discuss, realizing that it can be divided into many hundreds of parts and sub-parts other than those shown. The figure identifies the four lobes: the frontal, parietal, occipital, and the left and right temporal lobes. As the figure shows, the brain stem is continuous with the spinal cord, the main pathway for information connecting the brain and peripheral nervous system. The brain and spinal cord make up the central nervous system, and nerves running from them form the peripheral nervous system. The peripheral nervous system controls information flow between the brain and the rest of the body, and controls basic functions such as breathing, swallowing, and blood pressure.

Figure 20.1: Major Areas of the Brain

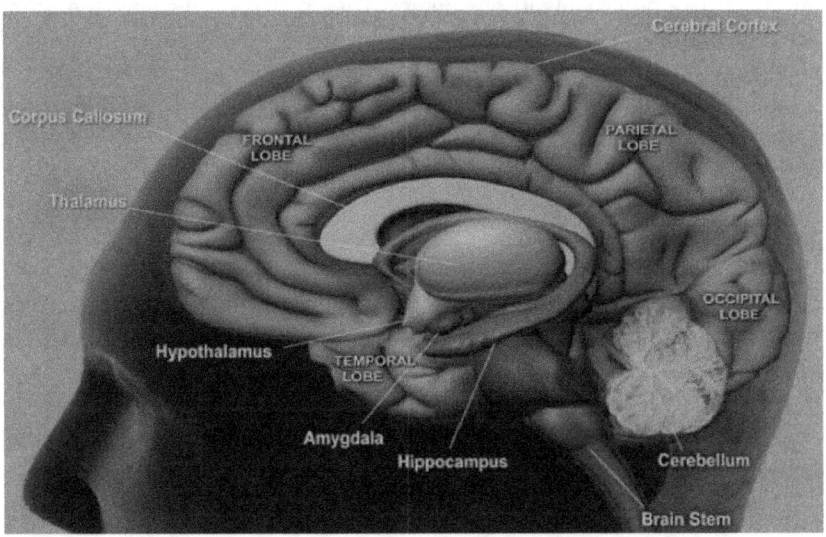

The brain stem has several sub-organs that carry sensory information to the thalamus, whose main function is to relay these signals to the cerebral cortex where most information processing occurs. Below the thalamus lies the hypothalamus, which plays a key role in connecting the hormonal system to the central nervous system. Wrapped around the thalamus is the hippocampus (one located on each hemisphere). The hippocampi consolidate short-term memory, learning, and emotion, and help to connect regions of the cerebral cortex with each other. Sitting atop the hippocampi at their endpoints are the amygdalae, whose main role is the processing of emotions. The amygdalae and hippocampi are among many other organs that comprise the limbic system, which is the emotional "feeling brain" as contrasted with the cerebral cortex, the "rational brain." The corpus callosum is a thick track of over 200 million nerve fibers connecting the brain's left and right hemispheres. Finally, we have the cerebellum, which is primarily concerned with motor control and balance. At the center of the brain lie four large fluid-filled spaces called the ventricular system where most of the neurons are produced during early development. The ventricular system is filled with cerebral spinal fluid, which is completely recycled several times per day. This system has many functions "including cushioning and protection of the brain, removal of waste material, and transport of hormones and other substances."[8]

Neuron Functioning

Figure 20.2 is an image of a neuron, an axon, multiple dendrites, and the process of synaptic connection. Neurons are the basic units of the brain, and are much like other cells in our bodies. They contain a nucleus and other constituents of somatic cells, but their uniqueness lies in their patterns of connectivity as members of a network of other neurons that convey conscious information. Neurons signal information to other neurons by converting environmental input into electrochemical impulses and transmitting them to other cells. Each neuron has one axon attached to the cell body, and numerous dendrites (some have thousands), which are branched extensions of the cell. Axons are coated by a myelin sheath made of cholesterol that acts like the insulation around electrical wiring, and in similar fashion, amplifies nerve impulses and protects the axon from short-circuiting. Myelinated axons transmit impulses about 100 times faster than unmyelinated axons, and are the brain's "white matter." Its "gray matter" consists of cell bodies and unmyelinated axons.[9] The number of branching dendrites varies among neurons, but each serve as a receiver, picking up impulses from neighboring neurons.

Neurons pass their information along the axon in the form of electric signals by the exchange of charged atoms (ions) in and out of the axon's permeable membrane (see Figure 20.2). The energy needed to perform the neuron's activities is provided by the same adenosine triphosphate (ATP) that drives all cells. Impulses travel down axons like neon lights flickering on and off until they reach the axon terminal point; the presynaptic knob. At this point, the message is changed from electrical to chemical in the form of neurotransmitters stored in tiny packages called vesicles. These vesicles open up and spill their contents out into the synaptic gap between the sending and receiving neurons. Neurotransmitters cross the synaptic gap to make contact with postsynaptic receptor sites where, depending upon the ratio of excitatory to inhibitory messages present, the message is translated back into electrical form, either for further transportation of the signal or its inhibition. The synaptic gap is about 20-30 nanometers wide. To put this in perspective, the width of a human hair is about 3,000 times bigger than that.[10] A neuron has the capacity to make many thousands of connections with other neurons, and all 100 billion of the brain's neurons "form over 100 trillion connections with each other—more than all of the Internet connections in the world!"[11]

Once neurotransmitters have passed on their messages, excess amounts are pumped back up into the presynaptic knob or degraded by enzymes. This activity takes place at a dazzling pace. A neurotransmitter remains in the synaptic gap for only 1/500 of a second before being transported back or destroyed to prevent signal confusion when the next signals arrive. The anatomical and chemical systems and circuits are interactive entities, each affecting and affected by the others.

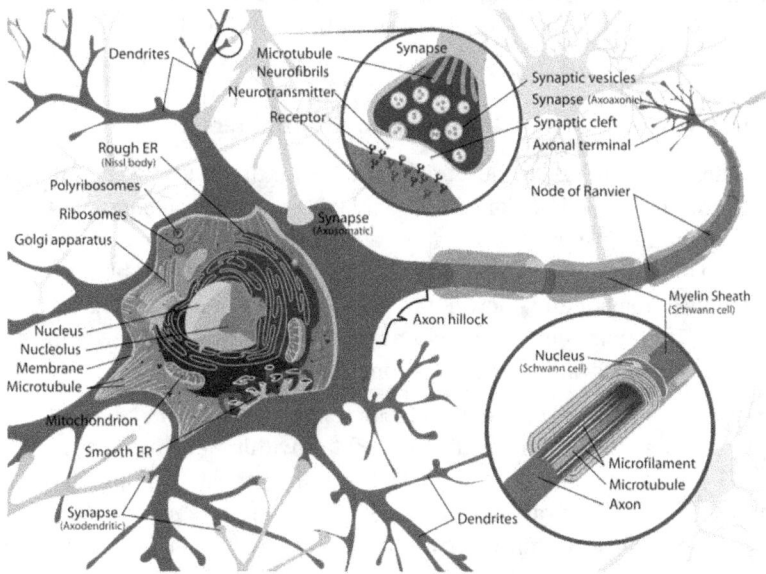

Figure 20.2: Neurons, Axons, Dendrites and Synapses

How much more efficiently does the brain operate compared to human efforts to mimic it? Neuroscientists try to follow the logic of the electrochemical processes in the brain using super-fast computers with high-performance neuronal network simulator software (NEST). Using NEST, a research team showed just how difficult it is to duplicate brain processes. The team created a huge artificial neural network of 1.73 billion cells connected by 10.4 trillion synapses. While this is very impressive, it is just a tiny fraction of the neurons and synapses the brain contains. It is no surprise that the neuroscientists were not able to simulate brain activity in real time. It was noted that: "It took 40 minutes with the combined muscle of 82,944 processors [small computer chips that receive input and provide appropriate output, just like a neuron] in K computer [a supercomputer in Kobe, Japan] to get just 1 second of biological brain processing time. While running, the simulation ate up about 1PB [1 petabyte (PB) is equal to 10^{15} bytes] of system memory as each synapse was modeled individually."[12]

Forty minutes, almost 83,000 processors, and a quadrillion bytes of memory just to get a mere second of the output that our billions of neurons get in real time day in and day out! Data such as these show why comparing brains to computers—while understandable in some sense since they both process information—is wrong. Neuroscientist Miguel Nicolelis considers all analogies of the brain with computers to be "a bunch of hot air." He remarks that "The brain is not computable and no engineering can reproduce it." Nicolelis' statement is based on the fact that brain activity is the result of unpredictable, nonlinear interactions among its multiple billions of cells, adding that "You could have all the computer chips ever in the world and you won't create a consciousness."[13]

The "Go-Stop" Neurotransmitters

There are many different neurotransmitters shunting information around in the brain, but we will be concerned only with behavior-balancing serotonin and dopamine. Just as the immune and blood clotting systems must be activated and deactivated appropriately, we must be motivated both to engage in adaptive behavior and know not to go beyond reasonable limits. Unlike the balanced immune and blood clotting systems, however, the behavioral "stop/go" system is under conscious control.

The behavioral stop/go system is facilitated by three interacting systems of emotional and behavioral regulation located within separate brain circuits fueled by dopamine and serotonin. The first is the behavioral approach system (BAS) designed to motivate us to seek reward. Figure 20.3 shows that dopamine is the major neurotransmitter that motivates and rewards us. Neurons whose principle neurotransmitter is dopamine are found primarily in the ventral tegmental area (VTA), the substantia nigra (SN), and the hypothalamus. Both the VTA and the SN are involved in reward via their connection with the nucleus accumbens, a major pleasure center. To assure that we do what is necessary for the continuation of life, God has provided us with a built-in reward system when we do what we must to survive and to be "fruitful and multiply." The reward is a shot of dopamine, which is why it is nicknamed the "happy hormone." The BAS is a general approach mechanism designed to solve the adaptive problem of desiring and attaining critical resources, such as food, water, and sex by rewarding us when we partake of them.[14]

In the pursuit of our goals we may encounter dangers that must be avoided. There is often no time to think about our options when our senses tell us to fight (defending self or others from attack), flee (running from attack or other dangers when fighting is not an option), or freeze (such as when a sexual assault victim stops struggling to avoid physical harm). We would have difficulty surviving if we lacked this visceral early warning system that psychologists label the fight/flee/freeze system (FFFS). The FFFS is part of the autonomic nervous system (ANS); a part of the peripheral nervous system associated with the neurotransmitter-hormones norepinephrine and epinephrine, and with brain structures such as the hypothalamus, amygdala, and prefrontal cortex (PFC).

Figure 20.3: Major Pathways of the "Stop-Go" Neurotransmitters

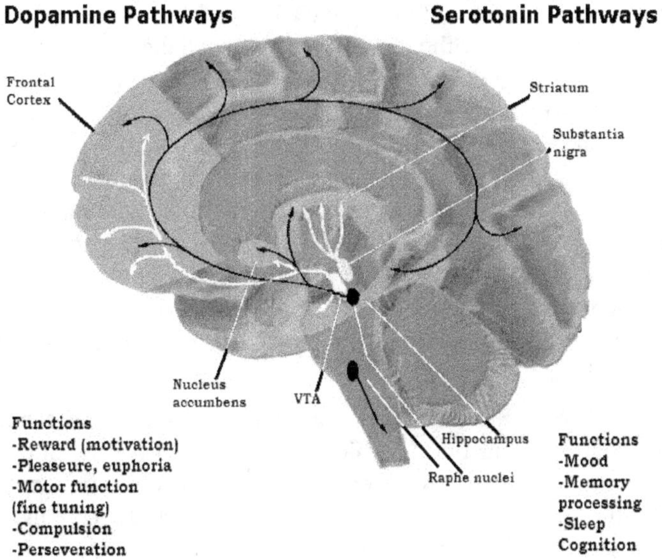

Dopamine Pathways   **Serotonin Pathways**

Frontal Cortex
Striatum
Substantia nigra
Nucleus accumbens
VTA
Hippocampus
Raphe nuclei

Functions
-Reward (motivation)
-Pleaseure, euphoria
-Motor function (fine tuning)
-Compulsion
-Perseveration

Functions
-Mood
-Memory processing
-Sleep
Cognition

Preventing the BAS from going too far in the pursuit of our pleasures is the behavioral inhibition system (BIS). The BIS is associated with serotonin and with limbic system structures such as the hippocampus and the amygdala that feed their memory circuits into the PFC.[15] Many of the effects of serotonin have to do with causing the levels of other neurotransmitters to increase or decrease at different brain regions. The BIS is activated by joint activation of the BAS and FFFS; that is, by conflicting approach-avoidance stimuli that produces anxiety, risk assessment, and caution. The BIS is thus a system of conflict detection and resolution that inhibits behavior until either the BAS or FFFS is most strongly engaged (do I go for the potential reward or not?). The BIS strives for the ideal between desire and prudence, and represents the emotional component of moral rules.[16]

Figure 20.3 identifies a large area called the frontal lobe. A part of the frontal lobe lying just above the eyes is the prefrontal cortex (PFC), which has been called "the most uniquely human of all brain structures."[17] The PFC is the terminal point for much of the emotional input from the limbic system, and the origin of our responses to it, adding rationality to emotion. Because of its many connections with other brain structures, it is considered to play the major integrative and supervisory role vital to the forming of moral judgments, mediating affect, and social cognition. As such, it is a major participant in the moral guidance of the BIS.

Why do we have a system urging us to exercise self-control when seeking our wants and needs? The short answer is that without the BIS the world would not be a very pleasant place. Without the BIS's stop signal the BAS would run wild, just as the immune and blood clotting systems would without their stop signals. The BIS enables us to navigate complex interactions with others in groups in a way beneficial to all. The BIS can be thought of as the so-called "social brain" that enables us to seek our goals with the good sense to know that our goals are best achieved by cooperating with others and enjoying a good reputation among them. Thanks to the BIS, most of us learn to seek our pleasures with temperance and prudence. If we are to abide by God's will to love our neighbors as ourselves, we must not act in impulsive ways when we seek our natural desires at the expense of the legitimate needs and safety concerns of others.

While our stop/go systems are material, once activated they come under the conscious control of our minds, which has the capacity to monitor its own inputs and outputs and to override temptation. God did not create us incapable of abusing the impulse to engage in immoral pleasures, and becoming addicted to sex, drugs, and alcohol. He created us capable of being moral creatures, but that necessitates the creation of humans capable of being immoral creatures. He does not force us in either direction because He wants us free to determine the direction of our lives ourselves. Like all bodily systems, He implanted our BIS/BAS/FFFS systems to help us navigate our earthly lives in accordance to His will; but it is everyone's free choice to do or not to do this.

Wiring the Brain: Synaptogenesis

Few of the "wiring" patterns of the brain are present at birth. Wiring needed for sheer survival, such as brain areas controlling heart rate, breathing, and body temperature are present at birth, and housed primarily in the brainstem and the cerebellum. This is the brain's "hardware." The wiring of its "software" is governed primarily by experience, and is called synaptogenesis—the creation of synaptic connections between neurons. At the peak of synapse creation, the cerebral cortex of healthy toddlers may create up to 2 million synapses per second, and by

2 years of age they will have approximately 100 trillion.[18] During the first few months of life, dendrites proliferate and glial cells wrap around axons to begin the process of myelination. Dendrite growth and axon myelination continues throughout life, but proceeds at an explosive rate during infancy and toddlerhood. The "lower" brain regions necessary for survival are the first to be myelinated, and some "higher" brain regions, especially the prefrontal cortex, are not fully myelinated until early adulthood.[19]

The brain has limited space, so synapses have to compete, and most will not survive the competition. The brain creates and eliminates synapses throughout life, but creation exceeds elimination in the first two years. After this period, there is a lot of elimination, after which production and elimination are roughly balanced until adolescence when elimination exceeds production again.[20] Just as gardeners prune their rose bushes to produce healthier roses, pruning synapses is required for the normal growth of the brain, and results in a "lean, mean, learning machine." In the earliest stages of synaptogenesis, axons send out their feelers looking for any available partners. Frequent couplings in response to environmental input establish functional connections between neurons like the establishment of a trail in the wilderness. The more often the trail is trodden, the more distinct it becomes from its surroundings, and the easier it is to follow. Brains are built to store experiences and build on circuit trails already there, which is why trail-blazing is easy in our early years but becomes more difficult as we age.

Experiences with strong emotional content are accompanied by especially strong electro-chemical impulses and become more sensitive and responsive to similar stimuli in the future.[21] Synapse retention is thus very much a use-dependent process, with those connections that are engaged most frequently and most strongly being most likely to be preserved. This process is summed up in neuroscience's pithy saying: "The neurons that fire together, wire together; those that don't, won't."[22] We obviously want "firing and wiring" patterns to establish neural tracks leading to a terminus signaling that the world is a loving and friendly place.

We have taken this brief journey into the most amazingly complex structure in the universe to again impress upon you the wonders of God's work. The brain's complexity is truly mind-boggling, because whatever mysteries lie within its folds, this grapefruit-sized object made of the same atoms that make everything else, has enabled humans to literally reach for the stars. It is so well calibrated to environmental circumstances that, in concert with the genome, the brain makes each and every one of us unique. Darwinists will tell you that this marvel of design arose from the same evolutionary process as everything else. They can tell us why a larger and more complex brain was "needed" and why it is useful, but not *how* it came to be. As noted earlier, it would require an unimaginable number of genetic mutations to arrive at a human brain with intelligent self-awareness. Moreover, because the nervous system controls all other bodily systems, the brain and the systems it controls would have to evolve simultaneously—a brain with nothing to control would be pretty useless, as would be any bodily mechanism without a brain to direct it. Given the waiting time problem for positive coordinated mutations, it would take a far longer time than the universe is old to get all those advantageous mutations to evolve concurrently. The most amazing thing about the brain, however, is that this particular configuration of dead atoms is the material facilitator of our immaterial conscious minds.

Endnotes

1. In Holt, J. 2012, p. 178.
2. Fincher, J. 1982, p. 23.
3. Mitchell, K. 2007.
4. Stiles, J. and Jernigan T. 2010, p. 337.
5. Mayr, E. 2001, p. 252.
6. Kolb, B., Mychasiuk, R., & Gibb, R. 2014.
7. Kuhn, J. 2012, p.44.
8. Stiles, J. and Jernigan, T. 2010, p. 330.
9. Fields, R. 2014.
10. Boulos, R. et al, 2013.
11. Weinberger, D., Elvevag, B. and Giedd, J. 2005, p. 5.
12. Scornavacchi, M. 2015 p. 14.
13. In Regalado, A. 2013, np.
14. Ktorupić, D., Gračanin, A. and Corr, P. 2016.
15. Hughes, K. et al, 2012.
16. Walsh, A. 2019.
17. Goldberg, E. 2001, p. 2.
18. U.S. Department of Health and Human Services, 2009.
19. Sowell, E., Thompson, P. & Toga, A. 2004.
20. Giedd, J. 2004.
21. Shi, S. et al., 2004.
22. Penn, A. 2001, p. 339.

CHAPTER TWENTY-ONE

Mind/Soul, Consciousness, and Love

> "To me, then, human consciousness lies outside science, and it is here that I seek the relationship between God and man."
> Nevill Mott: Nobel Laureate physicist

Consciousness and the Mind

The impression gained from the electro-chemical shunting around of molecules in the brain is that they fully determine human behavior in bottom-up fashion. This is not the case. If sex is biology's "queen of problems," consciousness has to be its "king of problems" (the origin of life is the "Grand Emperor"). No one has the slightest idea how anything material like the brain could be conscious of itself. Materialism relegates the conscious mind to brain activity with no independent existence; "A computer made of meat," as some neuroscientists like to call it. They say this because thought cannot occur without the brain, and because they can change brain states by stimulating neurons or administering drugs and other substances. The brain and the mind are co-dependent, but while we can have a brain without a mind, we cannot have a mind without a brain any more than we can see without eyes. This may be true, but there is a large and growing literature that affirms that the consciousness mind exists as an abstract entity separate from the brain, and perhaps it is another word for the soul in its earthly existence.

The idea that mind and body are two distinct things with different essential qualities is known as mind/body or mind/brain dualism. It is an ancient notion, but the most well-known version is credited to the 17th-century French physician, philosopher, and mathematician, Rene Descartes. According to Descartes, we can doubt anything except that we doubt, and doubting is thinking. Thus, the only thing we can be absolutely certain of is that we think, as his famous dictum asserts: "*Cogito ergo sum*" ("I think, therefore I am"). For Descartes, the body is an extended (it occupies space) material unthinking thing subject to mechanical laws, while the mind is an un-extended (it transcends space) immaterial thing that thinks, and is not subject to mechanical laws.[1] This is not a popular view among materialists, but David Chalmers, who has perhaps researched and written more on consciousness than any other scientist alive, argues that the conscious mind does not logically depend entirely on the brain, and thus cannot be reduced to it. Instead, he argues that consciousness supervenes on the brain in some unknown way. He writes: "I resisted mind-body dualism for a long time, but I have now come to the point where I accept it, not just as the only tenable view but as a satisfying view in its own right....I can comfortably say that I think dualism is very likely true."[2]

The configurations of our brain states occur in top-down fashion in response to our goals, beliefs, and desires in a way not captured by bottom-up mechanistic descriptions. Recall again our discussion of the brain chemistry of love. Reductionist accounts describe what goes on in the brain *after* people are in love; in no way is it a description of *why* they are in love. The relevance of the brain activity has for every lover is not mechanistic; the relevance is the information in their minds about what each represents to the other. I am thrilled with the windows into the mind that neuroscience has provided, but disturbed by the price we pay for transcribing the mind onto chemicals and tissue. Mental and physical events are intimately connected, but to say that the former is *nothing but* a function of the latter is illogical. Just as your car gets you to the next town without making you go there, so does your brain facilitate your mental journeys without making you take them.

Our brains allow us to function as free agents by providing the mechanisms by which we emote, think, believe, desire, plan, and worship, but it is our minds that instigate these things, and then the brain facilitates them. All human actions require a mind that forms intentions and an acting agent to carry them out. The contents of the mind form intentions, and physical brain states reflect those intentions. Minds exercise causal agency; we pick up a spade to dig a hole or strike a computer key to make a letter, all at the instigation of mind, and our extended bodies are the instruments by which we accomplish such tasks. That an immaterial mind has such power is not such an abstract idea that it may seem. God is disembodied mind, and since we are created in His image, our minds (souls?) are disembodied after death. The brain is a very complicated piece of biological machinery, but it cannot understand why that which it perceives gives rise to an intentional action; only a mind can do that.

Just as the lungs exist so that we may breathe, the brain exists so that we may "mind." Studies of London taxi drivers provide compelling evidence of the brain's ability to transform itself in response to mental activity. London cab drivers have to constantly learn new routes in a city of over 600 square miles with streets laid out in a spaghetti-like snarl. Using functional magnetic resonance (fMRI) machines, researchers find that cab drivers have significantly larger hippocampi (the organs of memory) than Londoners employed in other working-class occupations. Moreover, the longer they had been employed as cab-drivers the larger these structures are found to be.[3] If mental activity changes brain structures, perhaps it is not that the mind is "just the brain at work," but rather that the brain is "just the mind at work."

When we think of a mathematical equation, a musical score, or a philosophical argument, it is the meaning we ascribe to those thoughts that are fundamental, not the neural substrates recruited to make the meaning possible. It is the mind that organizes the bombardment of external stimuli that gives meaning to the brain's physical activity. The brain cannot bootstrap itself into meaningful activity by purely internal processes; it must interact with information content from outside itself in a reciprocal causal way. As professor of neurosurgery, Eban Alexander notes: "The brain serves more as a reducing valve or filter, limiting pre-existing consciousness down to the trickle of the illusory 'here and now' in which we find ourselves in this physical realm."[4] Alexander is saying that during our brief sojourn on Earth (his illusory "here and now"), we need a relief valve to limit our pre-existing consciousness to manageable levels, and the brain performs this filtering for us. This assumes that souls exist prior to a person's physical existence, but this is denied by almost all Christian denominations.

Mind and Information

It is useful to think of mental phenomena like we think of DNA—as information. Mental information is stored in the neural structures of the brain just as biological information is stored in DNA. In their packaged form, both are temporarily "material," or at least materially housed. When our thoughts are communicated to other parts of the brain, or outward to other minds, they are converted from their material substrate to a form of immaterial energy. When stored by self or other, they are again embodied in matter. Think of this as analogous to writing on a computer. When I write at my computer, what I intend to say precedes the electrical patterns engaged within the computer, and then my thoughts become physically manifested on the screen and are stored in the electrical entrails of the computer. The information content is obviously not determined by the electrical patterns within the computer, although I need the computer (or some other physical medium) to make my thoughts manifest. My thoughts precede these electrical patterns, and are thus not caused by those patterns. When I decide which keys to hit to make manifest the thoughts that "come to mind," a specific sequence of electrical activity is fired up within the computer; thus, my immaterial mind has acted causally on a material object.

Likewise, when I wrote this sentence I "fired up" a specific sequence of electrochemical activity in my brain, which reflected the neural correlates of my mind. If I change my mind to form a different sentence, a different sequence of neuronal firing takes place. This is top-down causation by which physical events in my brain are caused by my mind. Every mind state is also a brain state, but mental properties are not reducible to neural properties. Nobel Laureate neuroscientist and surgeon Sir John Eccles supports this view: "The more we discover scientifically about the brain the more clearly do we distinguish between the brain events and the mental phenomena and the more wonderful do the mental phenomena become. There is a fundamental mystery in my personal existence, transcending the biological account of the development of my body and my brain. That belief, of course, is in keeping with the religious concept of the soul and with its special creation by God."[5]

Likewise, Nobel Laureate physicist Eugene Wigner opines that quantum physics renders materialism logically inconsistent with the mind: "one may well wonder how materialism, the doctrine that 'life could be explained by sophisticated combinations of physical and chemical laws,' could so long be accepted by the majority of scientists. The reason is probably that it is an emotional necessity to exalt the problem to which one wants to devote a lifetime."[6] Writing of the awesome intellect of the great mathematicians, Wigner notes: "certainly it is hard to believe that our reasoning power was brought, by the Darwinian process of natural selection, to the perfection which it seems to possess."[7] Stressing the complexity of the mathematics involved in quantum mechanics, he adds: "It is difficult to avoid the impression that a miracle confronts us here quite comparable in its striking nature to the miracle that the human mind can string a thousand arguments together without getting itself into contradictions or the two miracles of the existence of the laws of nature and of the human mind's capacity to divine them."[8]

Getting to Mind

How did we acquire a conscious mind capable of such complex abstract reasoning? How did this lump of mushy tissue acquire the capacity to be aware that it exists, to inject meaning into its circuitry, and the intellectual power to probe itself? Consciousness is *the* most difficult problem with which psychologists and neuroscientists have to wrestle. But why should consciousness be such a perversely difficult problem when nothing is more obvious than the fact that we are conscious? As obvious as its existence is, consciousness is so difficult to define because it is abstract and operates mysteriously. As an immaterial abstraction, it is a part of reality that cannot be reduced to physics and chemistry. This is anathema to the materialist notion that only material exists, so the best way out is to consider consciousness an illusion, as Daniel Dennett (one of atheism's "Four Horsemen") has said: "It's the brain's 'user illusion' of itself."[9] Nobel Laureate physicist Robert Millikan, however, views consciousness as a gift of God: "The most amazing thing in all life, the greatest miracle there is, is the fact that a mind has got here at all, 'created out of the dust of the earth.' This is the Bible phrase, and science today can find no better way to describe it—a mind."[10]

Materialists offer no real alternative except to say this highest manifestation of life is another strange coincidence we must simply accept as a brute fact. Witness Stephen Jay Gould's statement: "Consciousness, vouchsafed only to our species in the history of life on earth, is the most god-awfully potent evolutionary invention ever developed. Although accidental and unpredictable, it has given *Homo sapiens* unprecedented power both over the history of our own species and the life of the entire contemporary biosphere"[11] How could Gould's accident possibly have occurred via evolution when Darwinists tell us that evolution's sole "purpose" is the continued existence of life? Natural selection is supposed to select only those alleles that contribute to the twin goals of survival and reproduction, so all humans would need are the instincts that further these goals, just like all other species.

Humans have these instincts, so why add a conscious mind to the human repertoire when the rest of creation does quite well without it? Adding mind is an impediment to seeking reproductive success, since humans generally seek it with moral rules that prevent them from seizing every mating opportunity that presents itself. It is difficult to see evolutionary pressures being exerted to gift us with such an impediment to biological fitness. Some neuroscientists try, believing that consciousness emerged from the increasing complexity of social life, and that the need to navigate more social relationships led to a bigger brain. However, the human brain is much bigger relative to body size than is necessary to fulfill the Darwinian imperative, and it is energetically costly to maintain. Much brain energy is expended appreciating abstract reasoning in mathematics and science, and in the creation and appreciation of music, art, and the beauty of the world. This is a wonderful gift as Wigner says, "we neither understand nor deserve," but it does not afford our species any fitness advantage in raw Darwinian terms, so we should not have such a marvelous "illogical" gift. Physicist John Polkinghorne likewise notes that: "Human powers of rational comprehension vastly exceed anything that could be simply an evolutionary necessity for survival, or plausibly construed of some sort of collateral spinoff from such a necessity."[12]

Darwin himself had a "horrid doubt" about what his theory meant for the human mind. In a letter to W. Graham in 1881, Darwin wrote: "Nevertheless you have expressed my inward conviction, though far more vividly and clearly than I could have done that the Universe is not the result of chance. But then with me the horrid doubt always arises whether the convictions of man's mind, which has been developed from the mind of the lower animals, are of any value or at all trustworthy."[13] If our thoughts are nothing but accidental molecules shunting around in brains made by the accidental concatenation of atoms formed in accidental universe, we may join Darwin in asking how we can believe them at all trustworthy. C.S. Lewis also wondered how such an accident would provide him with a correct account of all the other accidents, and wrote that: "It's like expecting that the accidental shape taken by the splash when you upset a milk jug should give you a correct account of how the jug was made and why it was upset."[14]

The Evidence from Near-Death and Pre-Death Experiences

There is a large amount of scientific data supporting the belief that our conscious minds (souls?) are disembodied after death in the form of studies of near-death experiences (NDEs). An NDE is one in which a patient is ruled clinically dead—no heartbeat or brain activity—and who is then resuscitated. About 10 percent of resuscitated patients report seeing things that suggest a transcendence of space, time, and perceptual boundaries. These visions include deceased relatives they never knew in life, and many other things they could not possibly have known. Many NDE patients report intense emotional feelings of joy, peace and love, and being drawn to a bright light. Almost all report coming closer to God by their experience and no longer fearing death. This is perhaps because: 'When an individual knows with a sense of unshakable certitude that he can exist outside of his own body, he intuitively understands that physical death is not an end."[15]

Oncologist Kenneth Long has collected thousands of cases of NDEs. He was once skeptical of NDEs, but now writes: "Every shred of evidence from near-death experience and a number of other related experiences all convincingly point to the conclusion that consciousness, that critical part of who we are, survives physical brain death." Long relates that many NDE patients did not want to return to their bodies because "they feel very intensely present, positive emotions in their near-death experience, more so typically than they ever knew on earth. They very strongly like this afterlife realm, this unearthly realm, which some call Heaven, and there's a sense of familiarity like they've been there before."[16] Remember, NDE individuals have no brain activity, so these perceptions had to come from something transcending the physical.

Neurosurgeon Eben Alexander was also a former skeptic who refused to look at the NDE data until he had an NDE himself. Given his experience, and after a long look at the data, he is now of the opinion that: "Those who assert that there is no evidence for phenomena indicative of extended consciousness, in spite of overwhelming evidence to the contrary, are willfully ignorant. They believe they know the truth without needing to look at the facts."[17] In his book *The Map of Heaven: How Science, Religion, and Ordinary People are Proving the Afterlife*, Alexander recounts hundreds of examples of NDEs and pre-death experiences in which people related the death of another person, whose death was unknown to others but later verified.

Psychiatrist Bruce Greyson also provides a number of examples of NDEs, including that of 9-year-old Eddie Cuomo who was in a feverish coma for 36 hours. When his fever broke, Eddie told his parents that he seen his deceased Grandpa (and other deceased relatives) whom he never saw in life. Eddie also revealed that he saw his sister Teresa, who told him he had to go back to the living. Edie's father discounted this because he had spoken with Teresa, who was attending college in another state, two nights before. Tragically, when Eddie's parents called the college, they were informed that Teresa had been killed in an automobile accident. The college had not been able to contact the Cuomos because they were at the hospital maintaining vigil at Eddie's bedside. This was a "clincher" instance of NDE for Greyson, who quotes a British physician who "concluded that the type of deathbed vision in which the dying 'appear to see and recognize some of their relatives of whose decease they were unaware, 'affords perhaps one of the strongest arguments in favour of [spiritual] survival.'"[18]

Some formerly skeptical physicists are also exploring NDEs using the tools of quantum mechanics, particularly quantum entanglement, a phenomenon Einstein described as "spooky action at a distance." As Contzen Pereira and Shashi Reddy explain quantum entanglement: "Physicists have experimentally demonstrated the entanglement of two particles no matter how far apart they are (even a billion miles apart, in theory), so a change in one particle instantly creates a simultaneous change in the other as if they were connected or in someway the same particle."[19] The basic idea of the "cell-soul pathway," as Pereira and Reddy call it, is that in life cell consciousness and soul consciousness are connected, and at death the body releases cell consciousness in the form of energy that is transferred to disembodied consciousness.

Mind and Language: What Purpose?

As noted, a conscious, self-aware, and intelligent, mind has no immediate Darwinian advantage, the proof being that no other species has been similarly blessed with it. It must, therefore, have some other purpose. Gould was right to say that it has given us "unprecedented power both over the history of our own species and the life of the entire contemporary biosphere," but I bet he didn't realize that his words agree with Genesis 1-28: "And God blessed them, and God said unto them, Be fruitful, and multiply, and fill the earth, and subdue it: and have dominion over the fish of the sea, and over the fowl of the air, and over every living thing that moves upon the earth."

Of course, all animals have a rudimentary form of awareness. They are aware of dangers, food sources, and mating opportunities, but they are not aware that they are aware. As purely instinctual beings, they are constrained to act in accordance with their genetic makeup. They are adapted to their environments, but they cannot create them. Because humans have a mind, they have escaped the captivity of mere instinctive responses to sensory perceptions. We enjoy an inner world that demands that we impose order on the content of our sensory perceptions and think above and beyond them. Humans actively create their environments, rather than merely adapting to them, through their ability to think rationally and to communicate their thoughts to others. Language is a powerful externalizer of our thoughts, and the words that give them voice can powerfully influence others, for good or ill. Animals are able to communicate their intentions through sounds and gestures, but such communication is a far cry from the fine nuances of language. Language is so central to our lives that we tend not to spend time thinking about it, and is yet another phenomenon that defies Darwinian explanations.

All the approximately 7000 languages linguists have cataloged contain a universal grammar capable of generating the rules of any specific language. There are so many factors involved in language that evolutionists have difficulty explaining it in a manner consistent with the accumulation of beneficial mutations because the distance between humans and other primates in communicative ability is vast. Neurolinguist Elizabeth Bates suggests that this vast distance implies two options for the existence of language are possible, neither of which is palatable to Darwinists: "either Universal Grammar was endowed to us directly by the Creator, or else our species has undergone a mutation of unprecedented magnitude, a cognitive equivalent of the Big Bang."[20] A linguistic Big Bang would mean that language, with all the structures that make it possible, exploded on the scene instantaneously, defying the Darwinian notion of the painfully slow gathering of beneficial mutations. All issues involved with mutation rates and waiting time discussed in Chapter 16 leads us unequivocally to the first option: we were endowed "directly by the Creator" with this great gift.

In 2014, a team of eight distinguished scientists in evolutionary biology, computer and information engineering, anthropology, and linguistics, reviewed efforts to understand language in evolutionary terms. They note that there are numerous ideas about the problem matched "by a poverty of evidence, with essentially no explanation of how and why our linguistic computations and representations evolved." The archaeological record and studies of animals have provided virtually no parallels to human language or any latent capacity for it, so we have no idea of what selective pressures could be involved. The researchers also note there is little hope of connecting genes to linguistic processes, and write that: "All modeling attempts have made unfounded assumptions, and have provided no empirical tests, thus leaving any insights into language's origins unverifiable. Based on the current state of evidence, we submit that the most fundamental questions about the origins and evolution of our linguistic capacity remain as mysterious as ever, with considerable uncertainty about the discovery of either relevant or conclusive evidence that can adjudicate among the many open hypotheses."[21]

For all the mystery involved, humans learn language effortlessly. "Learn" is perhaps not the right word. Children learn things like mathematics and reading with some difficulty, but they *develop* language like they develop muscle; they breathe it in as if by osmosis. They may even acquire two or more languages simultaneously if exposed to them long enough while their brains are most plastic; often without the accent of one being transferred to the other(s). Evolutionists posit that an increase of cells in the frontal and motor speech centers of the brain accounts for language, but this is little better than saying humans are taller than monkeys because of an increase of cells in the flesh and bones of our legs, or that a Rolls Royce is more reliable than a Ford because it's built better—true, but trivially so.

Evolutionists are clueless as to why evolution would select those additional cells that are irrelevant to the evolutionary goals of survival and reproduction. We see the great advantages of language once it is present, but this doesn't tell us how it came to be. Besides the brain, the lungs, diaphragm, larynx, trachea, tongue, and many more structures are uniquely configured for language. What were these language-facilitating organs doing before one person uttered a primitive word to another? Darwinists may be perplexed, but the Bible tell us that consciousness, mind, intellect, and language are gifts given by God so that we may hear, speak, and understand His word: "The Lord God hath given me the tongue of the learned, that I should know how to speak a word in season to him that is weary: he wakeneth morning by morning, he wakeneth mine ear to hear as the learned" (Isaiah 50:4).

The capacity to find meaning in the universe and to experience God is impossible without the faculties of an intelligent conscious mind and language. We think about God with our minds and speak to Him with our tongues. Communication is a two-way exercise, but can we expect an all-powerful God, so far above us, to bother communicating with us? After all, He is not only our divine master, He is also our maker. He is thus immeasurably greater than us, but is He wholly "other?" Christians believe in a personal God and that we are made in His image, so to many theologians this implies at least some essential similarity with humankind. Johnson and Potter view the gift of language as necessary for our I-Thou communion with God, and express the idea that this is a characteristic God shares with us that we may come to know Him:

God is conceived of as a communicator—whether in the form of the revealed word of God, the Ten Commandments, or his presence in religious experiences. Many theists have always believed that God created humanity in his image and that he desires ultimate communion with virtuous souls in some sort of afterlife existence. How would this be possible unless the terms of human thought—a language of thought—have some intrinsic similarity to divine thought? And how could there be meaningful communion without some medium for communication?"[22]

Fine-Tuning the Brain with Mother Love

Regardless of the immaterial nature of the mind, during our brief sojourn on Earth the brain is its home, and as such its priming for goodness and love is imperative. Because our experiences are captured in the process of synaptogenesis, experiencing love sets us on a firm path. Love is a mysteriously powerful force; the noblest, most beautiful, exquisite, and meaningful experience of humanity. By love we are born, through it we are sustained, and for it we may sacrifice life itself. Love insulates the child, brings joy to youth, and comfort and sustenance to the aged. Its boundless power cures the sick, raises the fallen, and comforts the tormented. Love is the magnetism between male and female; the unending devotion of parents and children, and the active concern for the well-being of our fellow man. God is love; if we are to share in His love, we must have a mind capable of understanding it. He made us able to receive His love, and when we grow in Christ we may spread and share that love with others. This is why He made us; to try to love as He does.

God is indeed love, but as an old Jewish proverb has it: "God cannot always be everywhere, and so created mothers." Mothers are our first environment; first in the womb and then at the breast. Mother love is not *eros*; the kind of love that depends upon the lovable qualities of the beloved, but rather *agape*, a selfless "giving" love rooted in the relationship between the lover and the loved. A mother's love for the issue of her womb is like God's love for us. God's love for his creatures is not rooted in our virtue or in our just desserts, but rather in God's nature as *agape* itself. Likewise, a mother's love is rooted in the knowledge that the creature in her arms is her child.

A father's love is somewhat different but complementary. In a general sense, mothers make their children *capable* of love, and fathers take this capacity and cultivate it through the inculcation of the ideals of humanity to make them *worthy* of love. If mother love reflects the loving nature of God, father love reflects the image of God as a just God who expects obedience to his word. Of course, mothers' agape and fathers' eros overlap; they are not mutually exclusive; they just signify different male or female tendencies that are both needed.

Humans are born with highly undeveloped brains relative to other species. Newborn horses, dogs, cats, etc., can walk within hours or days and can fend for themselves not long after. Human babies do not typically walk until about one-year-old, and they certainly cannot take care of themselves for years. This is why the experience of loving nurturing is required during the earliest periods of synaptogenesis. Just as the infant's immune system is immature at birth, so is its brain. Both rely on experience for maturity; the immune system requires exposure to antigens and the brain requires exposure to love.

It has been noted how breastfeeding and being cuddled sent signals through the infant's brain running to its pleasure centers telling it that all is right with its world. To the infant, who at this stage of life can only "think" with its skin, the mother is the fountainhead of all satisfactions. Affectionate touching, kissing, cuddling, hugging, and rocking are tangible assurances for the infant that it is loved and all is right with its world. A mother's touch communicates her loving emotions and sends signals of security to her infant. The neural tube is formed from the same embryonic tissue (the ectoderm) making the skin and the brain intimately connected.[23] We know from animal studies that the more they are picked up and stroked the more abundant are the synaptic connections.[24] This is why tactile stimulation is so important for an infant's brain development, why mothers have the desire to provide it, and why they receive joy in the process. The expansion of two individual natures by love is achieved in this fashion; each enriching the other, which was surely God's intention. In John 13:34, Jesus commanded us to love one another as He loves us. To love one another as Jesus loves us is a tall order out of human reach, but this does not mean that we should ever stop trying.

Endnotes

1. Descartes, R., de Spinoza, B., and Hutchins, R. 1952.
2. Chambers, D. 1996, p. 357.
3. Woollett, K., & Maguire, E. 2011.
4. Alexander III, E. 2015, p. 20.
5. In Brian, D. 1995, p. 371.
6. Wigner, E. 2013, p.177.
7. Wigner, E. 1990. p. 3.
8. Ibid., p. 7.
9. In Buckley, A. 2017.
10. Millikan, R. 1927, p. 69.
11. Gould, S. 1997, p. ix.
12. In Frankenberry, N. 2008, p. 345.

13. Darwin, C. 2005, p. 137.
14. Lewis, C. 1986, p. 27
15. Ring, K. 1984, p.110.
16. Long, K. Nd; np.
17. Alexander, E. III, Nd; np.
18. Greyson, B. 2010, p.169.
19. Pereira, C. and Shashi Kiran Reddy, J. 2016, p. 962.
20. In Johnson, J. and Potter, J. 2005, p. 87.
21. Hauser, M. et al, 2014, p. 1.
22. Johnson, J. and Potter, J. 2005, p. 92.
23. Dias, M., & Partington, M. 2015.
24. Mayford, M., Siegelbaum, S., and Kandel, E. 2012.

Chapter Twenty-Two
Free Will and Determinism

> "Our own consciousness tells us that our wills are free. And the information which that consciousness directly gives us is the last and highest exercise of our powers of understanding."
> Max Planck: Nobel Laureate physicist

An Age-Old Question

The issue of free will has bedeviled philosophers for millennia. In the 20th century we had psychologist B. F. Skinner insisting that we don't have free will but keep pretending that we do, and philosopher J.P. Sartre insisting that we have free will but keep pretending that we don't. Since the great minds across the ages have not come to a consensus on this matter, we can hardly claim to resolve it here, but as an issue central to Christianity, it must be addressed thoughtfully. Galatians 5:13 tells us we are free to choose: "You, my brothers and sisters, were called to be free. But do not use your freedom to indulge the flesh; rather, serve one another humbly in love." We have been endowed by God with the ability to make free choices, and those choices should be faithful to His commands, although He grants us the freedom to decide that they won't be. Atheism contends that free will is an illusion and that our thoughts and behavior are fully determined by our genes and our past experiences encoded in the wiring of our brains.

I define free will succinctly as the ability to choose a course of action independent of any outside influence, and determinism as the doctrine claiming that one's choice of action in any situation is the necessary result of a sequence of outside causes channeled through our genetic makeup and prior experiences. Determinism is not to be confused with necessity. It is not a case of given X, Y *will* occur, but rather, given X, Y has a certain *probability* of occurring. This is true in quantum mechanics, much less so in classical physics and chemistry, very much so in biology, and most true in the human sciences (economics, psychology, sociology, etc.).

Neither should we confuse free will with unpredictability, as Greek philosopher Epicurus did by deducing free will from Democritus' atomic theory within which the "swerve" of an atom can occur without cause. Roman philosopher Lucretius developed a theory of free will based on the same idea:

"The atoms do not move in straight or uniform lines; there is in their motion an incalculable declination or deviation, an elemental spontaneity that runs through all things and culminates in man's free will."[1] The modern version of this is based on Heisenberg's quantum mechanical principle of uncertainty, which has been used to affirm libertarian (unfettered) free will. Quantum effects are random, but statistically predictable. By definition we cannot control random events, and thus we cannot freely influence them, but would you want the kind of freedom that renders it impossible to probabilistically predict your behavior? The random firing of neurons is one of the defining features of schizophrenia, and being in that sad condition is not my idea of freedom. As Sociologist Max Weber said, this is "the privilege of the insane."[2]

Even those who defend free will most fiercely do not claim that the will is entirely free of causal chains. If free will is taken to mean action without a cause, all actions would be unpredictable and chaos would reign. Unpredictability may be a characteristic of quantum mechanics, but in human affairs it is a disaster. Everything in the universe has a cause (even quantum phenomena, despite their weirdness); we are of that universe, so what we think and do have causes. Agreeing to this does not commit us to strict determinism, since we are capable of causing our own thoughts and actions. If we are not responsible for own behavior, then praise and blame alike are pointless, as are concepts of theology such as human uniqueness (made in God's image), sin, faith, and love.

It might well be that a soft determinism is necessary for free will, because if I did not think that what I do produces meaningful consequences, why would I do anything? All rational action, coaching, training, tutelage, and guidance are deterministic in the sense that they are designed to produce effects. We preach, discuss, and write books and articles under the assumption that we can change the minds of others, whom we assume are free to accept or reject what we propose. I know that I am a free agent and that living according to that position is necessary, but I also know that my agency is constrained or enabled by my temperament, upbringing, knowledge, conscience, physical and cognitive abilities and disabilities, the size of my bank account, and the constraints imposed on me by others. But without a belief in free will or agency—our ability to shape our own worlds—our minds would be imprisoned in a deadly *que sera, sera* fatalism.

Some folks believe that any kind of causal talk about our behavior detracts from our freedom and dignity, which I find counterintuitive. Let us say that I know you have found a wallet containing a considerable sum of money and I predict that you will turn it in. By making that prediction, have I turned you into an automaton? I have made a prediction based on my knowledge of your moral character, but rather than insulting you I have praised you, and praiseworthiness and blameworthiness are the pillars that support the free will concept. If I said that although I have known you for several years, you are a free autonomous agent, and therefore I do not know if you will turn it in, I would be insulting your moral character by implying that you may decide to keep the wallet.

Although I accept soft determinism, I disavow hard determinism. If human action was just the result of pre-existing determinative forces, motive would translate into action with the lubricated ease of a Rolls Royce engine, but it does not. Choices are burdensome; they must often be made against the gravitational pull of impulse, habit, convention, sloth, and the presence of alternative options. We often make bad ones, recognized only as such by hindsight, when all we can do is to resolve not to make them again. That is what free will is; the ability to make both good and bad choices and to learn from them. I also disavow absolute (libertarian) free will because we are subject to natural law. Humans cannot have absolute free will, for only God is unencumbered by natural law, and is thus the only being that is absolutely self-determined.

Free Will and Free Mind

Returning to Romeo and Juliette one last time; both had the desire to mate hard-wired into their wills in a generalized form we call erotic desire. What was it about Juliette that attracted Romeo and led him to focus his desire exclusively on her? What was going on in his mind that made him love her so intensely that he would die for her (which in fact he did)? We can talk about the hormone of lust (testosterone), the neurochemicals of romantic love (dopamine, serotonin, and norepinephrine), and of attachment (oxytocin and vasopressin), but these are the biochemical tools love makes use of, and not the emotion that Shakespeare called "winged cupid painted blind." To say that Romeo had no choice but to fall in love with Juliette, and that love is reducible to, and determined by, neurochemistry, is to mock common sense. After all, he did choose her over Rosaline, and when he did his neurochemistry *responded* appropriately by translating the abstract language of his mind to the material language of his brain. All purposeful human actions evidence this pattern.[3]

German philosopher Arthur Schopenhauer's words lead us down a different path regarding free will: "Man can do what he wants, but he cannot will what he wills."[4] In other words, absent constraints, we are free to do whatever our wills desire, but we are not free to choose what they desire. Here "will" has a different meaning than its meaning in "free will," yet it stands closer in many ways to the notion that our wills initiate movement toward goals. Schopenhauer is saying that our conscious choices are driven by subconscious desires that emanate from our shared human nature, genetic endowment, and the learned habits of life. These are the things that lead us to "will" (desire) certain things and courses of action over others.

Many of us are slaves to our desires, but this does not mean that our desires are bad. They are part of what it means to be human, but we have veto power over them and must not follow them to excess. We are free to forego the exercise of natural will, as do priests, nuns, and monks who vow to forgo earthly pleasures to follow God. These people defied the most urgent human desires when they consciously and deliberately chose chastity and poverty over corporeal pleasures and resource acquisition. Does anyone believe that they could not have done otherwise? In the Schopenhauerian sense, it is not the will that is free, but the agent's mind that is free to choose how, why, or when, natural desires are sought, and to what extent. Conscious rational deliberation—mind—is the key to human freedom because it can break the causal chain initiated by the will at any point. If the will is to be free, it requires a free mind.

If we do not have free will (free mind) why do we all, including those who deny it, act as though we do? We acquire the notion that we are free agents, albeit with boundaries, through the everyday experience of our own volition. Surely, we could not fabricate the idea of free will without a basis for it in the reality of everyday experience. If this is not the case, why give medals to brave soldiers if they could not have done otherwise; why send criminals to prison if their actions were not their own, why praise writers, artists, musicians, or scientists if their creations were predetermined?

None of this philosophical talk matters to materialists, who believe that modern neuroscience has put the question beyond doubt: there is no such thing as free will. Citing "recent experiments in neuroscience," Stephen Hawking flatly asserts that: "It is hard to imagine how free will can operate if our behavior is determined by physical law, so it seems that we are no more than biological machines and that free will is just an illusion."[5] I wonder if he really believed that his brilliant body of work was determined by the physical laws he explored, and not at all attributable to his own brilliance, perseverance, courage, and agency. The courage he manifested in continuing to do science despite contracting amyotrophic lateral sclerosis—a progressive neurodegenerative disease—at an early age, surely evidences a very strong will. Hawkins toed the materialist line that asserts that when we sense we are making free choices it is simply a matter of becoming aware of what the brain has already decided what we should do, and we delude ourselves when we think that our free intentions caused those choices.

Hawking's appeal to "recent experiments in neuroscience," refers to a series of experiments showing that the brain registers decisions to make simple movements such as pressing buttons or moving a finger before subjects consciously decide to act. The most cited example is the famous 1983 experiment by Benjamin Libet. Libet had subjects perform simple tasks with EEG electrodes attached to their scalps while sitting in front of an oscilloscope (an instrument that displays and analyzes electronic wave signals). A clock with a fast circulating dot was created on the oscilloscope's screen. Subjects were asked to move a finger whenever they liked, and then report the position of the dot when they became conscious of the decision to do so. Subjects' reports, coupled with oscilloscope readings, indicated that unconscious brain activity was observable as EEG signals about half s second before they reported conscious awareness of their decision to move.[6] This finding was jumped on by atheists to paint humans as automatons adrift in a mindless universe, and it was declared a major problem for imbeciles who believe in spooky things like free will.

Neuroscientist Iain McGilchrist's take on all this is that it "is only a problem if one imagines that, for me to decide something, I have to have willed it with the conscious part of my mind. Perhaps my unconscious is every bit as much 'me.' In fact, it had better be, because so little of life is conscious at all."[7] In other words, the subconscious areas of our minds are still us, and thus associated with our wills. After all, we do not make conscious decisions to put one foot in front of the other when we walk, but our decision to walk from one place to another is ours nonetheless. To use such experiments to argue against free will ignores the fact that we have a mechanism designed to enable us to act without conscious thought called the autonomic nervous system (ANS), the seat of the "fight or flight" response. If someone throws a rock at you there is no time for conscious deliberation; you simply duck. You don't perceive the threat and chew it over in your mind before deciding that it is a good idea to duck.

Libet's readiness potential reflects subjects' awareness that they have to perform some task and are readying their brains for it, just as police officers ready their brains for rock throwers when responding to a riot. Likewise, Libet's observations reflect his subjects' anticipation of the experimental response rather than the actual conscious implementation of the movement. In other experiments, Libet instructed subjects to stop their intended actions once they became aware of the urge to carry them out. Subjects had a window of about 150 milliseconds after the subconscious urge to complete the task in which to consciously veto the urge, and no readiness potential was observed in this process. Libet called this abortive process "free won't;" our conscious minds at work every time we abort an automatic subconscious urge to light that cigarette, take another beer, or refuse an illicit relationship—it is freely taking control of our immediate desires. Those who interpret Libet's findings to mean that we are automatons ignore the fact that when decisions of importance are made in everyday life they are never as simple as "When will I move my finger." You don't automatically act when deciding such things as asking your girlfriend to marry you, buying a house, quitting a job, or where to go on vacation. These decisions require conscious thought, including thought relating to the constraints that hamper your decision. This is what free will is; willing something and then consciously deciding if that's really what you should do.

It is important to note that Libet did not interpret his results, or similar results obtained by others, the way Hawkins or other atheists did. On the contrary, Libet was a firm believer in free will: "My conclusion about free will, one genuinely free in the non-determined sense, is then that its existence is at least as good, if not a better, scientific option than is its denial by determinist theory... Such a view would at least allow us to proceed in a way that accepts and accommodates our own deep feeling that we do have free will. We would not need to view ourselves as machines that act in a manner completely controlled by the known physical laws."[8] Ironically, the man most often credited as the destroyer of free will was a firm believer in it.

What would a person be like if he or she really believed that free will is an illusion; I mean a person who was subjectively convinced of it, not just one parroting the words of famous atheists such as Hawking, Dawkins, or Sam Harris? The intuition that we are the masters of our own fate is an imperative if we are to live meaningful lives. Believing that one's lot in life is the result of someone or something else such as "society" only leads one into a dark sinkhole of victimhood and moral anarchy. A number of psychological experiments have shown that people who dismiss the idea of free will grant themselves the right to behave immorally. One such experiment involving college students found that students exposed to positive messages of determinism were significantly more likely to cheat on a variety of tasks than students exposed to positive messages of free will. The authors of one such study concluded that: "Much as thoughts of death and meaninglessness can induce existential angst that can lead to ignoble behaviors, doubting one's free will may undermine the sense of self as agent. Or, perhaps, denying free will simply provide the ultimate excuse to behave as one likes."[9] Criminologists have long identified the fatalistic belief that one is the slave of outside influences as one of the features of people occupying the lowest rungs of society, and that it is a belief that spawns hostility toward mainstream culture and promotes criminal behavior.

The Compatibilist Option

The position that Christians might adopt on the free will/determinism debate to counter atheist denials is to tie the notions together in a philosophical position called compatibilism. Compatibilism is a position that insists that free will and determinism can peacefully coexist because one can believe in both without being logically inconsistent. Compatibilism does not deny that human events have a chain of prior events leading to them, but avers that as long as we are free from external coercion, we have the freedom to decide how to respond to them. The criminal justice systems of all civilized societies are compatibilist in that they hold criminals responsible for their actions yet leave space for a variety of mitigating factors (mental disease or defect, coercion, etc.) when sentencing them for their crime.

St. Augustine used the concept of free will to account for the problem of evil (why a benevolent God allows evil to exist) in his early theology, but later placed increasing limitations on free will and became a compatibilist. Notice his Schopenhauer-like distinction between "free will' and "will" when he writes that sinners, "do not will to do what is right, whether because they do not know whether it is right or because they find no delight in it. For we will something with greater strength in proportion to the certainty of our knowledge of its goodness and the deep delight we find in it."[10] Augustine is saying people consent to some action according to the "delight" they find in it, and it is not the will that is free, but the agent's mind is free to contest the direction of the will ("fee won't"). Choice involves consent to follow one's "delights" or not. Jesse Couenhoven notes of Augustine that: "He repeatedly uses the expression 'the free choice of the will,' and in this way treats the relation between the two terms as correlative with motive and act. That is, 'will' is a way of speaking of what motivates our 'free choices.'"[11]

Regardless of where one sits on the free will-determinism issue, we can agree that all sane human beings engage in goal-directed behavior. A strict determinist might say that this is not indicative of free will because animals also engage in goal-directed behavior dictated by their natures, and we do not invest animals with free will. A compatibilist reply to this is that humans have the unique ability to take ownership of their natural desires and control them. Humans have the ability to reflect on their desires to form judgments concerning their desirability in light of moral and pragmatic considerations. This is another way of saying that choices are burdensome. Members of religious orders who have given up the most basic of human desires to follow God are really claiming ownership of their wills.

Strong adherents of both free will and determinism view them as logically inconsistent positions, and thus see compatibilism as evading the question. Compatibilists may counter this by appealing to Niels Bohr's principle of complementarity—the wave-particle dual nature of light. There was much initial resistance among physicists to this counterintuitive duality, but as it became more and more empirically endorsed it led to modern quantum theory. Albert Einstein and Leopold Infeld had the following to say about this supposed conundrum:

> But what is light really? Is it a wave or a shower of photons? There seems no likelihood for forming a consistent description of the phenomena of light by a choice of only one of the two languages. It seems as though we

must use sometimes the one theory and sometimes the other, while at times we may use either. We are faced with a new kind of difficulty. We have two contradictory pictures of reality; separately neither of them fully explains the phenomena of light, but together they do.[12]

Human beings exhibit this same complementary duality. Substitute human action for light, and free will and determinism for waves and particles, and we can likewise conclude that neither free will nor determinism alone can explain human actions; we need both concepts to do so. Just as there is no longer any paradox in the wave-particle duality of light, there should be no paradox in viewing humans as both free agents and determined. Soft determinism gives us the only kind of free will worth having. It is a free will/free mind that follows the reasoned dictates of our natures, but it also lays on our shoulders the responsibility of owning our actions.

Free Will and God's Omniscience

Belief in free will/free mind by which we can choose to come to Christ is essential to Christianity, but it is not true of all denominations. The 16th-century French theologian John Calvin denied free will, arguing that because God is omniscient, He knows everything you will do before you do it, and therefore your actions are predetermined. Central to Calvinism is the belief in predestination whereby some have been elected by God for salvation from all eternity, and all others are destined for eternal damnation. Calvin based this on Romans 9:15: "For he said to Moses, I will have mercy on whom I will have mercy, and I will have compassion on whom I will have compassion." Calvin interpreted this to mean that there is nothing people can do can save them from damnation if they are not among the elect. There is no hope of salvation in the act of choosing God, because God chose you before you were born for salvation or damnation. Calvin wrote: "Hence we maintain that, by his providence, not heaven and earth and inanimate creatures only, but also the counsels and wills of men are so governed as to move exactly in the course which he has destined."[13] If this is so, how can we be held accountable for sinful choices that flow out of our wills given that we are destined to move as God has destined? How can God be love in this view, and why was the atoning sacrifice of Jesus Christ even necessary?

Another tenet of Calvinism is that man is totally depraved by original sin, and to assert that we are free moral agents who can choose Christ as our savior is to deny our depravity. Calvinism allows for a free will of sorts, in that men may willfully sin. Man is not compelled by any outside force to do so; he cannot avoid doing so because of his corrupt nature. Thus, our choices are only between a greater or lesser evil. Calvinist theologian Loraine Boettner says that free-will is an empty concept, and adds: "Man is a free agent but he cannot originate the love of God in his heart. His will is free in the sense that it is not controlled by any force outside of himself. As the bird with a broken wing is 'free' to fly but not able, so the natural man is free to come to God but not able.[14]

Adding to this dismal picture, Calvin wrote: "Men do nothing save at the secret instigation of God, and do not discuss and deliberate on anything but what he has previously decreed with himself, and brings to pass by his secret direction." [15] This is a God-of-the-gaps argument raised to the Nth power. Since Calvin's God is the author of every sinful act we perform—God did it! We are not responsible for our actions; not because of our genes and environment, but because our deliberations and decisions are initiated by the "secret instigation of God." Such a fatalistic theology delivers Christianity into the hands of atheists; since God is the author and instigator of our evil acts, He is not worthy of our love and devotion.

How can we reconcile God's omniscience with human free will without traveling down the Calvinist road? Because an omniscient God knows what you will do does not mean that He made you do it. Humans exist in the arrow of time in which there is a before, a now, and an after. Suppose I predicted yesterday what you will do tomorrow in circumstance X (call this "time 1"or "before") and you do exactly that today ("time II" or "now"). I can say at "time III" or "after" that "I knew it," but I did not make you do it. I did not know it at time I; I simply made a prediction based on my knowledge of you. I had to wait until time II to see if my prediction was right. God, on the other hand, is outside of time such that there is no before or after; only an eternal now in which a human past, present, and future exist simultaneously: "I was, I am, I will be." God knew what you would do because he knows everything, but don't make God responsible for you doing it. Perhaps God has surrendered part of His power to foresee future events (the concept of *kenosis*) brought about by free human actions since He has gifted us with that freedom.

The Catechism of the Catholic Church makes this argument when it points out that God desired man to be "left in the hand of his own counsel...God created man a rational being, conferring on him the dignity of a person who can initiate and control his own actions. God willed that man should be 'left in the hand of his own counsel', so that he might of his own accord seek his Creator and freely attain his full and blessed perfection by cleaving to him."[16] This is a much more hopeful position which is in accordance with God as a loving God who sent his Son to atone for the sins of our depraved nature. Contra Boettner, man is both free in a compatibilist sense and able to come to God by choice.

Endnotes

1. In Durant, W., 1944, pp. 150-151.
2. In Eliaeson, 2002, p. 35.
3. Walsh, A. 2016.
4. In Tagliaferre, L. 2013, p. 123.
5. Hawking, S. and Mlodinow, L. 2010. p. 32.
6. Libet, B., Wright Jr, E. and Gleason, C. 1983.
7. McGilchrist, I. 2009, p. 187.
8. Libet, B. 1999, pp. 56-57.
9. Vohs, K. and Schooler, J. 2008, p. 54.
10. In Couenhoven, J. 2007, p. 284.
11. Ibid., p. 284.
12. Einstein, A. and Infeld, L. 1938, pp 262-263.
13. In Lane, A. 1981, p. 74.
14. Boettner, L. 2017, p. 30.
15. In Lane, A. 1981, p. 74.
16. Catechism of the Catholic Church, nd, np.

Chapter Twenty-Three

The Problem of Evil: Atheism's Best Argument

> "God created things which had free will. That means creatures which can go wrong or right. Some people think they can imagine a creature which was free but had no possibility of going wrong, but I can't. If a thing is free to be good it's also free to be bad."
> C. S. Lewis:
> philosopher, novelist, and lay theologian.

The "Rock of Atheism"

St. Thomas Aquinas believed that the existence of evil is the best argument atheists have to deny God. For this reason, the problem of evil is considered the "rock of atheism." Experiencing evil, either at the hands of others or by natural disasters, has probably done more to break religious faith than all atheist philosophical arguments combined. Who can fathom why an omnipotent and benevolent God permits murder, torture, rape, wars, slavery, disease, earthquakes, hurricanes, floods, and countless other evils that cause so much suffering? How can we square the existence of an all-loving and all-powerful God with the evil that exists in the world? The atheist argument is that an omnipotent God would prevent evil if He is benevolent. If He cannot, he is not omnipotent. If He could prevent evil and does not, He is not benevolent. Their argument boils down to saying that it is logically impossible for the Christian God to coexist in the same universe with evil, and since we know evil exists, it follows that God does not.

The area of theology that attempts to deal with this question is known as theodicy; a term derived from the Greek *theos* ("God") and *dike* ("justice"). Theodicy seeks to explain why it is possible for an omnipotent and benevolent God to exist despite the existence of appalling evil, and offers justifications for why God permits it. God needs no justification for what he does or does not permit any more than our parents need justification for their marriage and our existence. What does need justification, however, is our belief in the God of the Bible against atheist claims that the existence of evil flies in the face such beliefs. Christians must address the problem of evil if we are to uphold the rationality of our world view.

Atheists like to taunt Christians by asking why there is so much evil in the world if God exists. Christians should turn the tables and ask atheists why there is so much good in the world if he doesn't. The daily news is full of evil acts, both moral and natural, because they are atypical. For every mother who murders her child there are millions who are gently nurturing their children; for every flight that crashes there are millions that land safely; for every evil act there are millions of loving one's. Evil gets all the ink and pixels because, while evil is everywhere, it is not the norm—good is.

For Augustine, evil was the turning away from the light of God and has no objective reality: "For evil has no positive nature; but the loss of good has received the name 'evil.'"[1] Augustine did not argue that evil is not an experienced reality; he simply argued that it is the negation of the good spawned by the fall of man. The Christian Science Church also dismisses evil as an illusion because it views evil as logically inconsistent with a loving God. But if evil is an illusion with no ultimate reality, there is no moral distinction between obedience and disobedience to God's word, and Christ's atonement on the cross for our sins was futile. Evil is real, and its significance for the atonement is expressed insightfully by Dorothy Sayers: "For whatever reason God chose to make man as he is—limited and suffering and subject to sorrows and death—He had the honesty and the courage to take His own medicine. When He was a man, He played the man. He was born in poverty and died in disgrace and thought it well worthwhile."[2] Exactly! Suffering and evil are real, and Jesus experienced them more intensely than any man.

The idea that evil has no positive nature is not just an obscure metaphysical belief. Cold and darkness are also negations with no positive nature that we experience as real; their reality is revealed by their opposites that do possess positive natures. Heat has positive existence, it is generated by the kinetic motion of atoms transmitting energy from one thing to another. Cold has no such properties; it is a word invented to describe the subjective experience of low heat (low kinetic motion). Likewise, light has a positive existence in the form of streams of photons, while darkness is the negation of light with no positive existence. Physicists don't measure coldness or darkness; they measure temperature and light intensity. Cold is a term used to describe perceptions of low heat, and darkness is used to describe a space without light. We feel evil like we feel cold when little heat is present, we experience it like we experience darkness when photons are minimal, and we feel evil when the love of God is absent in the hearts of men.

Free Will and the Problem of Evil in *The Brothers Karamazov*

In Fyodor Dostoevsky's novel *The Brothers Karamazov*, there is a chapter called The Grand Inquisitor. This chapter provides us with the finest example in literature of the meaning of free will for Christians, and how it is tied to the problem of evil. Dostoevsky's premise is that without God humans cannot be free. He did not mean freedom immunized from conscience to do whatever one pleases, but rather the freedom to choose good or evil. That this freedom is restricted without God became a brutal truth in Dostoevsky's beloved Russia when it succumbed to socialism 37 years after his death. Dostoevsky believed that man must choose between Christ or atheism despite the evils of the world, a choice he faced himself and made in favor of Christ.

In the Grand Inquisitor, the intellectual Ivan Karamazov serves as a symbolic representative of all atheists who deny God because they are troubled by the problem of evil. He considered faith in a benevolent God impossible because of the evil in the world. Ivan relates a dream he had to his brother, Alyosha, a novice monk. Ivan meant it to be a devastating critique of theism by citing the most wanton evil done against defenseless children. Such evil convinced Ivan that even if God exists, he is a malicious and hostile God because he permits the evil that washes the world with tears.

The dream story begins at the height of the Spanish Inquisition with Jesus Christ appearing on the streets of Seville performing miracles. Jesus is recognized by the people, who begin to flock about Him, and by an old cardinal—the Grand Inquisitor—who orders His arrest. Jesus receives a visit from the cardinal that night in His cell, who reprimands Him for returning and hindering the work of the Church. The Grand Inquisitor accuses Christ of placing an intolerable burden of freedom upon man by expecting flawed humans to voluntarily choose to follow Him. He informs Jesus that the Church has rectified this by removing the awful burden of freedom, and in return it has provided the miracle, mystery, and authority that man craves. The cardinal further informs Jesus that man has willingly surrendered his freedom to the Church in exchange for happiness and security, and that He must not undo this work.

The cardinal also berated Christ for refusing the three temptations offered by Satan during Christ's 40 days in the desert: refusing to turn stones into bread (miracle); refusing to cast Himself from the pinnacle of the Temple to be saved by angels (mystery), and the offer of sovereignty over the Earth (authority). Had Jesus accepted these temptations, all people would have absolute certainty of his divinity and thus be forced to accept Him as Lord. According to the cardinal, Christ's rejection of Satan's offers left man's free will intact, but it took away his security, and it is security that men value more than freedom. Lacking certitude of Christ's divinity, the Church had to assume His power, since in Christ's way only the strong would achieve salvation. What then would become of the millions too weak to accept responsibly for their freedom of choice? Christ's insistence that man must freely choose to follow Him allows for the evils committed by those who choose not to. Christ does not want followers bought and paid for by miracle, mystery, and authority, but by reasoned faith in the revealed truth. Jesus remains silent throughout the cardinal's monologue, and kisses him on his bloodless lips at its end. The stunned cardinal then sets Him free with the admonition not to return.

In Antonio Malo's interpretation of The Grand Inquisitor, he says that Ivan paints an image of the Church and its flock that neither would recognize (as a devout Russian Orthodox Christian, Dostoevsky was often critical of Catholicism). Like many atheists, Ivan was fighting a Christianity that does not exist; a mere caricature. Malo also notes that the cardinal's view that Jesus is too elevated for man to successfully imitate is wrong: "the true Jesus, more than a model to imitate, is the origin of grace that allows everyone to follow in his footsteps, notwithstanding our sinful condition."[3] Jesus' symbolic kiss of forgiveness reveals that He, unlike the old cardinal, has faith in mankind and in the power of love.

If we ignore the caricature, The Grand Inquisitor is an insightful allegory. Alyosha did not take Ivan's tale the way Ivan had intended; he took it as eulogy, not vilification. For Alyosha, Jesus' silence signaled His patient confidence that all evil will eventually be rectified. The characters of Alyosha and Ivan illustrate differences in attitude toward suffering between the believer and the atheist. Both show concern for the suffering of others, but Ivan's concern is manifested only by intellectualizing it. He collects anecdotes of cruel evil from newspapers and books and uses them against God, but Alyosha goes out among the suffering to see what he can do to alleviate it. Alyosha's faith in God lends itself to an active concern for the well-being of others, kindness, and a morality based on the solid foundation of God's love. Ivan's doubt leads to the rejection of morality, coldness toward his fellow man, and a crippling existential despair that ended in madness; his soul lacerated by the absence of God's grace.

Evil and Soul-Making Theodicy

Some philosophers argue that God could have made human nature such that evil was not an option and still given us the ability to make choices, just not moral choices. Those who argue this position are seeking a perfect world on Earth, but they forget two things: (1) the perfect world is the next one, and (2) God wants moral choices, and that requires a freedom that necessarily entails the possibility of evil. A world in which free will exists is to be preferred over one that does not, regardless of the benefits that such a world may offer. If God coerced us into behaving morally at all times it would erase our freedom and stifle moral growth. We would also have difficulty knowing what goodness is without knowing evil. It is thus logically impossible for God to give us free choice without entailing the possibility of evil.

Augustinian theodicy absolves God of responsibility for evil, maintaining that God created finitely perfect human beings who fell from perfection by freely turning from God, and it was this act of rebellion that is the root of evil and suffering. Adam and Eve created evil by rebelling against God, and we all suffer as a consequence. John Hick, one of the most distinguished theologians of the 20th century, rejected Augustinian theodicy as relying on a too literal reading of the fall. Hick argues that if Adam and Eve had been created as perfect beings and living in infinite plenitude they would not have rebelled against God. After all, if they were perfect beings with everything they could possibly want, why would they have accepted the serpent's offer? Hick opts instead for the theodicy of the second century theologian, Irenaeus, which claims that God is responsible for evil (by giving us free will), but it is justified because of its benefits for human development. Irenaeus maintained that humans were not created in a perfected state, but is in an evolutionary process of development from morally imperfect beings to morally perfected beings. Irenaean theology thus avers that while humans are born in the *image* of God, they must develop their characters in the *likeness* of God.

Irenaean theodicy is premised on a truly omnibenevolent and omnipotent God, and the corollary of this is universal salvation. Only if all souls eventually achieve salvation does the suffering of humans throughout history make sense. Hick believes that God created the world to serve as a "vale of soul-making," or character building. This entails "human goodness slowly built up through personal histories of moral effort has a value in the eyes of the Creator which justifies even the long travail of the soul-making process."[4] For Hick the narrative of the soul's ascent to God is of an evolutionary journey toward perfection, beginning on Earth and continuing in the hereafter. Rather than Augustine's backward look to the fall, for which countless generations of humans bear no responsibility, Irenaean theodicy is forward-looking to future human perfection. This will be achieved by striving with faith in God and in love of humanity in a process that extends into a postmortem existence.

Universal Salvation, Spiritual Annihilation, or Eternal Damnation?

The idea of universal salvation contradicts Augustine's notion of eternal damnation for sinners. The idea of hell is often considered a major part of the problem of evil, since it conflicts with the Christian idea of a loving, forgiving, and just God. Many modern theologians note that the view that numerous souls will experience everlasting suffering is not embraced by those who believe that "God's love must be maximally extended and equally intense."⁵ The vast majority of people who ever lived were not Christians; they received whatever religious faith they had from their family and culture, and were not freely chosen. It is a denial of God's all-embracing love to believe that those who have never even heard of Jesus must burn in hell for eternity. John Hick asks: "Should we conclude that we who have been born within the reach of the gospel are God's chosen people, objects of a greater divine love than the rest of the human race? But then, on the other hand, do we not believe that God loves all God's creatures with an equal and unlimited love?"⁶

St. Gregory of Nyssa, believed in universal salvation and rejected the idea of eternal damnation, as did the majority of the founders of Christian orthodoxy. Origen Adamantius, martyr, and profound student of the Bible, viewed what we have come to call hell as a place of purifying, healing, and restoring (akin to Catholicism's Purgatory), not of torturing and destroying. I find his more intellectually and spiritually satisfying than Augustine's idea of hell, for does not Paul tell us in Ephesians 3:11 that God had a definite purpose in mind before He created us? Given this, and given our belief in an omnibenevolent God, Gerald Bray asks "if the non-elect have no hope of salvation and God does not want them to suffer unduly, why were they ever created in the first place?"⁷ Our image of God cannot conceive of Him as one who would create beings just to have them suffer eternally. The notion of universal inclusiveness is seen in Gregory's words: "For it is evident that God will in truth be all in all when there shall be no evil in existence, when every created being is at harmony with itself and every tongue shall confess that Jesus Christ is Lord; when every creature shall have been made one body."⁸

Many people cringe at the thought of people who lived decent lives that die with a few peccadillos blotting their souls suffering the same fate as the monstrously evil. How could souls in Heaven ever rest content in the knowledge that loved ones are suffering the eternal fires of hell, and perhaps for transgressions that would get them a few days in jail from an earthly judge? The belief that hell is a place of everlasting torment creates a religion of fear, coercing people to worship; sincerely or not. What then of free will? The Church of England's Doctrine Commission expressed its horror of the traditional view of hell and offered its own view of the afterlife: "Christians have professed appalling theologies which made God into a sadistic monster. ... Hell is not eternal torment, but it is the final and irrevocable choosing of that which is opposed to God so completely and so absolutely that the only end is total non-being."⁹

Just as we may be alarmed at ordinary sinners being consigned to everlasting torment, we may also be chagrined to think of the truly evil—a Hitler, Stalin, or Mao—sharing an eternal heavenly existence with the saintly. But according to the doctrine of conditional salvation, this will not be the ultimate fate of such evil people, for God is a just God who punishes people commensurate with the evil they have done. Gregory, Origen, and Irenaeus taught that the truly evil among us will suffer destruction of their souls in a form of spiritual capital punishment called annihilation—their souls will simply cease to be just as their earthly lives ceased to be upon corporeal death.[10] Hell is the second death, and eternal punishment is eternal destruction. This is the Church of England's "total non-being."

Traditionalists may consider this blasphemous, but theologian Mark Scott avers that while several biblical passages imply universal salvation: "Only Matt. 25:41, 46 explicitly affirms eternal damnation, and it might reflect later theological sensibilities, not the original teachings of Jesus."[11] Lawyer and theologian Edward Fudge augments Scott, noting that most of the Church's founding fathers denied eternal damnation as contrary to a just and loving God: "Although not canonical, the writings of those fathers of the church are worth our reading, for these are men who were taught by the apostles, or by those whom the apostles had taught. Their writings offer a window into some of the thinking of some leaders of that early generation of believers. If we find among the apostolic fathers some unanimity of opinion, it is not to be taken for granted."[12] These words are similar to those written by American Constitutional originalists relevant to the founding fathers against those who consider the Constitution a "living" document; if you want to know what the meaning of the Constitution actually is, there is no better source than those who lived closest to it.

The essence of Hick's theodicy is that morality and character can only be built by experiencing evil and responding to it positively in the spirit of "That which does not kill me strengthens me." An extreme example of this is given by Holocaust survivor and psychiatrist Viktor Frankl who writes of his experiences of suffering and the opportunity for growth it provided in the Auschwitz concentration camp:

> The way in which a man accepts his fate and all the suffering it entails, the way in which he takes up his cross, gives him ample opportunity – even under the most difficult circumstances to add a deeper meaning to his life. It may remain brave, dignified and unselfish. Or in the bitter fight for self-preservation he may forget his human dignity and become no more than an animal. Here lies the chance for a man either to make use of or to forgo the opportunities of attaining the moral values that a difficult situation may afford him. And this decides whether he is worthy of his sufferings or not.[13]

However, most victims of this atrocity were murdered, so what purpose is there in that? They had no opportunity to build up their characters. Hick's response to this is that the suffering of others serves a purpose for observers to develop sympathy, empathy, and compassion, which may help present and future victims of evil. A world in which humans cannot harm others would also be a world in which they cannot make the moral choice to help them. While logical, I find this to be immoral on its face; should not humans be treated as valued ends in themselves and not as a means to the end of another's character building? What right do we have to expect the suffering of another to serve our moral benefit? I am, however, partial to Irenaean theodicy because of its universalism—all except the truly evil will eventually be saved. It has the taste of "the best of all possible worlds" (and worlds to come) and is emotionally satisfying. That is, God continues to draw all the unsaved to Himself even after death. Of course, this does not mean that soul making is the reason that God permits moral evil, or that hell is not what Augustine says it is, for humans cannot know the mind of God with certainty.

The Problem of Natural Evil

Free will and soul-making theodicies apply only to moral evil; to the choices made by free agents. Natural evils—hurricanes, tornadoes, earthquakes, volcanic eruptions, diseases, famines, and other such occurrences—cannot be justified this way because humans have no hand in most of these occurrences. Because of this, the problem of natural evil is said to pose a greater threat to belief in a benevolent God than moral evil. Richard Swinburne attempts to fall back on a soul-making theodicy to account for natural evil, saying that natural disasters provide people with additional opportunities to develop sympathy and to give aid and comfort to victims: "The pain makes possible those choices that would not otherwise exist."[14] With all due respect to Swinburne's sterling reputation as a theologian, I believe this to be a desperately weak argument, but I also believe that a natural evil theodicy is easier to formulate and defend than is a moral evil theodicy.

The natural evils atheists trot out as proof of God's non-existence are necessary for life to exist. Despite the necessity of plate tectonics for life, it sometimes produces natural evils in the form of earthquakes and other natural disasters. Natural evils provide arguments against an omnipotent and benevolent God because they cause approximately 68,000 deaths each year.[15] These things are indeed horrendous, but it is forgotten that without plate tectonics there would be no life at all—no landmass, no magnetic shield, no atmosphere, nor anything else necessary for life. There is not one natural disaster that does not play an important role in the Earth's ecology. As terrible and heart-rendering as the loss of life is, it pales in comparison to the 400,000 and 500,000 people murdered ("moral evil") worldwide each year according to the UN.[16] Natural disasters are the inevitable result of the laws of God's design for life just as moral evil is the inevitable result of our gift of free will.

God has created a world that functions according to natural laws, and once they are in place, He does not disturb them. Natural "evils" are consequences of these laws. Natural laws make our world "just right" for human life, but there are sometimes disastrous trade-offs. God could prevent natural evils caused by plate tectonics by stopping the rotation of the iron core, but if He did so there would be no sentient beings to experience good or evil. God cannot suspend His laws of nature when people make their home near fault lines, volcanos, or flood plains without becoming Newton's tinkerer. The world is not perfect, but professor of science and theology, John Haught, is repelled by the very idea of a perfect world. He says that such a world would be "dead on delivery. Since it would already be perfect, it would also be finished; and if finished, it would have no future, no room for freedom...Such a world would be devoid of the contingency, indeterminacy, freedom, vitality, and futurity apart from which there could be no truly dramatic creative process."[17]

Natural laws ensure regularity, consistency, and predictability in a world of cause and effect without which we could never discover them. If cause A resulted in effect B only when no humans were present, we would be living in a world of constant miracles, which are by definition the suspension of natural law. A world in which natural laws guarantee consistency permits intentional action, moral deliberation, and scientific investigation. A world of miracles in which a man falls from a 10-story building and gets up unscathed, is inimical to both moral and rational activity because it would be the Grand Inquisitor's world of miracles, mystery, and authority. "The universe declares the power and glory of God" (Psalm 19:1), but a miracle world would declare it so obviously that we would have no choice but to believe. God wants us to take responsibility for ourselves in a world of predictable cause and effect. Physicist Don Page notes that "God may grieve over unpleasant consequences of elegant laws of physics and might even directly experience all of them Himself (as symbolized by the terrible suffering He experienced in the Crucifixion), but there may be that inevitable tradeoff that God takes into account in maximizing total goodness."[18] We see in these words a reiteration of the argument made in Chapter 19 that all design involves compromises in some feature so that the whole may be optimized.

This is my feeble attempt at a theodicy of natural evil, but all theodicies ultimately fail because they are the attempts of humans presuming to know the mind of God and seeking to justify Him by human standards of good and evil, which cannot be done. Humans are attempting to answer questions that only God can answer. It is therefore wise to heed St. Augustine's advice when confronted with different interpretations of Scripture used to support a particular theodicy:

> In matters that are obscure and far beyond our vision, even in such as we may find treated in Holy Scripture, different Interpretations are sometimes possible without prejudice to the faith we have received. In such a case, we should not rush in headlong and so firmly take our stand on one side that, if further progress in the search of truth justly undermines this position, we too fall with it. That would be to battle not for the teaching of Holy Scripture but for our own, wishing its teaching to conform to ours, whereas we ought to wish ours to conform to that of Sacred Scripture.[19]

Augustine's note of caution is the "scientific" way to look at scriptural evidence, and thus I could not agree more. However, I believe that the best answer to the problem of evil is not answered in any learned theodicy, but in the promise of Revelation 21:4: "And God shall wipe away all tears from their eyes; and there shall be no more death, neither sorrow, nor crying, neither shall there be any more pain: for the former things are passed away." **AMEN!**

Endnotes

1. In Geisler, N. and Corduan, W. 2002, p. 322.
2. Sayers, D. 1994, p. 4.
3. Malo, A. 2017, p. 268.
4. Hick, J., 2007, p. 256.
5. Jordan, J. 2012, p. 53.
6. Hick, J. 1998, p. 26.
7. Bray, G. 1992, p. 23.
8. Pearson, C. 2009, p. 28.
9. Church of England's Doctrine Commission, 1995, p. 199.
10. Bray, G., 1992, p.21.
11. Scott, M. 2010 p. 321.
12. Fudge, E., 2012, p. 353.
13. Frankl, V., 1985, p. 88.
14. Swinburne, R., 2010, p. 95.
15. Centre for Research on the Epidemiology of Disasters, 2015.
16. United Nations Office on Drugs and Crime, 2014.
17. Haught, J. 2015, p. 132.
18. Page, D. 2014, pp. 12-13.
19. In Grant, E. 2004, p. 222.

Chapter Twenty-Four
Christianity, Atheism, and Morality

> "In view of such harmony in the cosmos which I, with my limited human mind, am able to recognize, there are yet people who say here is no God. But what really makes me angry is that they quote me for the support of such views."
> Albert Einstein Nobel Laureate physicist

Christianity and Morality

We all want to live in a good society next to neighbors on whom we can rely not to hurt us. We have a whole science—criminology—devoted to exploring what makes people want to hurt others, but not one that explores what makes people want to help others. We call people good when they value us and have an active concern about our well-being; this is moral behavior. Moral people obviously act in more virtuous and principled ways than immoral people, which benefits both themselves and society, and when society benefits, everyone benefits, including atheists. The 18th-century French philosopher and skeptic, Voltaire, knew the value of another's religion to himself: "I want my attorney, my tailor, my servants, even my wife to believe in God…because then I shall be robbed and cuckolded less often."[1]

A free society is glued together by the voluntary obedience of its citizens to laws, which are moral rules. We get these rules from parents, teachers, and others with our best interests in mind, but ultimately they come from God's commandments ("Thou shalt not steal," "Love thy neighbor as thyself;" "Do unto others as you would have them do onto you," and so forth). These are the rules of any decent society that we sense intuitively, and it was this moral intuition that moved British philosopher and novelist C.S. Lewis, to abandon atheism for Christianity. Lewis wrote that "human beings, all over the earth, have this curious idea that they ought to behave in a certain way, and cannot really get rid of it."[2] He argued that because there is this pressing moral law there must be a moral lawgiver. True morality stems from the nature of God because only He can provide a coherent foundation for objective moral values. Without God we do not have an objective foundation for moral judgments and are left only with moral relativism asserting that what is "good" and "bad" depends on one's culture, or on whom you ask.

Sam Harris says that we can find the foundation for objective moral values in science, but Albert Einstein would say that Harris got it backwards: "You are right in speaking of the moral foundations of science, but you cannot turn round and speak of the scientific foundations of morality...every attempt to reduce ethics to scientific formulae must fail."[3] While many atheists subscribe to the universality of human rights and are moral people who care about others, care about justice, value good, and denounce evil, they cannot provide an ultimate justification for a universal objective morality by which good and evil are clearly delineated and understood. You simply cannot ground morality in an atheist worldview.

Most people know intuitively that belief in God is central to morality, and that harmful and immoral acts are far more likely to issue from an atheist worldview than a Christian one. Psychologist Will Gervias tested these intuitions in five experiments consisting of 1,152 participants who were asked to judge a variety of immoral acts such as serial murder, incest, and cannibalism as representative of Christianity or atheism. Participants, even atheist participants, judged each immoral act as more likely to be committed by atheists than by Christians. Gervias concluded that: "The present findings, combined with previous research...suggest that it is specifically immoral negative actions that are seen as representative of atheists, consistent with other evidence suggesting that many view belief in God as a prerequisite for morality." [4]

When atheists prove to be of good moral character, they are obeying C. S. Lewis' moral intuition. Romans 2:14-15 put this best many centuries before Lewis: "For when Gentiles, who do not have the law, by nature do the things in the law, these, although not having the law, are a law to themselves, who show the work of the law written in their hearts, their conscience also bearing witness, and between themselves their thoughts accusing or else excusing them." These verses inform us that those who do not have the *written* law have the law of conscience to morally direct them. Romans 2:13 points out that persons without the law who follow their conscience are better persons than those who have the law but do not follow the law: "For it is not the hearers of the law who are righteous before God, but the doers of the law who will be justified [declared righteous]."

If God is Dead, all is Permissible

The reason that the atheist worldview provides no basis for affirming objective moral values is because it is nihilistic. Nihilism is a cynical and pessimistic outlook that views human beings as mere byproducts of pitiless nature that have evolved on an insignificant rocky planet located in one of billions of galaxies swirling around in a hostile, mindless, and accidental universe. Richard Dawkins put the atheistic view of human worth honestly when he says, "there is at bottom no design, no purpose, no evil, no good, nothing but pointless indifference...We are machines for propagating DNA... It is every living object's sole reason for being."[5] Why in this view should we think that human beings have any more worth than cockroaches? If Dawkins' depressing words describe reality, and many believe they do, then all is permissible.

Philosopher Joel Marks agrees with Dawkins' assessment, and sees it as pointing to the conclusion that there is no objective morality: "The long and the short of it is that I became convinced that atheism implies amorality; and since I am an atheist, I must therefore embrace amorality." He notes that what he calls "soft atheists" believe that a person can be an atheist and still believe in morality, and he views atheists such as Sam Harris as softies for holding that view. Marks continues: "So was I [a softy], until I experienced my shocking epiphany that the religious fundamentalists are correct: without God, there is no morality. But they are incorrect, I still believe, about there being a God. Hence, I believe, there is no morality."[6]

The most notable person to proclaim that if God is dead, then so is morality, was the 19th-century arch-atheist German philosopher Friedrich Nietzsche. Nietzsche was a strange creature who thought love "the greatest danger" and morality humanity's greatest weakness. He taught that people can only be free if they reject God, arguing that good and evil are simply social constructs the weak have imposed on the moral neutrality of nature. Good and evil are inventions that provoke the development of a conscience, and that inhibits the greatness of the strong. Nietzsche maintained that to rid ourselves of morality we must abandon Christianity. The "will to power" of his "superior men" (*Übermensch*) will fill the vacuum. For Nietzsche, atheism is a moral, rather than an intellectual revolt, because morality stands in the way of fulfilling one's desires. He wrote: "All superior men who were irresistibly drawn to throw off the yoke of any kind of morality and to frame new laws."[7] If God is dead, meaning, morality, and reason die with Him, and all that is left are the relentless drive for power of the *Übermensch:*

> When one gives up Christian belief one thereby deprives oneself of the right to Christian morality... Christianity is a system, a consistently thought out and complete view of things. If one breaks out of it a fundamental idea, the belief in God, one thereby breaks the whole thing to pieces: one has nothing of any consequence left in one's hand... Christian morality is a command: its origin is transcendental... it possesses truth only if God is truth – *it stands or falls with the belief in God* [my emphasis].[8]

Unlike Nietzsche, Sam Harris believes morality is a good thing, the purpose of which is to advance human well-being, a position with which I fully agree.

He defines well-being as a state of happiness and satisfaction with life. I agree with this too, but he trashed Christianity as inimical to these goals. He doesn't want to acknowledge any purpose for our search for happiness other than maximizing pleasure and minimizing pain. He doesn't want to be judged for his actions in this pursuit, so thinks it best we get rid of the Judge. With Christopher Hitchens, Harris believes that "religion poisons everything," and that its practice leads to poor health and impedes social progress. I interrogate these absurd claims in the remaining chapters. If atheist "morality" is superior to Christian morality, we should expect atheism to lead to good, prosperous, and free societies full of happy individuals and families, and Christianity to produce the opposite.

Putting Atheist Morality to the Test

In a YouTube talk titled *The Closing of the Modern Mind*, psychologist Jonathan Haidt says that: "members of religious communities are simply better citizens; they give more, not just to their religious communities but to their society in a variety of ways...its participation in a religious community that has an effect that reigns in selfishness and draws them out into community." We will contrast this with the destruction of the moral impulse under atheism; made as plain as day in the 20th century in the former Soviet Union, Red China, and NAZI Germany. Atheism splattered the 20th century with blood and tears. Its legacy has left an indelible blotch on the soul of humanity, resulting in the death of over 100 million people around the globe. Many more millions are plodding through a life of meaninglessness, hopelessness, and degradation as Dawkins' "survival machines" waiting only for death and oblivion. Militant atheists turn a blind eye to the historical record and the cruel legacy of their creed. They do so because it is a gigantic black hole in the center of their worldview that they know will suck in and annihilate their arguments if they orbit too close to it.

The Soviet Union

Karl Marx knew that the success of socialism depends on spiritually disarming the people by ridding a country of its two epicenters of morality—religion and the family. He wrote: "Once the earthly family is discovered to be the secret of the heavenly family, the former must be destroyed in theory and in practice."[9] After the Russian Revolution, the Soviet Union destroyed thousands of churches and killed or imprisoned thousands of priests. It declared the state officially atheist and then went after Marx's earthly family. The Soviets passed laws legitimizing the offspring of the unmarried, made divorce available on demand, and encouraged free love as the "essence of communist living."[10] They threw bourgeois morality out of the window, and Ivans abandoned their Natashas from Riga to Vladivostok, leading to millions of father-absent children roaming the streets, and forming criminal gangs that ran rampant across the country.[11] After 10 years of chaos, the government turned 180 degrees and began praising marriage and the family, condemning divorce, and it restored the legal concepts of legitimacy and illegitimacy.

This about-turn did not stop the rot because atheism remained the official state religion. Peter Hitchens has harrowing tales to tell of his time as a journalist in Moscow before the fall of the Soviet Union in his book *The Rage Against God*. He wrote of bribes to receive proper medical treatment, riots over cancelled vodka rations, abortion as a means of birth control, mistrust of neighbors, two families living in squalid apartments meant for one, theft of everything not nailed down, and a level of incivility that made a New Yorker sound like an English gentleman. Experiencing this Godless society led this former socialist and atheist (and brother of Christopher Hitchens) to embrace God.[12]

In his 1983 Templeton Prize Lecture, Nobel Prize winner for literature, Aleksandr Solzhenitsyn, revealed the reason why Russia became a dystopian nightmare. He noted that as a child he heard people say that Russia was experiencing vile oppression and great disasters because it had abandoned God. He remarked that if he "were asked today to formulate as concisely as possible the main cause of the ruinous Revolution that swallowed up some sixty million of our people, I could not put it more accurately than to repeat: Men have forgotten God; that's why all this has happened."[13]

Russian president, Vladimir Putin, agrees. In his 2014 state of the nation address, Putin said: "Many Euro-Atlantic countries have moved away from their roots, including Christian values...Policies are being pursued that place on the same level a multi-child family and a same-sex partnership, a faith in God and a belief in Satan. This is the path to degradation." He further said that social and religious conservatism is the only key to preventing the world from slipping into chaotic darkness. Patriarch Kirill I, head of the Russian Orthodox Church, also noted the spiritual disarmament of the people in Western countries: "The general political direction of the elite bears, without doubt, an anti-Christian and anti-religious character...We have been through an epoch of atheism, and we know what it is to live without God...We want to shout to the whole world, 'Stop!'"[14] Putin and Kirill know better than most how important Christianity is to a decent society because both lived most of their lives in a Godless Russia and experienced its social and spiritual malaise. The Russian government has rebuilt 30,000 churches since communism fell, and it actively promotes Christianity at all levels of society.[15] Unfortunately, there seems to be an emerging collusion of church and state in Russia, but that sad fact does not detract from the truth of Putin's and Kirill's words.

China

There were similar efforts to kill Christianity and the family during the Chinese Cultural Revolution; an insane decade between 1966 and 1976. Mao Zedong harvested the naiveté and fervor of young people and used them to destroy the old culture. These youthful Red Guards destroyed hundreds of churches and imprisoned or killed many thousands of priests, ministers, and practicing Christians. Qingbo Xu explains that atheism was proclaimed the official state religion and, following Marx's mandate, the traditional family was to be destroyed: "People were told that they had no right to love their children, parents, and other relatives, and friendship networks had to be dismantled because 'they ruined one's subsistence and well-being.'"[16] The idea of the Cultural Revolution was to empty the old culture and spiritually impoverish the people so that a new culture could be built according to Mao's communist ideal.

Christianity experienced widespread popularity after Mao Zedong's death. Scholar of religion, Fenggang Yang, estimates that if the current rate of growth continues, by 2030 China will have the largest Christian congregation in the world at about 247 million.[17] However, Chinese politician Zhu Weiqun has stated that the Communist Party should "unambiguously promote Marxist atheism to society," describing it as "the nations' mainstream ideology," and the party should seek to "strengthen propaganda education about a scientific worldview, including atheism."[18] Since Weiqun's 2015 pronouncement, China has initiated one of harshest crackdowns on religion in years. There are numerous reports of rural village chiefs forcing Christians to denounce their religion on pain of losing state welfare benefits. But Bob Fu, founder of the religious rights group ChinaAid, said that the church would survive and even thrive: "I have hope for the future, these campaigns were done in Roman times, under Stalin and under Mao, and none succeeded. It will only have the opposite effect, and if Communist party cadres studied history, they would see this. Crackdowns will cause the church to grow faster, and help church be more united."[19] Attacking Christianity will again undermine the nation's morality, which it desperately needs since: "Today's China is plagued by widespread mistrust and loneliness, as well as pervasive corruption and greed."[20]

Nazi Germany

Although never officially atheist, Nazi Germany is another example of its evils. The Jewish Holocaust is well-known, but little is known about Nazi attacks on Christianity. Much of the evidence of the Nazi's plans for Christianity comes from the record of the 1945 Nuremberg war crimes trials compiled by William Donovan of the OSS, the precursor of today's CIA. There is much evidence that Hitler hated Christianity, but he was anything but ignorant of psychology. When he was struggling for power in Weimar Germany in the 1920s, his arch enemies were communists, so it was politically expedient to paint himself as a "defender of the faith" and traditional Western values against Godless communism. However, when Hitler became chancellor of German in 1933, he began to do whatever he could to de-Christianize Germany and move it toward a pagan world of "a religion of race and blood." In one of his Table Talks with high-ranking Nazis, he said that "the only way of getting rid of Christianity is to allow it to die little by little."[21] During Hitler's dictatorship, more than 6,000 pastors, priests, monks, and nuns were imprisoned or executed in Germany because Christian morality stood in the way of his evil plans, and the same measures were taken in all occupied countries.[22]

Hitler could not overly antagonize Christian Germans during the war, but Nazi documents show that after their expected victory, they would attack Christianity as they had attacked Judaism. The Nuremberg war crimes indictment reads in part: "The Nazi's persecution of Catholic and Protestant churches was a classic power struggle. At first, the Party followed Hitler's politically pragmatic policy of placating the Churches while undermining their political power. Once political dominance had been achieved, the churches were no longer needed, and the Party turned to a more radical and aggressive policy spearheaded by Himmler."[23] According to other Nuremberg war-crimes documents: "The Nazi conspirators, by promoting beliefs and practices incompatible with Christian teaching, sought to subvert the influence of the Churches over the people and in particular over the youth of Germany. They avowed their aim to eliminate the Christian Churches in Germany and sought to substitute Nazi institutions and Nazi beliefs and pursued a programme of persecution of priests, clergy and members of monastic orders whom they deemed opposed to their purposes and confiscated Church property."[24]

It is hardly surprising that all three nations attacked Christianity, for they were all Marxist, and Marxism and atheism are fraternal twins. Make no mistake; the National Socialist Worker's Party (NAZI) was socialist; after all, that's what they called themselves. The go-to man for the truth of this is Adolf Hitler himself who, in 1941 speech, stated that "basically National Socialism and Marxism are the same."[25] He even acknowledged his debt to Marxism when he said: "I have learned a great deal from Marxism as I do not hesitate to admit…The whole of National Socialism is based on it."[26]

I don't think anyone who lived in Russia, China, or Germany during these periods believes that atheism builds moral foundations for a good society. A story told by philosopher of science and theologian William Lane Craig is instructive. He tells of meeting a Russian cosmologist, Andrei Grib, in St. Petersburg and asking him why there was a great upsurge of Christianity following the collapse of communism. Grib replied, "Well, in mathematics we have something called 'proof by the opposite.' You can prove something is true by showing its opposite to be false. For seventy years we have tried Marxist atheism in this country, and it didn't work. So everyone figured the opposite must be true."[27]

Christianity and the Economic Benefits to Society

The Christian worldview provided impetus for science; science begets technology, and technology creates jobs and wealth. One reason for this is given by economist Kelly Mua, who notes that the link between Christianity and prosperity exists "mainly by fostering religious beliefs that influence individual traits such as honesty, work ethic, thrift, and openness to strangers."[28] These traits, of course, are moral ones. In another study, a team of economists used data from 66 countries encompassing the years 1981 through 1997 and concluded: "We find that on average, religious beliefs are associated with 'good' economic attitudes, where 'good' is defined as conducive to higher per capita income and growth...Overall, we find that Christian religions are more positively associated with attitudes conducive to economic growth."[29]

Some scholars have attempted to put a dollar amount on the benefits of Christianity to America. In his book *America's Blessings: How Religion Benefits Everyone, Including Atheists,* Rodney Stark estimates that Christianity benefits the American economy to the tune of $2.6 trillion per year. He arrived at that figure by asking: "What would it cost if America suddenly were transformed into a fully secularized society?"[30] That is, if there were no practicing Christians in America behaving according to Christian morality, what might the costs to the U.S. economy be? Because practicing Christians are less likely to be involved in crime and substance abuse, there are massive savings for the criminal justice system. Happier intact homes mean less divorce, illegitimacy, and parental abuse and neglect, leading to huge savings in welfare and less criminal behavior. Religious schools not only provide superior moral training; they also produce better-educated children who are more likely to graduate. This means better job prospects, less unemployment, fewer deadbeats, and lower unemployment costs. Regular religious attenders give more money to charity and volunteers their time much more than non-attenders, and more charitable giving and volunteering saves many millions of dollars in what otherwise would be a burden on the taxpayer. Finally, better physical and mental health among practicing Christians saves billions in health care costs.

In a massive study filled with voluminous statistics, economists Brian and Melissa Grim arrived at three estimates of the value of faith-based businesses and institutions to the American economy. Their most conservative estimate, taking into account only the revenues of faith-based organizations, is $378 billion annually. Their second estimate, taking into account the market value of goods and services provided by religious organizations and the contribution of faith-based businesses, puts the value at over $1 trillion. Their third estimate includes those behaviors Stark includes in his analysis: "Our third, higher-end estimate recognizes that people of faith conduct their affairs to some extent (however imperfectly) inspired and guided by their faith ideals. This higher-end estimate...places the value of faith to U.S. society at $4.8 trillion annually, or the equivalent of nearly a third of America's gross domestic product (GDP)."[31]

Of course, the secular benefits accruing from Christian beliefs should not be the reason for allowing God into our hearts. Some theologians might dismiss these pragmatic factors as a secular version of Pascal's Wager (it is wise to accept God since you have much to gain if you do and much to lose if you don't). While I agree that belief in God comes from the depths of the heart, and one must voluntarily place Him there, and not because it is prudent or self-serving to do so, this is not an argument that an atheist would find at all credible. I am simply pointing out how incredibly naïve atheist claims are that Christianity is detrimental for individuals and for society. One wonders how atheists can make the patently false claim that religion "poisons everything" when confronted with massive evidence to the contrary.

Endnotes

1. In Wilson, E., 1993, p. 219.
2. Lewis, C., 2001, p. 8.
3. In Lennox, J., 2011, p. 99.
4. Gervais, W. 2014, P. 7.
5. In Craig, W. and Meister, C., 2010, p. 18.
6. Marks, J. 2012, p.2.
7. Nietzsche, F., 1997, p. 14.
8. Nietzsche, F., 1990, pp. 80-81.
9. In Jal. M., 2010, p. 136.
10. Hazard, J., Butler, W. & Maggs, P., 1977, p. 470.
11. Hosking, G., 1985, p. 213.
12. Hitchens, P., 2010.
13. Solzhenitsyn, A., 2006, p. 577.
14. In Bennetts, M., 2014.
15. Caldwell, Z. 2018.
16. Xu, Q., 2014, p., 142.
17. In Phillips, T., 2014.
18. In Hewitt, T., 2016.
19. In Hass, B. 2018.

20. Melchior, J. 2014.
21. Hitler, A. 2007. p. 49.
22. Overy, R., 2004, p. 281.
23. Tatara, C., pp. 43-44.
24. Hulme, C. and Salter, M., 2001, p. 5.
25. In Pipes, R. 2011, p. 259.
26. In Muravchik, J. 2003, p. 164.
27. Craig, W., 2010, p. 29.
28. Mua, K., 2016.
29. Guiso, L., Sapienza, P., and Zingales, L., 2003.
30. Stark, R., 2012, p. 163.
31. Grim, B., and Grim, M., 2016, p. 2.

CHAPTER TWENTY-FIVE

Christianity and the Healthy, Happy, Loving Life

> "Let us leave sadness to the devil and his angels. As for us, what can we be but rejoicing and glad?"
> St. Francis of Assisi

Individual Happiness: Atheism v. Christianity

Although he doesn't acknowledge it, Richard Dawkins is surely aware of the pain, misery, fear, and squalor in counties that embrace his atheistic worldview, yet he wants to turn his own country in the same blighted direction. He was instrumental in a £100,000 campaign to litter London buses with the slogan "There probably is no God, so relax and enjoy life," The message within the message is that people are happier without God. If Dawkins is right, having divested themselves from the bugaboos of religion, atheists should be the happiest of people, but this is far from the case. Commenting on the bus campaign, Mary Kenny wrote: "Far from relaxing and enjoying life, most atheists I have encountered are gloomy blighters with a depressing and nihilistic message that there is no purpose to life so where's the point of anything?"[1]

The bus campaign doesn't seem to have had much impact on people who ride them. Figures published by Britain's Office for National Statistics 2016 Well-Being research program found that atheists report lower levels of happiness, life satisfaction, and feeling of self-worth than religious people, with practicing Christians and Jews topping the list on all indicators of well-being.[2] A study comparing British evangelical Christians with national norms on a number of health and well-being indicators found that evangelicals score higher on all of them. The study concluded that: "In general, the faith and values and the disciplined lifestyles reported by evangelicals appear to be beneficial to health and well-being. This could well be enhanced by their strong sense of purpose and belonging to God, stable families, and caring faith communities"[3] Another British study concluded: "Religious people seem to have a greater purpose in life, which is why they are happier. Looking at the research evidence, it seems that those who celebrate the Christian meaning of Christmas are on the whole likely to be happier."[4]

On this side of the Atlantic, the Pew Research Center conducted more than 35,000 telephone interviews and found that "highly religious people" (those who pray every day and attend religious services each week) "are more engaged with their extended families, more likely to volunteer, more involved in their communities and generally happier with the way things are going in their lives." The study also finds that almost half of highly religious people visit extended family at least once or twice a month compared with only three-in-ten who are less religious. Sixty-five percent of the highly religious donated money, time or goods to help the poor compared with 41 percent of the non-religious, and 40 percent of the highly religious said they were "very happy" compared to 26 percent of the non-religious.[5] A further study from the General Social Survey and the World Values Survey concluded that: "Results from both data sets support prior research by showing a positive association between happiness and both political conservatism and religiosity. Importantly, it was found that political conservatism and religiosity interact in predicting happiness levels."[6]

Unhappiness and a sense of the ultimate meaninglessness of life may lead to clinical depression. In her book on depression and suicide, psychiatrist Julia Kristeva agreed with Mary Kenny that atheists are gloomy blighters; writing that: "The depressed person is a radical, sullen atheist."[7] Being depressed may lead to suicidal thoughts and to suicide itself. Kristeva was right on the money; according to a study in the *American Journal of Psychiatry*:

> Religiously unaffiliated subjects had significantly more lifetime suicide attempts and more first-degree relatives who committed suicide than subjects who endorsed a religious affiliation...Furthermore, subjects with no religious affiliation perceived fewer reasons for living, particularly fewer moral objections to suicide. In terms of clinical characteristics, religiously unaffiliated subjects had more lifetime impulsivity, aggression, and past substance use disorder.[8]

Another study examining the relationship between spiritual values/church attendance and suicidal thoughts in a nationwide Canadian Community Health Survey concluded: "Results suggest that religious attendance is associated with decreased suicide attempts in the general population and in those with a mental illness independent of the effects of social supports." [9]

It is understandable that atheists who reflect on their atheism have a good chance at being depressed by its nihilism. Blaise Pascal, the brilliant 17th-century French mathematician and physicist, saw atheism as the root of unhappiness. He wrote in his insightful book *Pensees* ("thoughts"): "There are only three types of people; those who have found God and serve him; those who have not found God and seek him, and those who live not seeking, or finding him. The first are rational and happy; the second unhappy and rational, and the third foolish and unhappy."[10] A "gloomy blighter" is not a pleasant person to be around. Whichever comes first—atheist or gloomy blighter—it is undeniable that people who wish others a Merry Christmas in the true spirit of that day are far happier eating their Christmas plum pudding than atheists mumbling their politically correct "Happy Holidays."

I previously differentiated between atheists who peacefully coexist with theism, and antitheists who seek to destroy it. As we might expect, there are profound personality differences between these two groups. Christopher Silver examined personality dimensions of 1,153 members of atheist/skeptical/freethinking groups. He divided them into six groups according to the depths of their non-belief, beginning with agnostics and ending with antitheists. Compared to other non-believers, antitheists scored lowest on all measures of personal growth and highest on all measures of maladjustment. Antitheists were found to be highly narcissistic, and to have few trusting relationships with others. They have the highest anger when their views are challenged, and were lowest on agreeableness (unfriendly, uncooperative, and lacking in compassion). Likening them to New Atheists such as Dawkins, Silver stated: "certainly this profile here moves well beyond the protesting and socially assertive atheist. In a sense, they are socially distant, opinionated, closed-minded people who believe they are better than others."[11]

Bertrand Russell's search for happiness and ultimate meaning is a good place to end our discussion of happiness. Russell relates how he felt when he had occasion to visit the sick wife of a colleague: "Suddenly the ground seemed to give away beneath me, and I found myself in quite another region. Within five minutes I went through some such reflections as the following: the loneliness of the human soul is unendurable; nothing can penetrate it except the highest intensity of the sort of love that religious teachers have preached; whatever does not spring from this motive is harmful, or at best useless." [12] This is a remarkably frank statement from the man who said God didn't give him enough evidence to believe.

The Family, Religion, and Antisocial Behavior

In *God is Not Great*, Christopher Hitchens claimed that Christianity encourages antisocial behavior. Citing Christian claims that Christianity promotes prosocial behavior, he writes: "In fact, if a proper statistical inquiry could ever be made, I am sure the evidence would be the other way [i.e., atheists commit fewer antisocial acts than Christians]."[13] Hitchens obviously did not do his homework. There are literally hundreds of "proper statistical inquiries" on the matter that prove him egregiously wrong in yet another of his outrageous claims. A review of 113 studies linking regular church attendance to criminal offending and illegal drug use concluded that: "attending religious services is the best documented correlate of [the prevention of] crime."[14]

A later review of 270 studies concluded: "our updated systematic review suggests the beneficial relationship between religion and crime is not simply a function of religion's constraining function or what it discourages—opposing drug use, violence, or delinquent behavior—but also through what it encourages—promoting prosocial behaviors."[15] A study of almost 184,000 mostly young adults remarked that: "Results indicate that the protective relationship between religiosity and criminal behaviors such as drug selling and theft is consistent across gender as well as across the developmental periods of adolescence and young adulthood...the protective effect of religiosity on criminal behavior was consistently observed across important sociodemographic differences."[16] Finally, a study of 17,266 adults from 13 nations found that religious people are more law-abiding than the non-religious, and that: "This relationship was more pronounced in the case of 'overt' aspects of religiosity (especially church attendance and membership) than in the case of any specific religious beliefs."[17]

Religious families monitor their children's educational and extra-curricular behavior, providing them with moral guidance in the context of warmth and caring, and parents and children in religious households rate the quality of their relationships with one another significantly higher than do parents and children in non-religious households. A review of the relevant research on this noted that: "The tendency of religious beliefs to place great value on children increases parental motivation to spend time and energy on their children. Not only are religious parents less likely to abuse or yell at their children but they are also more likely to hug and praise them often and to display better parent functioning."[18]

Unwed motherhood is a result of the decline in morality in an increasingly secular United States. The plague of illegitimacy is an engine driving all kinds of antisocial behavior. Growing up in a fatherless home increases the probability of immoral rearing by single, poor, and often isolated mothers. Kendall and Tamura analyzed crime and out-of-wedlock birth rates in the United States and concluded, "we find that an increase of 10 nonmarital births per 1,000 live births is associated with an increase in future murder and property crime rates of between 2.5 and 5 percent."[19] Likewise, a large study of rural communities in four U.S. states found that "a 10 percent increase in female-headed households was associated with a 73 to 100 percent higher rate of arrest for all offenses except homicide [it was associated with a 33% increase in homicide]."[20] A Christian upbringing helps to prevent unmarried childbirth. According to a Fragile Families Research Brief, women who attend religious services frequently are "73 percent more likely to be married at childbirth than mothers who attend services infrequently or not at all."[21] The sad legacy of growing up in a fatherless home is vividly related below:

> Eighty-five percent of youth in prison have an absent father, 71% of high school dropouts are fatherless, 90% of homeless and runaway children have an absent father and fatherless children and youth exhibit higher levels of depression and suicide, delinquency, promiscuity and teen pregnancy, behavioral problems and illicit and licit substance abuse, diminished self-concepts, and are more likely to be victims of exploitation and abuse.[22]

A 2018 *Child Trends* report reveals that illegitimate births have increased from just 5 percent in 1960 (when prayer was still in our schools) to 40 percent in 2016. The report notes illegitimate birth rates of 70 percent for African Americans, 53 percent for Hispanics, 29 percent for Anglos, and 17 percent for Asian Americans.[23] It is thus no surprise that minority children are vastly overrepresented in prisons given their high rate of illegitimate birth. It is politically incorrect to pass moral judgement on birth status, but not doing so has hurt generations of children, since illegitimacy has almost become normalized. Studies abound showing the benefits of traditional two-parent families. The U.S. Department of Health and Human Services noted that "children from two-parent families live longer and enjoy overall better health than children from single-parent families or whose parents divorced in childhood."[24] A huge Swedish study of 986,342 children also found that children living in a single-parent household had greatly elevated risk of psychiatric issues, suicide or attempted suicide, and drug and alcohol addiction.[25]

Christianity and Loving Families

When digesting these statistics, it is helpful to recall the saying: "The family that prays together stays together." Yet Richard Dawkins has infamously claimed that teaching children Christian values amounts to child abuse. Perhaps we should imprison parents for reading the Bible to children as they do in the atheist countries that he believes provide a superior upbringing. Belying Dawkins' silly and hateful claim is a study of children that concluded: "Children who were more spiritual were happier. Spirituality accounted for between 3 and 26% of the unique variance in children's happiness depending on the measures...The personal (i.e., meaning and value in one's own life) and communal (quality and depth of inter-personal relationships) domains of spirituality were particularly good predictors of children's happiness."[26]

Marriages in which both spouses attend church regularly are 2.4 times less likely to end in divorce than marriages in which neither spouse attends,[27] and more than six decades of studies have shown religious commitment to be the most important predictor of marriage stability and happiness.[28] Not only are religious families more likely to stay together, they are happier than non-religious families, and a stable marriage is positively associated with better physical, mental, and emotional health of both parents and children. A study of the marital well-being of 354 couples found that religiousness is related to marital well-being through "relational virtues" (commitment, forgiveness, emotional closeness, and sacrifice).[29] Another study found that the more couples practiced their faith the more satisfied they were with their marriages. Sixty- percent of couples who attended a place of worship at least once a month rated their marriages "very satisfactory" compared with 43 percent who attended less often or not at all.[30]

Drawing on studies composed of over 30,000 respondents, Bradford Wilcox found that the more husbands attended religious services the happier their wives were with the level of affection, understanding, and quality time spent with them.[31] A national study of over 9,000 respondents showed that adults who frequently attend religious services as children were significantly more likely to provide assistance to parents in their old age, and reported higher quality relationships and more frequent contact with parents: "It appears that the influence of religion in fostering early parent–child ties noted in prior research extends throughout the life course, influencing ties between adult children and their parents."[32]

Christianity and Physical and Mental Health

A survey of physicians in the *Journal of Family Practice* found that 96 percent agreed that spiritual well-being is important for physical and mental health.[33] Karl Marx, on the other hand, viewed religion is the opiate of the masses; a childish fairy tale whose soothing comfort prevented believers from embracing a communist paradise on earth. Sigmund Freud also saw religion as a sign of psychological immaturity, but then, Freud saw all people as psychologically immature. Contrary to these naysayers, Christianity leads to personal enlightenment, greater happiness, psychological maturity, and better physical and mental health. Take Robert Hummer and colleagues' conclusion about longevity, for instance. Based on National Health Interview Survey data, they write: "We showed that religious involvement is strongly associated with adult mortality in a graded fashion. Those who never attend services exhibit the highest risk of death, and those who attend more than once a week exhibit the lowest risk."[34] The study found that the risk of untimely death is almost twice as great for those who never attend services compared with those who attend once a week or more. This translates to a difference in life expectancy of seven years at age 20. Another study followed an initial group of 5,286 adults for 28 years and found that, controlling for age, individuals who attended church one or more times a week were 23 percent less likely to die in follow-up periods than non-attenders.[35]

There are few things more harmful to our overall health than hypertension and related cardiovascular problems, but religion is a spiritual anti-body. Physician Daniel Hall reviewed a number of studies comparing things such as exercise, diet, and medication that lower the risk of heart disease, and concluded: "The real-world, practical significance of regular religious attendance is comparable to commonly recommended therapies, and rough estimates even suggest that religious attendance may be more cost-effective than statins [cholesterol lowering medication]"[36] A study of 36,000 Norwegians found that people who regularly attended religious services had significantly lower blood pressure than non-attenders, with increasing attendance associated with decreasing blood pressure in gradient fashion.[37] One of my own studies, controlling for other health-related factors, age, sex, and various demographic and psychological measures, found that 30.9 percent of non-church attenders were hypertensive versus only 7 percent of regular church attenders.[38]

Finally, a comprehensive review of the religiosity and physical and mental health literature found that daily spiritual experiences (DSE) are associated with a wide range of healthy physical (increased energy), psychological (less depression and greater feelings of self-efficacy), and behavioral (volunteering, charitable giving) outcomes. It was also noted that: "DSE is also associated with greater pain tolerance, fewer days spent in long-term care after hospitalization, better self-rated health, and better physical health perceptions in those with chronic illness. Furthermore, changes in DSE during treatment for alcohol and drug disorders have been associated with a greater likelihood of abstinence, increased prosocial behaviors, and reduced narcissistic behaviors."[39]

Charitable Giving

In Christian thought, the highest form of morality is love, as St. Paul's noted in I Corinthians 13:13: "And now abideth faith, hope, love, these three; but the greatest of these is love." The love of man for God must be manifested in an active concern for others in the form of charitable giving of time, money, and blood. No other religious or secular body has been more forceful in pushing the idea of helping the unfortunate as Christianity. As American educator Carlton Hayes put it: "From the wellspring of Christian compassion, our Western civilization has drawn its inspiration, and its sense of duty, for feeding the hungry, giving drink to the thirsty, looking after the homeless, clothing the naked, tending the sick and visiting the prisoner."[40] The historical record shows that many pagan cultures saw helping the sick and needy as a sign of weakness while Christians helped them in the belief that serving the sick and needy also serves God. Historian Alvin Schmidt writes that: "Christianity filled the pagan void that largely ignored the sick and dying, especially during pestilences. In so doing, it 'established the principle that to help the sick and needy is a sign of strength not weakness.'"[41]

Christian denominations run many charitable organizations. According to the Pontifical Council for Pastoral Assistance to Health Care Workers: "the Catholic Church manages 26 percent of health care facilities in the world... the Church has 117,000 health care facilities, including hospitals, clinics, orphanages, as well as 18,000 pharmacies and 512 centers for the care of those with leprosy."[42] Catholic Charities USA is the largest private provider of welfare services of all people in the nation, serving millions of people each year regardless of their religion, sex, race, sexual orientation, or anything else. Many other Christian and Jewish denominations cater indiscriminately to all manner of people who are in need. When was the last time you heard of any such atheist-based or humanist-based programs? Probably never. This is exactly what D.J. Taylor said upon noting that Norwich, England, topped the table for "no religious belief" in the British census. Taylor wrote: "my godless city is full of Christian charity. But it would be difficult to deny that, first, the church is still making its presence felt here in the Great Eastern Land, and, second, were you to take it out of Norwich, or any other similarly sized city, you would create a gaping hole that no government agency—and certainly not the British Humanist Association—would ever be able to fill."[43]

It is no surprise to find that practicing Christians live up to the "Love thy neighbor" command far more than atheists. An American national sample of 12,100 adults indicated that: "more frequent church attendance and greater participation in religious activities increase the levels and likelihood both religious and/or secular giving"[44] Another study of 30,000 people from 50 different sites across the U.S. found that people who attend church once or more per week were 25 percent more likely to donate money, and 23 percent more likely to volunteer time than their secular counterparts.[45] Yet another study across 29 states found that practicing Christians give far more time and money, and far more often, to both religious and secular causes than secularists.[46] Journalist George Will summarized it bluntly: "America is largely divided between religious givers and secular nongivers."[47]

A five-year study by David Campbell and Robert Putnam resulted in a sweeping look at contemporary American religion in a book titled *American Grace*. Campbell and Putman provide a number of reasons why practicing Christians make better neighbors than their secular counterparts. For instance, 40 percent of church attenders regularly volunteered to help the poor and elderly, compared with 15 percent of their secular counterparts. This was true for all kinds of services; giving money, donating blood, extending help to acquaintances experiencing hard times, such as unemployment, depression, or ill-health.[48] The evidence that Christianity, far from "poisoning everything," is the antidote to so much poison in the world's bloodstream, is so overwhelming that it is scientific malfeasance for Dawkins and his kind to act as though it does not. They are too smart to plead ignorance of this evidence. They simply ignore it or look at it with such jaundiced eyes that their brains become so befuddled that they get things backwards. I have challenged atheist colleagues to produce one, just one, published peer-reviewed article showing that Christians are less happy, more depressed, less physically and mentally healthy, more criminal, and less charitable than atheists—I'm still waiting.

But perhaps some religions do "poison everything," and this should remind us not to conflate religion with God. Victor Stenger used the tragedy of 9/11 to take a cheap shot, writing that: "Science flies you to the moon; religion flies you into buildings."[49] Stenger should listen to John Glen, who actually flew to the moon. Glen reported that his experience strengthened his belief in God: "To look out at this kind of creation and not believe in God is to me impossible."[50] In fact, almost all the 29 astronauts who visited the moon were practicing Christians, with many serving as church leaders in their congregations. Aboard the first lunar landing, the crew took turns reading from the Book of Genesis, and Buzz Aldrin, an elder in the Presbyterian Church, took communion. Aldrin also said: "Just before I partook of the elements, I read the words which I had chosen to indicate our trust that as man probes into space we are in fact acting in Christ." Aldrin also pointed out that: "There are many of us in the NASA program who do trust that what we are doing is part of God's eternal plan for man."[51]

Richard Dawkins also uses the attacks of 9/11 (which he says gave birth to his militant atheism) to trash all religion, but why blame Christianity for 9/11? Horrible crimes have been committed in the *name* of Christianity in the past, but not in *conformity* with Christianity. Why put Islam and Christianity in the same box since the same cannot be said of Islam? Matt Slick points out a profound difference between the two faiths: "The final words of Muhammad were: 'O Lord, perish the Jews and Christians. They made churches of the graves of their prophets. There shall be no two faiths in Arabia' (Hadith Malik 511:1588). And the last words of Jesus where His enemies were concerned? 'Father, forgive them; for they do not know what they are doing' (Luke 23:34)."[52] As far as I am concerned, Islamic terrorism has provided militant Western atheists with an opportunity to attack their real target—Christianity.

A caveat to keep in mind when reading this chapter is that religious belief overlaps with conservative Republicans, just as atheism overlaps with liberal Democrats. According to the Pew Research Center, 69 percent of Americans who identify as atheists identify politically as liberal Democrats, while just 10 percent identify as conservative Republicans.[53] Studies consistently show a conservative-

Republican advantage over liberal-Democratic individuals on all indicators of well-being noted in this chapter, and to almost identical degrees. That is, conservative Republicans are happier, more likely to be married, enjoy better physical and mental health, and give more to of themselves in terms of money, blood, and volunteering than their liberal Democratic counterparts.[54] Do we see this because most conservative Republicans are Christians, or because most Christians are conservative Republicans? Perhaps conservativism and Christianity are both indicators of the increasing wisdom humans acquire once they leave behind their youthful rebellious attachment to socialism and atheism. There are numerous examples of famous individuals who were formerly strong liberals or socialists who became conservatives with maturity. Internet searches will reveal long lists of them, but you won't find lists of famous conservatives who became liberals or socialists. I would love to see a plausible explanation for this from a committed leftist.

Endnotes

1. Kenny, M., 2008.
2. Bingham, J., 2016.
3. Smith, G., 2017, p. 11.
4. University of Warwick, News and Events, 2003.
5. Pew Research Center, 2016.
6. Bixter, M. 2015, p. 7.
7. Kristeva, J. 989, p. 5.
8. Dervic, K., et al, 2004, p. 2303.
9. Rasic, D., et al, 2009, p. 32.
10. Pascal, B., 1958, p. 257.
11. Silver, C. 2013, p. 197.
12. In Sullivan, A. 2018.
13. Hitchens, C. 2007, p. 13.
14. Ellis, L. and Walsh, A., 2000, p. 205.
15. Johnson, B. and Jang, S. 2011, p. 129.
16. Salas-Wright, C. Vaughn, M. and Maynard, B. 2014, p. 673.
17. Ellis, L. and Peterson, J. 1996, p. 761.
18. Dollahite, D. and Thatcher, J., 2005, p. 5.
19. Kendall, T. & Tamura, R. 2010, p. 213.
20. Osgood, D. & Chambers, J. 2003, p. 6.
21. Fragile Families Research Brief, 2005, p. 2.
22. Kruk, E., 2012, p. 49.

23. Child Trends. 2018.
24. Wood, R., Goesling, B., & Avellar, S., 2007, p. 48.
25. Weitoft, G., Hjern, A., Haglund, B. and Rosén, M., 2003, p. 289.
26. Holder, M., Coleman, and Wallace, J. 2010. p.131
27. Call, V. and Heaton, T., 1997.
28. Wilcox, B., 2004.
29. Day, R., and Acock, A., 2013.
30. Bahr, H. and Chadwick, B., 1985.
31. Wilcox, B., 2004.
32. King, V., Ledwell, M., and Pearce-Morris, J., 2013.
33. Ellis, M., Vinson, D., and Ewigman, B., 1999.
34. Hummer, R., Rogers, R. Nam, C. and Ellison, C., 1999.
35. Strawbridge, W., Cohen, R., Shema, S., and Kaplan, G., 1997.
36. Hall. D., 2006.
37. Sørensen, T., Danbolt, L., Lien, L., Koenig, H., and Holmen, J., 2011.
38. Walsh, A., 1998.
39. Koenig, H., Pearce, M., Nelson, B., & Erkanli, A. 2016, p. 1764.
40. In Schmidt, A., 2004, pp. 147-148.
41. Schmidt, A. 2001, p.153.
42. Catholic News Agency. 2017.
43. Taylor, D. 2012.
44. Forbes, K., & Zampelli, E., 2013, p. 2487.
45. Brooks, A., 2003.
46. Brooks, A., 2004.
47. Will, G. 2008.
48. Campbell, D. & Putnam, R. (2010).
49. Stenger, V. 2009, p. 59.
50. In Zauzmer, J. 2016.
51. Schratz, P. 2019.
52. Slick, M. 2013.
53. Lipka, M. 2016
54. Walsh, A. 2017.

Chapter Twenty-Six

Christianity, Freedom, Democracy, and Human Rights

> "No Human society has ever been able to maintain both order and freedom, both cohesiveness and liberty apart from the moral precepts of the Christian religion."
> John Jay:
> first Chief Justice of the U. S. Supreme Court

Christianity and the March to Human Rights

It is not only science, capitalism, prosperity, and objective morality that we owe to Christianity. Christianity has also given the world freedom, democracy, and human rights. It is the only religion that is a standard-bearer for the value and dignity of the individual, as any historian will attest. Only Christianity champions the individual conscience against the collective will, which has, over the centuries, paved the way for human rights and democracy. The stirring words of the Declaration of Independence bear witness to the Christian origins of the freedoms found in the document and to whom we should give thanks: "We hold these truths to be self-evident, that all men are created equal; that they are endowed by their Creator with certain unalienable rights; that among these are Life, Liberty, and the pursuit of Happiness." The rights laid out in that document and in the United States Constitution, particularly the First Amendment of the Bill of Rights, while given voice by human hands, were not dreamt up in the human mind. Rather, as Alexander Hamilton wrote in 1775, they are rights planted in our hearts by the Creator: "The sacred rights of mankind are not to be rummaged for, among old parchments, or musty records. They are written, as with a sunbeam, in the whole volume of human nature, by the hand of the divinity itself; and can never be erased or obscured by mortal power."[1]

Should anyone doubt that Christianity is the fountainhead of freedom, point them to Freedom House's annual survey of freedom in the world. In its 2018 *Freedom in the World Index*, 85 of the world's 210 nations (40.5%) designated as "free," all but 4 (Japan, Mongolia, South Korea, and Taiwan) are Christian nations. Japan and South Korea have been heavily influenced by American political ideals, and perhaps the same can be said of Taiwan. Some Christian nations such as corruption-riddled Mexico are designated "partly free," and the only nations with majority Christian population designated "unfree" are those scarred by socialism, such as Cuba and Venezuela.[2] That 81 out of 85 (95.3%) of the world's free nations are Christian speaks volumes for the thesis that Christianity is the major force behind social, civil, and political freedom in the world.

Further indicative of the thesis is the fact that the 1948 United Nations Declaration of Human Rights is so patently based on Judeo/Christian principles that many Muslim nations refused to sign it on the grounds that it conflicted with sharia law (some of the worst violators of human rights, such as Iran, did sign it with a wink and a nod).[3] Sharia law is completely devoid of any notion of human rights. Its most problematic aspects include the absolute subjugation of women, inhuman punishment for crimes, many of which are not even crimes in the West (such as apostasy, sorcery, and witchcraft; the latter two being catchall offenses), and the lack of anything remotely resembling religious freedom. Atheist philosopher Jurgen Habermas acknowledges that human rights took root *only* in the Judeo/Christian tradition:

> Egalitarian universalism, from which sprang the ideas of freedom and social solidarity, of an autonomous conduct of life and emancipation, of the individual morality of conscience, human rights and democracy, is the direct heir of the Judaic ethic of justice and the Christian ethic of love...To this day, there is no alternative to it. And in light of the current challenges of a postnational constellation, we continue to draw on the substance of this heritage. Everything else is just idle postmodern talk.[4]

None of these benefits were envisioned by early Christians, but values and beliefs have latent as well as manifest consequences. The belief that God is a God of reason who wants us to get to know him through his creation, shapes values; values shape motives, and motives produce beliefs and actions that evolve over time. Only when the effects of these actions and beliefs become manifest can we trace them back to motives, and then back to earlier values and beliefs. The idea of the equality of all people as central to democracy is stated by Paul in Galatians 3:28: "There is neither Jew nor Greek, there is neither bond nor free, there is neither male nor female: for ye are all one in Christ Jesus."

Paul did not mean this as a call for voting booths and bicameral parliaments; nor did he mean that secular distinctions among people should be disregarded. He meant that whatever secular distinctions exist, all people are God's creations, and equally deserving of His love. All humans share the same origin, the same nature, and the same ultimate destiny regardless of their rank on

Earth, but except in the eyes of God, true quality does not, and cannot, exist as long as people possess different talents and ambitions. It would be centuries before it was recognized that all people are equally entitled to all political and civil rights regardless of these talents and ambitions. These rights were first recognized in Christian societies and are yet to be seen in the vast majority of non-Christian societies.

Severe injustices existed at the dawn of Christianity, including the practice of slavery. Even after the Emperor Constantine converted Rome to Christianity, slavery still existed throughout Europe. The slavery of Christians in Europe was ended by the end of the first millennium, although it was replaced by the less onerous institution of serfdom. In the 18th and 19th centuries, anti-slavery activism blossomed in Europe and America founded on morality and the idea of the equal love of God the brotherhood of all humanity. These activists continually invoked the Golden Rule: "All things whatsoever ye would that men should do to you, do ye even so to them" (Matthew 7:12). There were no abolitionist movements anywhere in the world that were not founded on Christian convictions; not one.[5]

Christian nations such as the U.S. and the U.K. engaged in slavery and the slave trade, and Charles Darwin rightly rebuked both countries for doing so. However, he also noted that they sacrificed more than any other country to end it: "It makes one's blood boil, yet heart tremble, to think that Englishmen and our American descendants, with their boastful cry of liberty, have been and are so guilty: but it is a consolation to reflect that we have made a greater sacrifice than ever made by any nation to expiate our sin."[6] This was written shortly after Britain freed all colonial slaves at a cost of £20 million, which was 37 percent of the government's revenue in 1831.[7] According to the Bank of England inflation calculator, this is the equivalent of £2,199,603,960 in 2018 pounds. In addition to this expense, there was the immense cost in lives and money associated with the 50-year-long patrol of Royal Navy warships. In combatting the slave trade, these ships intercepted over 1,600 slave ships and liberated over 150,000 slaves.[8] The United States paid an even greater price in lives and resources to end slavery in America.

With its great ethic of justice, Judaism is also conducive to democracy, but democracy is incompatible with Islam, the other Hebraic faith. While Judaism and Christianity are sometimes dogmatic about religion, they generally avoid political, civil, and social dogmatism. Islam, on the other hand, restricts freedom in all spheres of life. Norman Graebner observes: "Muhammad had crammed the Koran with political maxims, criminal and civil laws, and even theories of science; the biblical Gospel, on the other hand, imposed no demands on faith beyond the establishment of a proper relationship with God and men and men with each other. It was this quality in Christianity, among others, that permitted it to exist in a cultivated, democratic civilization."[9]

Alfred Stephan concurs, and writes that most Western scholars, and all Islamic scholars, find Islam to be incompatible with democracy: "According to this view, allowing free elections in Islamic countries would bring to power governments that would use these democratic freedoms to destroy democracy itself"[10] This was written in 2000, and was remarkably prescient. We witnessed the chaos and brutality that befell many Islamic states after the "Arab Spring" in 2011 that many naïve Westerners thought would bring democracy to the Middle East. What the Arab Spring actually did was to clear the way for radical Islamist takeover, so much so that the debate about Islam's compatibility with democracy is effectively over. The French Muslim intellectual Salem Ben Ammar's harsh opinion of democracy is typical of Islamic scholars: "To hell with democracy! Long live Islam! These two competing political systems are antithetical to each other. You can't be democratic and be a Muslim or a Muslim and be a democrat."[11]

The Reformation and the Slow and Bloody Road to Democracy

There is no real separation of mosque and state in any Islamic country, where the law privileges Muslims over non-Muslims in all things. In many of these counties, laws passed by secular politicians must pass scrutiny by religious authorities, and there is even a religious police force in Saudi Arabia to make sure that all toe the sharia line. It cannot be denied, however, that a similar situation existed in Christian Europe for centuries in which a cozy relationship between prince and priest existed. Religious dissenters had to tread lightly when the Church of Rome reigned supreme. This began to change after Martin Luther nailed his Ninety-Five Theses to the door of Wittenberg Castle Church in 1515, setting the Protestant Reformation in motion.

Luther translated the Bible into the vernacular and taught that it, not the Pope and his priests, was the source of divinely revealed knowledge. Luther's views found fertile soil because they came in the midst of the Renaissance, a cultural movement that profoundly affected intellectual life in Europe. The Renaissance was an age in which political, philosophical, social, and scientific orthodoxy were challenged. The primary impetus for this was the revival of ancient Greek philosophy, brought into Italy by Greek scholars fleeing Constantinople after it fell to the Ottoman Turks. The questioning of received orthodoxy changed the intellectual landscape and encouraged many Church reformers to rebellion. The dissemination of Luther's ideas was aided by the mass production of Bibles made possible by Johannes Gutenberg's invention of the printing press.

The Protestant Reformation led to conflicts between Lutherans and Catholics for state sponsorship, igniting numerous civil and interstate wars that devastated much of the continent. To end religious strife, Europe's nations and principalities signed a treaty in 1555 known as the Peace of Augsburg, and settled on the principle of *cuius regio, eius religio* ("Whose realm, his religion."). In essence, this principle meant that all individuals living in a given state must follow the religion of the state sovereign, and resulted in the *de facto* division of Western Europe into Lutheran and Catholic domains.[12]

The Peace of Augsburg still affirmed the unity of church and state and failed to address the issue of religious pluralism, which was becoming widespread among Protestants. The practice of non-established religions was forbidden, and those who practiced them were branded heretics and faced banishment, or worse, if they did not repent. The rise of Calvinism as a Protestant rival to Lutheranism gave rise to the Thirty Years War that engulfed Western Europe from 1618 to 1648. This series of wars ended with a number of treaties collectively known as the Peace of Westphalia. The Peace of Westphalia was a profound turning point in the history of Western civilization marking the dim beginnings of the modern democratic nation state. The Peace of Westphalia abrogated the Treaty of Augsburg, ending the sovereign's right to control matters of religious faith in their territories, and provided a model for dealing with religious disagreements in a peaceful manner. At long last, rulers no longer had the right to foist their religious convictions on their subjects or to interfere with their religious conscience.[13]

Atheists take these religious wars to make more absurd claims. Richard Dawkins claims that: "Religious wars really are fought in the name of religion, and they have been horribly frequent in history."[14] Likewise, Sam Harris boldly states that religion is "the most prolific source of violence in our history."[15] These chaps may be decent scientists, but they are terrible (or dishonest) historians. Robin Schumacher's examination of Philip and Axelrod's three-volume *Encyclopedia of Wars* decisively refutes such claims. Schumacher reports that the *Encyclopedia* lists 1,763 recorded wars waged over the course of human history. "Of those wars, the authors categorize 123, as being religious in nature, which is an astonishingly low. However, when one subtracts out those waged in the name of Islam (66), the percentage is cut by more than half to 3.23%."[16] Yet another New Atheist claim is skewered on the sword of truth.

Nevertheless, Christians must honestly admit the violence committed in the name of Christ, while soundly condemning it as the very antithesis of His message. As Matthew 5:9 says: "Blessed are the peacemakers for they shall be called sons of God." Christ does not wish to rule the Earth by military conquest; there is no universal caliphate to be won by violence in Christian theology. Modern Christianity seeks no sway over Caesar, and is more than content to possess the ultimate truth rather than the satisfaction of state power. As Jesus said to Pontius Pilate: "My kingdom is not of this world: if my kingdom were of this world, then would my servants fight." (John 18:36). The religious wars for denominational supremacy in Christendom fought in God's name were contrary His commands and a serious affront to Him.

Although there are many examples of powerful Christian rulers forcing conquered subjects to accept Christianity, Jesus and His disciples won converts through loving words and deeds, and never counseled war and violence to force people to accept Him. Forced conversion is anathema to Christianity because God wants people to accept Him voluntarily. Jesus was the great peacemaker who counseled His followers to turn the other cheek and to win over non-believers with love and rational debate. Muhammad, on the other hand, was a great general who marinated in warfare. His counsel regarding non-believers in the Qur'an (2:190-191) is: "Kill them wherever you come upon them and drive them out of the places from which they have driven you out. For persecution is far worse than killing. And do not fight them at the Sacred Mosque unless they attack you there. If they do so, then fight them—that is the reward of the disbelievers."

What about the crusades? Atheists delight to emphasize the massacres of Muslims and Jews by rampaging crusaders, and there is no justification for such barbarity. Yet, they ignore or downplay the primacy, longevity, and barbarity of the on-going Muslim "crusades." The last Christian crusade ended 725 years ago, but Muslims launched their own crusades (jihads) against the West 365 years earlier than the first Christian crusade launched in 1095. Islamic armies began the conquest of Persia in 634, just two years after Mohammed's death, and went on to subjugate over half of what used to be Christendom from Egypt to Spain. They plundered the Churches of St. Peter and St. Paul in Rome in 846, a full 249 years before the first crusade.[17] The crusades were "counter jihads" waged to stop further Muslim expansion across Christendom. Historian Bernard Lewis makes it clear that the purpose of the crusades was to recover lost Christian lands. He notes that: "it may be recalled that when the Crusaders arrived in the Levant not much more than four centuries had passed since the Arab Muslim conquerors had wrested these lands from Christendom—less than half the time from the Crusades to the present day—and that a substantial proportion of the population of these lands, perhaps even a majority, was still Christian."[18] I mention this to show how wrong atheists are to equate the relatively few wars waged in the name of Christianity, which were contrary to Christian doctrine, with wars waged in the name of Islam, which are consistent with its doctrine.

Early English Views of the Christianity/Human Rights Relationship

Americans are justly proud of their founding documents, but to understand them we must understand their antecedent documents, all drenched in Christian ideals. The English Petition of Rights (1628) and the Bill of Rights (1689) are particularly important. In common with Alexander Hamilton, Sir Edward Coke, Chief Justice of England in the early 17th century, and considered the father of Anglo/American common law, viewed the law of England "written with the finger of God in the human heart." He pointed out that:

> In nature, we see the infinite distinction of things proceed from some unity...as many rivers from one fountain, many arteries in the body of man from one heart, many veins from one liver, and many sinews from the brain: so without question *Lex orta est cum mente divina* [The law was written in the mind of God], and this admirable unity and consent in such diversity of things, proceeds only from God the fountain and founder of all good laws and constitutions.[19]

Two centuries later, Lord John Acton, a prominent British politician and intellectual, noted that democracy cannot long endure without God, because without Him society is coarsened and morality mocked. In an address delivered to the members of the Bridgnorth Institute in 1877, Lord Acton made plain his views on the mutually sustaining influences of Christianity and democracy, remarking that:

> The idea that religious liberty is the generating principle of civil liberty, and that civil liberty was a discovery reserved for the seventeenth century...That great political idea, sanctifying freedom and consecrating it to God, teaching men to treasure the liberties of others as their own, and to defend them for the love of justice and charity, more than as a claim of right, has been the soul of what is great and good in the progress of the last two hundred years. The cause of religion, even under the unregenerate influence of worldly passion, had as much to do as any clear notions of policy in making this country the foremost of the free.[20]

Why Christianity is Important to Freedom and Democracy

Freedom requires a moral foundation that only Christianity can supply, although the American Humanist Association begs to differ. Its motto, "Good without God," assumes that a secular value system can supply the moral foundations for democracy, but we have seen that this is an egregious falsehood. "Good without God" can only last as long as society has a cadre of people nourished on Judeo-Christian values remaining on Earth; after they have gone and we are left with citizens who have been raised in a society that has given God the boot, watch out! Robert Winthrop, one-time speaker of the House of Representatives, put it cogently: "Men...must necessarily be controlled, either by a power within them, or by a power without them; either by the Word of God, or by the strong arm of man; either by the Bible, or by the bayonet."[21] When we are ruled by the bayonet, as so many non-Christian societies are, our freedoms are gone. George Washington stressed this in his farewell address to the nation upon leaving his second term as President of the United States:

> Of all the dispositions and habits which lead to political prosperity, Religion and Morality are indispensable supports. In vain would that man claim the tribute of Patriotism who should labor to subvert these great Pillars of human happiness-these firmest props of the duties of Men and citizens...Let it simply be asked, Where is the security for property, for reputation, for life, if the sense of religious obligation desert the oaths, which are the instruments of investigation in Courts of Justice? And let us with caution indulge the supposition that morality can be maintained without religion. Whatever may be conceded to the influence of refined education on minds of peculiar structure, reason and experience both forbid us to expect that National morality can prevail in exclusion of religious principle.[22]

Only in a mature democracy with political checks and balances and the rule of law does everyone have a chance to have their say without fear of state reprisal. People in democracies can criticize their government with impunity, which the vast bulk of humankind could not do throughout history, nor cannot do today. The freedoms of religion, speech, press, and association enshrined in the First Amendment of the Bill of Rights recognizes that in the domain of moral conscience there is a power higher than the state that grants us these inalienable rights. The state does not grant us these rights, but it is expected to protect them. This is why socialist states know that they must purge God from society.

The great British philosopher, novelist, and lay theologian, G. K. Chesterton, tells us that it is difficult to exercise our freedoms if there is no higher power than the state: "But the truth is that it is only by believing in God that we can ever criticise the Government. Once abolish the God, and the Government becomes the God." He goes on to show how this was the case in the old Soviet Union with its commandment that it alone should be worshipped and obeyed. Chesterton adds: "The truth is that Irreligion is the opium of the people. Wherever the people do not believe in something beyond the world, they will worship the world. But, above all, they will worship the strongest thing in the world."[23] And that will be the state; it alone will govern your conduct and impose its vision of what is permissible and impermissible.

Alexis de Tocqueville, the great French philosopher of American life, noted in his masterpiece, *Democracy in America*, that America's freedoms and bounty rest on its religion: "there is no country in the world where the Christian religion retains a greater influence over the souls of men than in America; and there can be no greater proof of its utility and of its conformity to human nature than that its influence is powerfully felt over the most enlightened and free nation of the earth."[24] Even agnostics such as historians Will and Ariel Durant developed a humble respect for religion. They knew the value of religion to society and how it provides comfort to the suffering and the bereaved, how it teaches morality, and how it confers meaning and dignity on members of the flock. Their conclusion is most instructive: "through its sacraments has made for stability by transforming human covenants into solemn relationships with God...*There is no significant example in history, before our time, of a society successfully maintaining moral life without the aid of religion*" [my emphasis][25]

In an earlier work, Will Durant echoed George Washington's fears about a nation's morality when it abandons religion. It begins, he says, with the intellectual classes first abandoning religion, then the moral code it undergirds, and finally, literature and philosophy become not only non-religious, but anti-religious. This they view as liberating:

> The movement of liberation rises to an exuberant worship of reason, and falls to a paralyzing disillusionment with every dogma and every idea. Conduct, deprived of its religious supports, deteriorates into epicurean chaos; and life itself, shorn of consoling faith, becomes a burden alike to conscious poverty and to weary wealth. In the end a society and its religion tend to fall together, like body and soul, in a harmonious death.[26]

The ultimate proof of the value of Christianity for freedom, human rights, and democracy is that with the extremely rare exceptions noted earlier, these blessings are found only in nations founded on Christian principles. Proof-positive exists in the fact that the complete subjugation of the individual to the state that occurs in atheistic countries must be preceded by destroying Christianity, as was noted in the cases of the Soviet Union, Red China, and Nazi Germany. There can only be one ultimate master: God or the state. If one abandons God, life becomes shorn of any real purpose or ultimate meaning, and one may then turn to the worship of Big Government. When enough people do this, it will not be long before Western societies lose the freedoms and rights bequeathed to us by our Christian forefathers.

Endnotes

1. Hamilton, A. 1788, np.
2. Freedom House. 2018.
3. Mayer, A. 1993.

4. Habermas, J. 2006, p. 150-151.
5. Drescher, S. and Engerman, S., 1998.
6. In Richerson, P. & Boyd, R., 2010, p. 565
7. Blackburn, R. 2000.
8. Sherwood, M., 2007.
9. Graebner, N., 1976, p. 264.
10. Stephan, A., 2000, p. 48.
11. In Taylor, B. 2015.
12. Straumann, B., 2008.
13. Ibid.
14. Dawkins, R., 2006, p. 316.
15. Harris, S. 2005, p. 27.
16. Schmacher, R., 2012.
17. Durant, W. and Durant, A. 1950, p. 290.
18. Lewis, B. 1993, p. 12.
19. In Zimmerman, A. 2013, p. 92.
20. Acton, J. 1877.
21. Winthrop, R., 1852.
22. Washington, G., 1796.
23. Chesterton, G., 2001, p. 57.
24. de Tocqueville, A., 1994, p.199.
25. Durant, W. and Durant, A., 1968, p. 43 and p. 51.
26. Durant, W., 1935, p. 71.

CHAPTER TWENTY-SEVEN

The Modern American Assault on Christianity

> "Without God there is a coarsening of the society. Without God democracy will not and cannot long endure. If we ever forget that we are one nation under God, then we will be a nation gone forever."
> Ronald Reagan: 40th president of the United States

Slouching toward Gomorrah

The United States emerged from World War II the wealthiest, freest, and most powerful nation on the planet, and the 1940s through the 1960s was America's golden age. It was a time of prosperity and social conservatism; most families were intact and attended places of worship in their Sunday best; Billy Graham Crusades drew millions; drugs were something you got from the pharmacy when you were sick, and a sexually transmitted disease (STD) was an affliction only prostitutes and their clientele suffered. The post-war culture that emphasized self-control, discipline, and God and country has slowly been replaced by a leftist secular culture trumpeting moral relativism and "Doing your own thing." In the late 1960s we saw the rejection of the values of conventional society, previously limited to a few exotic beatniks, spill over to a flood of hippies, and then go on to infect mainstream culture. Since then, American culture has been, in Judge Robert Bork's telling phrase, "slouching toward Gomorrah."

The motto of hippie subculture was: "If it feels good, do it, and don't get hung up on guilt." Hippies swallowed, sniffed, and injected everything in sight, as if sobriety was a difficult state for them to tolerate. Such folks were relatively few, but today 22 percent of young adults are users of illicit drugs according to a 2015 Substance Abuse and Mental Health Services report.[1] The hippy-inspired sexual revolution made it inevitable that STDs were no longer suffered only by prostitutes and their sorry clients. A 2015 American Sexual Health Association report reveals that more than half of all Americans will suffer that penalty at some point in their lives, and that 19.7 million new STD cases are reported every year in the U.S., costing the economy $17.2 billion annually. The report also notes that 25 percent of teenagers contract an STD each year and that half of all sexually active persons will contract an STD by age 25.[2] The deadliest of these STDs is AIDS. According to a 2014 U.S. Department of Health & Human Services report, men who engage in homosexual behavior represent about 4 percent of the male population, but in 2010 they were responsible for 78 percent of new HIV infections among males, and for 54 percent of all male and female HIV infections.[3]

Popular Music and Entertainment: 1950s and 2000s

As President Reagan noted in the epigraph: "Without God there is a coarsening of the society." Our society has become very coarse over the last 70 years. Ever since Plato, it has been noted that popular music and entertainment express a culture's values better than any other single factor. In the 1950s music featured wholesome songs, such as Pat Boone, hand over his heart, crooning to his sweetheart that "Every star's a wishing star that shines for you," in *April Love*, or Elvis expressing tender love to his special girl in *Love me Tender*. Fast forward to the present and we find Lil Wayne, hand on his crotch, whining in *Every Girl*: "I wish I could f**k every girl in the world," and the equally obnoxious Pitbull howling that his "ho" has "got an a** like a donkey, with a monkey [vagina], look like King Kong," in *I know you want me*.

Not to be outdone, 21st century TV is also wallowing in muck. One of the most wildly popular shows today is the toilet humor of the cartoon show, *South Park*. Nothing is sacred in *South Park*. Songs that are supposed to celebrate the birth of Christ are turned into the most blasphemous filth imaginable, such as *Merry F**king Christmas*. The most obnoxiously depraved "Christmas" song of all, *Howdy Ho!*, features the Virgin Mary engaging in oral sex with the three wise men.[4]

Mocking Christianity is also fashionable in the arts. Andres Serrano's notorious photograph, *Piss Christ*, depicts a crucifix submerged in a glass of urine. Serrano received $5,000 from National Endowment for the Arts to fashion this loathsome "creation." Then we have the deeply offensive painting by Chris Ofili called *The Holy Virgin Mary* spattered with elephant dung and liberally collaged with pornography. Hillary Clinton, and other paragons of left-wing virtue, strongly defended government support for this garbage. Leftists feel that it is just fine to use tax dollars to display the work of degenerates in the public square while at the same time demanding tax dollars to remove crucifixions, crèches, and the Ten Commandments from it. The Supreme Court is their ally, issuing ruling after ruling aimed at outlawing anything smacking of Christianity from public view while defending filth as "freedom of expression."

Perhaps showcasing society's degeneracy most vividly are the annual Exxxotica Expos. These glittering expos feature pornography, sex toys, lots of lewd conduct, and draw huge crowds around the country annually. Billed as a "celebration of sex," it features scantily clad women (and some men) singing the praises of loveless sex, and even a dungeon where visitors are invited to "experiment" with sadomasochism. I wonder what manner of STDs are floating around there in the muck. The state of today's culture makes laughable Sam Harris' claim that atheism can supply morality, or that we can be "Good without God." Do such people ever stop to examine their claims, or are they so contemptuous of God that evidence doesn't matter?

The Spiritual Disarming of America's Schools

Abe Lincoln once wrote: "The philosophy of the school room in one generation will be the philosophy of government in the next. The only assurance of our nation's safety is to lay our foundation in morality and religion."[5] We must ask ourselves is whether we want to instill in our children the values of compassion, decency, hard work, and self-reliance, or do we want them animated by a hatred for the values that forged America, a deep desire to be free of irksome responsibilities and to outsource their lives to big government. The importance of teaching Christian values in school was so important to the Founding Fathers that they made it a part of the Northwest Ordinance, enacted in 1787 by the last Congress of the Confederation. The document made plain what the purpose of schools was to be: "Religion, morality, and knowledge, being necessary to good government and the happiness of mankind, schools and the means of education shall forever be encouraged."[6]

Early in America's history, the Bible was used in schools as a text for instilling moral principles. The Bible was eventually replaced by Noah Webster's *Primer*, which was filled with Bible verses designed to "enlighten the minds of youth in religious and moral principles and restrain some of the common vices of our country."[7] The Bible is banned in schools today, and Webster's *Primer* would be mocked as pious twaddle. The spiritual disarming of America's schoolchildren has had dire consequences. The 2000 *Congressional Record* notes the top seven disciplinary problems in schools in 1940 as ranked by school teachers: talking out of turn, chewing gum, making noise, running in the hall, cutting in line, dress code violations, and littering. In 1990, teachers listed the top seven disciplinary problems as drug and alcohol abuse, pregnancy, suicide, rape, robbery, and assault.[8] School problems have thus devolved from simple acts of youthful exuberance to immoral, self-destructive, and criminal behaviors.

The Supreme Court of the United States (SCOTUS) has been instrumental in this moral free-fall by removing God from our schools. SCOTUS did this with its "discovery" of "tension" between the two religious' clauses of the First Amendment. The wording that SCOTUS finds so difficult to decipher is: "*Congress shall make no law respecting the establishment of religion, or restricting the free exercise thereof.*" These rights, unlike others enumerated in the Bill of Rights, are singularly absolute, and need no interpretation. Both clauses are negative restraints on the government, and are thus pulling in the same direction. The establishment clause bars Congress from establishing a state religion and the Free Exercise Clause forbids it to inhibit the "free exercise thereof." Both Clauses are restraints on government, not on religion. Yet SCOTUS has used the establishment clause to effectively eviscerate the free exercise clause by reading the former so broadly that the slightest hint of government involvement amounts to Congress establishing a Church of The United States. Even if there is conflict between the clauses, there are no rules for deciding which should prevail. The establishment clause is the easiest to comply with because it can be violated by just one thing—the establishment of a state religion—whereas the free exercise clause may be violated in many ways. If there is to be a hierarchy of clauses, it is thus more reasonable that the free exercise clause should have priority.

The first case leading to the demise of Christianity in public schools was *Engel v. Vitale* (1962). The case involved the Board of Regents of New York State writing a non-sectarian prayer for recitation at the start of each school day. The Board viewed moral and spiritual training as part of a school's mandate and a tool for the development of moral character and good citizenship. The prayer reads: "Almighty God, we acknowledge our dependence upon Thee, and beg Thy blessings upon us, our teachers, and our country. Amen." Although recitation was voluntary, opposition from ten parents in the New Hyde Park school district resulted in a lawsuit to have it banned. After failing in district court and in the New York Court of Appeals, the parents, joined by the ultra-leftist ACLU, appealed to SCOTUS. SCOTUS ruled the prayer in violation of the establishment clause, and in doing so conflated children praying for their teachers, their country, and themselves, with Congress establishing a religion. This ruling slanders James Madison, the author of the religious clauses. Madison affirmed that we owe a duty to the Creator, and that: "This duty is precedent, both in order of time and in degree of obligation, to the claims of Civil Society. Before any man can be considered as a member of Civil Society, he must be considered as a subject of the Governour of the Universe."[9]

Perhaps we should also shield students from *The Star-Spangled Banner*, which contains some of the same sentiments found in the New York Regent's prayer. Part of the third verse of our national anthem reads: "Praise the Power that hath made and preserved us a nation! Then conquer we must, when our cause it is just, and this be our motto: In God is our trust." If we buy the twisted logic of *Engel*, singing *The Star-Spangled Banner* in state schools also establishes a religion and is unconstitutional. Justice Potter Stewart made these common-sense comments on the ruling that school prayer amounted to New York State establishing a national religion:

The Court does not hold, nor could it, that New York has interfered with the free exercise of anybody's religion...But the Court says that, in permitting school children to say this simple prayer, the New York authorities have established "an official religion." I cannot see how an "official religion" is established by letting those who want to say a prayer say it. On the contrary, I think that to deny the wish of these school children to join in reciting this prayer is to deny them the opportunity of sharing in the spiritual heritage of our Nation.[10]

The next case was *Abington School District v. Schempp* (1963), which involved Bible reading in Pennsylvania and Maryland schools. The Abington School District argued that Bible reading promoted moral values and American institutions; sentiment that are consistent with the words of the first SCOTUS Chief Justice, John Jay, who wrote: "The Bible is the best of all books, for it is the word of God and teaches us the way to be happy in this world and in the next. Continue therefore to read it and to regulate your life by its precepts."[11] The Schempp family disagreed, and filed suit. Although students could be excused if parents objected, this did not satisfy them. SCOTUS ruled that Bible reading in schools is unconstitutional. Justice Clark wrote in his opinion that although the majority of Americans would like Christian morality taught in public schools, their wishes must be sacrificed to the wishes of a tiny minority of atheists and malcontents who do not. Justice Stewart wrote in his dissent that: "a refusal to permit religious exercises thus is seen not as the realization of state neutrality, but rather as the establishment of a religion of secularism, or, at the least, as government support of the beliefs of those who think that religious exercises should be conducted only in private."[12]

After SCOTUS had prayers and the Bible removed from schools, it went after the Ten Commandment in *Stone v. Graham* (1980). The State of Kentucky required the Ten Commandments be posted on the walls of every classroom in the state. SCOTUS ruled that displaying them could give students the impression that the state was promoting the beliefs they represent (so, the state should not promote Thou shalt not kill, steal, lie, or covet!). The Court's wording in a *per curium* decision (no author identified) is quite amazing: "If the posted copies of the Ten Commandments are to have any effect at all, it will be to induce the schoolchildren to read, meditate upon, perhaps to venerate and obey, the Commandments."[13] Is moral guidance no longer a legitimate state interest? Apparently not; let our schoolchildren meditate on obscene rap lyrics, *South Park*, Exxxotica, lust, greed, and envy instead. Heaven forbid that they actually read, meditate upon, venerate, and obey God's laws!

In *Wallace v. Jaffree* (1985), SCOTUS ruled that schoolchildren could not even meditate on God. Before the Court was a 1981 Alabama statute authorizing one-minute silence in public schools "for meditation or voluntary prayer," and a 1982 statute authorizing teachers to lead "willing students" in a prayer to "Almighty God...the Creator and Supreme Judge of the World." SCOTUS ruled the statutes unconstitutional. In Justice Warren Burger's dissent, he noted that "The notion that the Alabama statute is a step toward creating an established church borders on, if it does not trespass into, the ridiculous. The statute... affirmatively furthers the values of religious freedom and tolerance that the Establishment Clause was designed to protect." Burger also pointed out the Court's hostility to religion: "To suggest that a moment-of-silence statute that includes the word "prayer" unconstitutionally endorses religion, while one that simply provides for a moment of silence does not, manifests not neutrality but hostility toward religion."[14] I do not see how the ruling could possibly be interpreted any other way.

SCOTUS has made the free exercise clause subordinate to the establishment clause at every turn. By doing so it implies that those who wrote the clauses were crude unlettered men, unable to put together a sentence that was not self-contradictory, or that meant exactly what it said. The clauses are only in conflict if one believes that anything remotely religious in the public sphere amounts to Congress establishing a religion. The free exercise clause simply forbids Congress from prohibiting the free exercise of religion; it does not say Congress must not prohibit free exercise "except in the public square."

In Justice William Rehnquist's biting dissent, he pointed out how alien the Court's ruling would have been to previous generations of Americans, especially the generation that wrote the Declaration of Independence, the Constitution, and the amendments to it in the Bill of Rights:

> It would come as much of a shock to those who drafted the Bill of Rights as it will to a large number of thoughtful Americans today to learn that the Constitution, as construed by the majority, prohibits the Alabama Legislature from "endorsing" prayer. George Washington himself, at the request of the very Congress which passed the Bill of Rights, proclaimed a day of "public thanksgiving and prayer, to be observed by acknowledging with grateful hearts the many and signal favors of Almighty God." History must judge whether it was the Father of his Country in 1789, or a majority of the Court today, which has strayed from the meaning of the Establishment Clause.[15]

If the State Alabama was behaving in unconstitutionally, so was George Washington and his contemporaries who endorsed prayer and promoted religion. As far as the anointed ones of SCOTUS are concerned, all individuals and institutions in America prior to their mysterious finding of tension between the religious clauses were in violation of the Constitution. This is how far the Courts have taken us toward making this nation not just non-Christian, but anti-Christian.

In its relentless battle against religious freedom, SCOTUS delivered another devastating blow in *Lee v. Weisman* (1992). SCOTUS interpreted the establishment clause in this case so broadly that its boundaries could not be seen, but somewhere in that vast expanse the free exercise clause was buried. The case involved Providence, Rhode Island, school's request that a rabbi deliver a non-sectarian prayer at its graduation ceremony in 1989. The parents of Deborah Weisman objected, and requested an injunction barring the prayer. A district judge denied the Weisman's motion, and the rabbi delivered the prayer. However, the Weissmans continued their objections and won a victory in a federal appeals court. The school district then appealed to SCOTUS, arguing that the prayer is nonsectarian and that if Deborah objected, she was free to remain seated during the prayer, and free not to attend the ceremony. But SCOTUS ruled that Deborah was "psychologically coerced" into bearing the "burden" of having to listen to the prayer, and if she took the option of absenting herself, she would forfeit the intangible benefit of the graduation ceremony. Thus, millions of high school graduates across the nation cannot engage in what could be a moving spiritual experience because one student was (or at least her parents were) offended, and SCOTUS bought it.

In Justice Scalia's dissent, he wrote mockingly that the Court has gone beyond what it knows it's doing and that "interior decorating is a rock-hard science compared to psychology practiced by amateurs." He further added that: "In holding that the Establishment Clause prohibits invocations and benedictions at public school graduation ceremonies, the Court...lays waste a tradition that is as old as public school graduation ceremonies themselves, and that is a component of an even more longstanding American tradition of nonsectarian prayer to God at public celebrations generally."[16]

The next case in this anti-Christian crusade was *Santa Fe Independent School District v. Jane Doe* (2000). In this case, SCOTUS ruled that even the practice of opening high school football games with a prayer, *initiated by students,* is unconstitutional. The complainants in this case could not argue that coercion to participate was involved as in *Lee v. Weisman,* since attendance at a football game is an entirely voluntary extra-curricular activity. Yet, Justice Stevens said students are faced with the "choice between whether to attend these games or to risk facing a personally offensive religious ritual...The Constitution, moreover, demands that the school may not force this difficult choice upon these students."[17] Here we have the image of someone facing the horrible "risk" of exposure to a community-affirming prayer described by Stevens as an "offensive religious ritual." Students voluntarily praying were surely surprised to find that SCOTUS considered them bullying agents of the state, or that they were forcing a "difficult choice" on other students. The Court ruling inhibits the free speech as well as free exercise of religion rights of students, and tells them that their beliefs and desires are subordinate to the beliefs and desires of one or two malcontents.

Chief Justice Rehnquist's dissent noted the hostility of the Court to Christianity, to America's heritage, and to the plain meaning of the establishment clause:

> But even more disturbing than its holding is the tone of the Court's opinion; it bristles with hostility to all things religious in public life. Neither the holding nor the tone of the opinion is faithful to the meaning of the Establishment Clause, when it is recalled that George Washington himself, at the request of the very Congress which passed the Bill of Rights, proclaimed a day of "public thanksgiving and prayer, to be observed by acknowledging with grateful hearts the many and signal favors of Almighty God."[18]

Anti-Christianity in Higher Education

Anti-Christian animus also reaches into higher education. In *Christian Legal Society v. Martinez* (2011), SCOTUS ruled against a challenge by the Hastings chapter of the Christian Legal Society (CLS) to the policy of the University of California Hastings College governing official recognition of student groups. Potential members of CLS were required to sign a "Statement of Faith" to live in accordance with Christian beliefs, and disqualified anyone who engaged in "acts of sexual conduct outside of God's design for marriage between one man and one woman." The CLS was denied registered student organization (RSO) status because of these rules. The CLS complained that it was the only group denied the right to participate in the life of the law school community on equal terms with other groups because of its faith-based nature, and that "To require a religious group like CLS to admit nonbelievers is a severe burden on its freedom of association." CLS is the only group ever to be denied RSO status at Hastings.

Justice Alito wrote in his dissenting opinion that the bylaws of other RSOs had provisions for denying membership to students disagreeing with their views and objectives, and that this is an essential part of any group's identity and message. The CLS was thus uniquely burdened solely because of its Christian views. Alito noted that the Court ruling turned a blind eye to three Constitutional rights denied the CLS by Hastings: (1) Freedom of religious expression (its bylaws were deemed unacceptable, so they could not express them); (2) freedom of speech (CLS was denied access to campus bulletin boards, facilities, and the school newspaper), and (3) freedom of association (unlike other RSOs, CLS was not permitted the freedom to choose with whom it wants to associate). It is important to note that the CLS did not deny membership based on sexual orientation, but rather on sexual conduct. It is as unlikely that a practicing homosexual would want to join the CLS as an African American is to want to join the KKK, or a Jew to join the Nazi Party.

The upshot of all this anti-Christian animosity is that SCOTUS is slowly obliterating a rich tradition of religious freedom in the hearts of the young people of America. This has led: "to a society that lacks a unifying or common narrative. When religion is exorcised from the public schools, we raise generations after generation of citizens with no clear sense concerning the foundations of moral values, ethical behavior, or interpersonal conduct. As a result, many Americans struggle with issues of worth and identity, lacking a clear sense of either self or community."[19] This is the tragedy shaped by an unaccountable legal priesthood sitting on the highest court in the land that thinks it knows the Constitution better than its authors, and who, much to the delight of atheists, is hell-bent on outlawing America's Christian heritage.

The Congressional Prayer Caucus Foundation paints a disturbing picture of what has happened since God was purged from our public schools. The cited figures are since 1962, the year the Bible was removed from our schools, and are adjusted for population increases:

> Criminal arrest of teens is up 150% according to the US Bureau of Census; teen suicides in ages 15-19 years up 450% according to the National Center of Health Services; illegal drug activity is up 6000% according to the National Institute of Drug Abuse; child abuse cases up 2300% according to the US Department of Health and Human Services; divorce up 350% according to the US Department of Commerce, and SAT scores fell 10% even though the SAT questions have been revamped to be easier to answer. Violent crime has risen 350%, national morality figures have plummeted, and teen pregnancy escalated dramatically after prayer and the Bible were removed from the schools.[20]

We may add to this laundry list of woes the horrendous escalation in mass shootings since SCOTUS's assault on Christianity. In the decade of the 1950s there was not one mass shooting; in the 60s there were 6; 13 in the 70s, 32 in the 80s, and 42 in the 90s, and they keep coming almost weekly.[21] The legal assault on Christianity is applauded by leftists, who know that Karl Marx was right when he reminded us that "communism begins where atheism begins."[22] A free society is glued together by religion, family, and other civic institutions that stand between individuals and their government. When these nongovernmental institutions are weakened, big government steps in to fill the void. This has been the dream of leftists ever since the totalitarian ideals of Plato's *Republic*. That is, every aspect of social life must be determined by a self-anointed elite that knows what we need better than we do. Chesterton and Winthrop were right—it's God or the State; the Bible or the bayonet.

Endnotes

1. Substance Abuse and Mental Health Services Administration, 2016.
2. American Sexual Health Association, 2015.
3. U.S. Department of Health & Human Services, 2014.
4. Walsh, A. 2018.
5. Federer, W. 1994, p. 392.
6. Yale Law School, nd.
7. In Sampson, R. 2005, p. 473.
8. Congressional Record, vol. 146, p. H8894.
9. Madison, J. 1785.
10. Stewart, P. 1962, *Engel v. Vitale*, 370 U.S. 421.
11. Jay, J., 1784.
12. Stewart, P. 1961, *Abington School District v. Schempp*, 374 U.S. 203.
13. *Per curium* decision. *Stone v. Graham*, 449 U.S. 39, 1980.
14. Burger, W. 1985. *Wallace v. Jaffree*, 472 U.S. 38.
15. Rehnquist, W. 1985, *Wallace v. Jaffree*, 472 U.S. 38, 1985.
16. Scalia, A. 1992, *Lee v. Weisman*, 505 U.S. 577.
17. Stevens, J. 2000, *Santa Fe Independent School Dist. v. Doe*, 530 U.S. 290.
18. Rehnquist, W. 2000, *Santa Fe Independent School Dist. v. Doe*, 530 U.S. 290.
19. Shrader, D. 2005, p. 5.
20. Congressional Prayer Caucus Foundation, 2013.
21. Prager, D. 2019.
22. In Sheen, F. 1948, p. 69.

References

ABCScience (1998). Molecular basis for evolution. *Elsevier Science Channel*, November 27th.
http://www.abc.net.au/science/articles/1998/11/27/17476.htm

Abel, D. (2011). Is life unique? *Life, 2*:106-134.

Acton, Lord J. (1877). The history of freedom in Christianity. The Acton Institute. https://acton.org/research/history-freedom-christianity

Aczel, A. (1998). *Probability 1.* San Diego: Harcourt Brace Jovanovich.

Alberts, B., Johnson, A., Lewis, J., Walter, P., Morgan, D., Raff, M., Roberts, K., & Walter, P. (2015). *Molecular Biology of the Cell,* 6th Edition. New York: Garland.

Andrews, E. (2017). *Is the Bible really the word of God? Is Christianity the One True Faith?* Cambridge, OH: Christian Publishing House.

Alexander, E. (nd). Dr. Eben Alexander on his near-death experience—and what he's learned about consciousness. *Goop Wellness.*
https://goop.com/wellness/spirituality/dr-eben-alexander-near-death-experience-hes-learned-consciousness/

Alexander, E. (2015). Near-death experiences, The mind-body debate & the nature of reality. *Missouri Medicine, 112*: 17-21.

American Sexual Health Association (2015). Statistics.
http://www.ashasexualhealth.org/stdsstis/statistics/

Appolloni, S. (2011). " Repugnant"," not repugnant at all": How the respective epistemic attitudes of Georges Lemaitre and Sir Arthur Eddington influenced how each approached the idea of a beginning of the universe. *Scientific Journal of International Black Sea University, 5:* 19-44.

Aquilino, W. (1999). Two views of one relationship: Comparing parents' and young adult children's reports of the quality of intergenerational relations. *Journal of Marriage and the Family,* 61:858-870.

Astronomy Essentials (2015). How long to orbit Milky Way's center? November 28th. http://earthsky.org/astronomy-essentials/milky-way-rotation.

Aviezer, N. (2010). Intelligent design versus evolution. *Rambam Maimonides Medical Journal, 1:*1-9.

Ayala, F., Cicerone, R., Clegg, M., Dalrymple, G., Dickerson, R., Gould, S., Herschbach, D., Kennedy, D., McInerney, J. & Moore, J. (1999). *Science and creationism: A view from the National Academy of Sciences.* Washington, DC: National Academy of Sciences.

Bacon, F. (1889). *Essays civil and moral.* London: Cassell & Company.

Bahr, H. & Chadwick, B. (1985). Religion and family in Middletown, USA. *Journal of Marriage and the Family,* 47:407-414.

Bailey, D. (2018). What are the cosmic coincidences? *Science meets religion.* http://www.sciencemeetsreligion.org/physics/cosmic.php

Ball, P. (2016). There's no space for young Einsteins today. *The Guardian*, February 12th. https://www.theguardian.com/commentisfree/2016/feb/12/einstein-gravitational-waves-physics

Barnes, L. (2012). The fine-tuning of the universe for intelligent life. *Publications of the Astronomical Society of Australia, 29*: 529-564.

Barnes, R. (2017). Tidal locking of habitable exoplanets. *Celestial Mechanics and Dynamical Astronomy, 129:* 509-536.

Barrow, J. & Tipler, F. (1986). *The Anthropic Cosmological Principle*, New York: Oxford University Press.

Bartha, I., Di Lulio, J., Venter, J. & Telenti, A. (2018). Human gene essentiality. *Nature Reviews Genetics.* 19**:** 51–62.

Basu, D., & Kulkarni, R. (2014). Overview of blood components and their preparation. *Indian journal of anaesthesia, 58*(5), 529.

Batygin, K., & Laughlin, G. (2015). Jupiter's decisive role in the inner Solar System's early evolution. *Proceedings of the National Academy of Sciences*, 112: 4214-4217.

Bauchau, V. (2006). Emergence and reductionism: From the game of life to science of life. In

Beaulieu, J., Bennett, D., Fouqué, P., Williams, A., Dominik, M., Jørgensen, U., Kubas, D., Cassan, A., Coutures, C., Greenhill, J. and Hill, K. (2006). Discovery of a cool planet of 5.5 Earth masses through gravitational microlensing. *Nature, 439*:437-440.

Bechley, G. (nd). Gunter Bechley: Fossils vs Darwin. https://gbechly.jimdo.com/

Behe, M. (1996). *Darwin's black box: The biochemical challenge to evolution.* New York: Simon and Schuster.

Behe, M. (2001). Reply to my critics: A response to reviews of *Darwin's Black Box*: the biochemical challenge to evolution. *Biology and Philosophy, 16:* 683-707.

Beleza, S., Santos, A., McEvoy, B., Alves, I., Martinho, C., Cameron, E., Shriver, M., Parra, E. & Rocha, J. (2012). The timing of pigmentation lightening in Europeans. *Molecular biology and evolution, 30*(1), pp.24-35.

Bellissent-Funel, M., Hassanali, A., Havenith, M., Henchman, R., Pohl, P., Sterpone, F., van der Spoel, D., Xu, Y. & Garcia, A., (2016). Water determines the structure and dynamics of proteins. *Chemical Reviews, 116:* 7673-7697.

Benner, S. (2014). Paradoxes in the origin of life. *Origins of Life and Evolution of Biospheres, 44*: 339-343.

Bennetts, M. (2014). Who's 'godless' now? Russia says it's U.S. *Washington Times*, January 28. http://www.washingtontimes.com/news/2014/jan/28/whos-godless-now-russia-says-its-us/?page=all.

Benzmüller, C., & Paleo, B. (2014). Automating Gödel's ontological proof of God's existence with higher-order automated theorem provers. In *Proceedings of the Twenty-first European Conference on Artificial Intelligence* (pp. 93-98). IOS Press.

Berger, A., Loutre, M. & Mélice, J. (2006). Equatorial insolation: from precession harmonics to eccentricity frequencies. *Climate of the Past, 2:* 131-136.

Bernhardt, H. (2012). The RNA world hypothesis: the worst theory of the early evolution of life (except for all the others). *Biology Direct, 7: I-10.*

Bingham, J. (2016). Religion can make you happier, official figures suggest. *Daily Telegraph*, February 2nd.

Bixter, M. (2015). Happiness, political orientation, and religiosity. *Personality and Individual Differences, 72:* 7-11.

Blackburn, R. (2000). *The overthrow of colonial slavery: 1776-1848*. London: Verso.

Blankenship, R. (2010). Early evolution of photosynthesis. *Plant physiology, 154*: 434-438.

Boettner, L. (2017). *The reformed doctrine of predestination*. Woodstock, Ontario: Devoted Publishing

Bohannon, J. (2012). DNA: The ultimate hard drive. *Science*, August 16th. https://www.sciencemag.org/news/2012/08/dna-ultimate-hard-drive.

Bonhoeffer, D. (1972). *Letters and papers from prison*. New York: Macmillan.

Borwein, J, & Bailey, D. (2014). When science and philosophy collide in a 'fine-tuned' universe. Physics.Org. https://phys.org/news/2014-04-science-philosophy-collide-fine-tuned- universe.html#jCp.

Boulos, R., Eroglu, E., Chen, X., Scaffidi, A., Edwards, B., Toster, J., & Raston, C. (2013). Unravelling the structure and function of human hair. *Green Chemistry, 15:* 1268-1273.

Bray, G. (1992). Hell: Eternal punishment of total annihilation? *Theology Evangel*, Summer: 19-24.

Brian, D. (1995). The voice of genius: Conversations with Nobel scientists and other luminaries. Cambridge, MA: Perseus Publishing.

Bromm, V., & Larson, R. (2004). The first stars. *Annual. Review of Astronomy and Astrophysics, 42:*79-118.

Brooks, A. (2003). Religious faith and charitable giving. *Policy Review*, October & November, 1-5.

Brooks, A. (2004). Faith, secularism, and charity. *Faith and Economics*, 43: 1-8.

Bryson, B. (2003). *A short history of nearly everything*. New York: Broadway Books.

Buber, M. (2002). *The Martin Buber reader*. Asher Biemann (Ed.). New York: Macmillan.

Buckley, A. (2017). Is consciousness just an illusion? BBC News. https://www.bbc.com/news/science-environment-39482345

Busby, C., & Howard, C. (2017). Re-thinking biology—I. Maxwell's Demon and the spontaneous origin of life. *Advances in Biological Chemistry, 7:*170-181.

Bussey, P. (2016). *Signposts to God: How modern physics and astronomy point the way to belief*. Downers Grove: IL: Intervarsity Press.

Bruteau, B. (1997). *God's ecstasy: The creation of a self-creating world*. New York: Crossroad Publishing Company.

Caldwell, Z. (2018). Russia's Orthodox Church has opened 30,000 places of worship in last 30 years. *Aleteia*, January 2nd. https://aleteia.org/2018/01/02/russias-orthodox-church-has-opened-30000-places-of-worship-in-last-30-years/

Call, V. & Heaton, T. (1997). Religious influence on marital stability. *Journal for the Scientific Study of Religion*, 36:382-392.

Campbell, D. & Putnam, R. (2010). Religious people are 'better neighbors.' *USA TODAY*. November 15th. http://www.usatoday.com/news/opinion/forum/2010-11-15-column15_ST_N.htm

Cammaert, E, (1937). *The laughing prophet: The seven virtues and G. K. Chesterton*, Methuen, York: England.

Carey, N. (2015). *Junk DNA: A journey through the dark matter of the genome.* Columbia University Press.

Carr, B. (2013). Lemaître's prescience: the beginning and end of the cosmos. In R. Holder and S. Mitton (eds.), *Georges Lemaître: Life, Science and Legacy* (pp. 145-172). Berlin, Heidelberg: Springer.

Carter, B. (1974). Large Number Coincidences and the Anthropic Principle in Cosmology. IAU 63, *Confrontation of cosmological theories with observational data*, 63:291–298.

Cassé, M. (2003). *Stellar alchemy: the celestial origin of atoms.* Cambridge: Cambridge University Press.

Catechism of the Catholic Church (nd). Article 3: Man's freedom, 1730. http://archeparchy.ca/wcm-docs/docs/catechism-of-the-catholic-church.pdf

Catholic News Agency (2017). Catholic hospitals comprise one quarter of world's healthcare, council reports. February 10th. https://www.catholicnewsagency.com/news/catholic_hospitals_represent_26_percent_of_worlds_health_facilities_reports_pontifical_council.

Centre for Research on the Epidemiology of Disasters (2015). *The human cost of natural disasters: A global perspective.* New York: United Nations.

Chalmers, D. (1996). *The conscious mind: In search of a fundamental theory.* New York: Oxford University press.

Chesterton, G. (2001). The Collected Works of G.K. Chesterton, Vol. 20. San Francisco: Ignatius Press.

Child Trends. (2018). *Births to unmarried women.* Bethesda, MD. https://www.childtrends.org/indicators/births-to-unmarried-women.

Choroser, M. (2013). Where Thomas Nagel went wrong. *The Chronicle of Higher Education*, May 13th. https://www.chronicle.com/article/Where-Thomas-Nagel-Went-Wrong/139129

Christian, J. (2011). *Philosophy: An introduction to the art of wondering.* Boston, Wadsworth.

Chung, W., Wadhawan, S., Szklarczyk, R., Pond, S., & Nekrutenko, A. (2007). A first look at ARFome: Dual-coding genes in mammalian genomes. *PLoS computational biology*, 3:e91.

Church of England (1995). *The mystery of salvation: The doctrine Commission of the General Synod.* London: Church House Publishing.

Clark, D. & Pazdernik, N. (2009) *Biotechnology: applying the genetic revolution.* Amsterdam: Elsevier.

Clark, R. (1985). *The life of Ernst Chain: Penicillin and beyond,* London: Weidenfeld & Nicolson.

Cleaver, G. (2006). Before the Big Bang: String theory, God, & the origin of the universe. *Perspectives on science and religion,* June 3-7, Philadelphia, PA: Mexanexus Institute.

Cliff, H. (2013). Could the Higgs Nobel be the end of particle physics? *Scientific American.* October 8th. https://www.scientificamerican.com/article/could-the-higgs-nobel-be-the-end-of-particle-physics/

Clinton, W. (2000). Remarks of the President. Office of the Press Secretary, the White House. https://clintonwhitehouse3.archives.gov/WH/EOP/OSTP/html/00628_2.html

Cockell, C., Bush, T., Bryce, C., Direito, S., Fox-Powell, M., Harrison, J., Lammer, H., Landenmark, H., Martin-Torres, J., Nicholson, N. and Noack, L., (2016). Habitability: A review. *Astrobiology, 16*:89-117.

Coghlan, A. (2014). Massive 'ocean' discovered towards Earth's core, *New Scientist,* June, 12th. https://www.newscientist.com/article/dn25723-massive-ocean-discovered-towards-earths-core/

Coghlan, A. (21017). Planet Earth makes its own water from scratch deep in the mantle. *New Scientist,* January 27th. https://www.newscientist.com/article/2119475-planet-earth-makes-its-own-water-from-scratch-deep-in-the-mantle/.

Collins, F. (2006). *The Language of God: A scientist presents evidence for belief.* New York: Free Press.

Collins, F. (2007). Collins: Why this scientist believes in God. *CNN News.* http://www.cnn.com/2007/US/04/03/collins.commentary/index.html

Collins, F. S. (2006). *The language of God: A scientist presents evidence for belief.* New York: Simon and Schuster.

Collins, R. (2007). The multiverse hypothesis: A theistic perspective. In Carr, B. (ed), *Universe or multiverse? pp. 459-480.* Cambridge: Cambridge University Press.

Collini, E., Wong, C., Wilk, K., Curmi, P., Brumer, P., & Scholes, G. (2010). Coherently wired light-harvesting in photosynthetic marine algae at ambient temperature. *Nature, 463*: 644-647.

Coming Untrue (2015). *Insulting our intelligence.* October, 15th https://www.cominguntrue.com/2015/10/insulting-our-intelligence.html?m=0]

Congressional Prayer Caucus Foundation (2013). The effects of removing prayer and the Bible from the schools in 1962. December 11th. https://cpcfoundation.wordpress.com/2013/12/11/the-effects-of-removing-prayer-and-the-bible-from-the-schools-in-1962/.

Congressional Record (2000). Congressional Record, vol. 146. Washington, DC: United States Printing Office.

Copithorne, W. (1971). The worlds of Wallace Pratt, *The Lamp, 53:11-14*

Corey, M. (2001). *The God hypothesis: Discovering design in our just right Goldilocks universe.* Lanham, MD: Rowman & Littlefield.

Couenhoven, J. (2007). Augustine's rejection of the free-will defence: An overview of the late Augustine's theodicy. *Religious Studies, 43*: 279-298.

Coyne, G. V., & Heller, M. (2008). *A comprehensible universe: The interplay of science and theology.* New York: Springer-Verlag.

Coyne, J. (2009). *Why evolution is true.* New York: Viking.

Craig, W. (2008). *Reasonable Faith: Christian Truth and Apologetics.* Wheaton, IL: Crossway.

Craig, W. (2010). *On guard: Defending your faith with reason and precision.* Colorado Spring, CO: David C Cook.

Craig, W., & Meister, C. (2010). *God is great, God is good: why believing in God is reasonable and responsible.* Downers Grove, IL: InterVarsity Press.

Craig, W. (2013). The God particle. *Christian Research Institute*, July 20th. http://www.equip.org/PDF/JAF3356.pdf.

Crick, F. (1994). The astonishing hypothesis. New York: Scribner's.

Curry, E. (2013). Do the polls show that science leads to atheism? *Perspectives on Science and Christian Faith, 65:* 75-77.

Dantzer, R. (2017). Neuroimmune interactions: from the brain to the immune system and vice versa. *Physiological Reviews, 98*:477-504.

Darwin, C. (1982). *The origin of species.* London: Penguin.

Darwin, C. (1879). To John Fordyce, 7 May 1879. Cambridge University Darwin Correspondence Project. ghttps://www.darwinproject.ac.uk/letter/DCP-LETT-12041.xml

Darwin, C. (1892). *Charles Darwin: his life told in an autobiographical chapter, and in a selected series of his published letters* (Vol. 5, edited by F. Darwin. London: John Murray.

Darwin, C. (1903). *More letters of Charles Darwin.* (F. Darwin, ed). New York: D. Appleton.

Darwin, C., Burkhardt, F., & Smith, S. (1991). *The Correspondence of Charles Darwin: 1858-1859* (Vol. 7). Cambridge: Cambridge University Press.

Darwin, F. (2005). Charles Darwin to W. Graham, July 3, 1881, in *The Life and Letters of Charles Darwin*, ed. F. Darwin, Boston: Elibron.

Davies, P. (1982). *The accidental universe.* Cambridge: Cambridge University Press.

Davies, P. (1983). *God and the New Physics.*, New York Penguin.

Davies, P. (1984). Superforce: The search for a grand unified theory of nature. New York: Simon & Schuster.

Davies, P. (2003). *The Origin of Life.* London: Penguin Books.

Davies, P. (2007). Taking science on faith. *New York Times*, November 24th. http://www.nytimes.com/2007/11/24/opinion/24davies.html?pagewanted=all

Dawkins, R., 1986. *The Blind Watchmaker.* New York: W.W. Norton and Company.

Dawkins, R. (1989). *The selfish gene.* Oxford: Oxford University Press.

Dawkins, R. (2006). *The God delusion.* New York: Houghton Mifflin.

Dawkins, R. (nd). Debate with John Lennox. www.dawkinslennoxdebate.com.
Day, R., & Acock, A. (2013). Marital well-being and religiousness as mediated by relational virtue and equality. *Journal of Marriage and Family, 75:* 164-177.
de Castro Fonseca, M., Aguiar, C., da Rocha Franco, J., Gingold, R., & Leite, M. (2016). GPR91: expanding the frontiers of Krebs cycle intermediates. *Cell Communication and Signaling, 14:* 3.
Dembski, W. (1998). *Mere creation: Science, faith & intelligent design*. Downers Grove, IL: InterVarsity Press.
Dembski, W. (2004). *The design revolution: Answering the toughest questions about intelligent design*. Westmont, IL: InterVarsity Press.
Dembski, W. A. (2006). *No free lunch: Why specified complexity cannot be purchased without intelligence*. Lanham, MD:Rowman & Littlefield.
Dembski, W. (2008). In defense of intelligent design. *The Oxford Handbook of Religion and Science, Oxford Handbooks in Religion and Theology, pp. 715-731*. Oxford: Oxford University Press.
Denton, M., Marshall, C., & Legge, M. (2002). The protein folds as platonic forms: new support for the pre-Darwinian conception of evolution by natural law. *Journal of Theoretical Biology, 219*: 325-342.
Dervic, K., Oquendo, M., Grunebaum, M., Ellis, S., Burke, A., & Mann, J. (2004). Religious affiliation and suicide attempt. *American Journal of Psychiatry, 161*: 2303-2308.
Descartes, R., de Spinoza, B., & Hutchins, R. (1952). *Great books of the western world*. London: Encyclopaedia Britannica.
de Tocqueville, A. (1994). *Democracy in America*, Bradley, P. (ed.). New York: Alfred A. Knopf.
de Vries, G. & P. Södersten (2009). Sex differences in the brain: The relation between structure and function. *Hormones and Behavior*. 55: 589-596.
Dias, M., & Partington, M. (2015). Congenital brain and spinal cord malformations and their associated cutaneous markers. *Pediatrics, 136:* e1105-e1119.
Dimitrov, T. (2010). 50 Nobel Laureates who believe in GOD. *Scientific GOD Journal*, 1:166-182.
Dingle, H. (1972) *Science at the crossroads*, London: Martin Brian & O'Keefe.
Dobzhansky, T. (1973). Nothing in biology makes sense except in light of evolution. *The American Biology Teacher,* 35:125-129.
Dollahite, D. & Thatcher, J. (2005). How family religious involvement benefits adults, youth, and children and strengthens families. Salt Lake City: *The Sutherland Institute*.
Donoghue, J. (2016). The multiverse and particle physics. *Annual Review of Nuclear and Particle Science, 66*:1-36.
Drescher, S. & Engerman, S. eds. (1998). *A historical guide to world slavery*. New York: Oxford University Press.
Drummond, H. (1894). *Ascent of man*. New York: Pott & Co.
Duck, M., & Duck, E. (2014). *Waters of creativity: Navigating the straits between science and theology to find the source of one's beginning*. Lake Mary, FL: Charisma Media.

Durant, W. (1935). *Our Oriental Heritage*. New York: Simon & Schuster.

Durant, W. (1944). *Caesar and Christ*. New York: Simon and Schuster.

Durant, W. & Durant, A. (1950). *The Age of Faith*. New York: Simon & Schuster.

Durant, W. & Durant, A. (1968). *The Lessons of History*, New York: Simon and Schuster.

Dyson, F. (1979). *Disturbing the universe*, New York: Harper and Row.

Dyson, L., Kleban, M., & Susskind, L. (2002). Disturbing implications of a cosmological constant. *Journal of High Energy Physics*, 10: 1-26.

Easton, J. (2005). Survey on physicians' religious beliefs shows majority faithful. *Medical Center Public Affairs, University of Chicago Chronicle*. July 14.

Ebifegha, M. (2009). *The Darwinian delusion: The scientific myth of evolutionism*. Bloomington, IN: AuthorHouse. P. 36

Ecklund, E. & Park, J. (2009). Conflict between religion and science among academic scientists? *Journal for the Scientific Study of Religion*, 48: 276-292.

Einstein, A. (1923). *Sidelights on Relativity (Geometry and Experience)*. New York: P. Dutton.

Einstein, A. (1940/1994). *Ideas and Opinions*. Ed. Carl Seelig. New York: Modern Library.

Einstein, A. (1941). Science, philosophy and religion, a symposium. In *Conference on Science, Philosophy and Religion Their Relation to The Democratic Way of Life. New York*. http://www.sacred-texts.com/aor/einstein/einsci.htm.

Einstein, A. & L. Infeld (1938). *The evolution of physics*. Cambridge: Cambridge University Press.

Ekström, S., Coc, A., Descouvemont, P., Meynet, G., Olive, K., Uzan, J., & Vangioni, E. (2010). Effects of the variation of fundamental constants on Population III stellar evolution. *Astronomy & Astrophysics*, 514, A62.

Ellis, G. (2011). The untestable multiverse. *Nature*, 469:294-295.

Ellis, G., & Silk, J. (2014). Scientific method: Defend the integrity of physics. *Nature, 516:* 321-323.

Ellis, L., & Peterson, J. (1996). Crime and religion: An international comparison among thirteen industrial nations. *Personality and Individual Differences, 20:* 761-768.

Ellis, L. & Walsh, A. (2000). *Criminology: A global perspective*. Boston: Allyn & Bacon.

Ellis, M., Vinson, D., & Ewigman, B. (1999). Addressing spiritual concerns of patients. *Journal of Family Practice, 48:* 105-106.

Esch, T. & G. Stefano (2005). Love promotes health. *Neuroendocrinology Letters*, 3:264-267.

Everard, D. (2012). *The wiser-mouse legacy: On being an authentic Christian!* Bloomington, IN: WestBow.

Ewart, P. (2009). The necessity of chance: Randomness, purpose and the sovereignty of God. *Science & Christian Belief, 21: 111-131*.

Feltz, B., Crommelinck, M., & Goujon, P., pp. 29-40, *Self-organization and emergence in life sciences*. Dordrecht, The Netherlands: Springer.

Ferreira, S. & Potgieter, M. (2004). Galactic cosmic rays in the heliosphere. *Advances in Space Research, 34*:115-125.

Fields, R. (2014). Myelin—more than insulation. *Science, 344*: 264-266.

Federer, W. (1994). *America's God and Country: Encyclopedia of Quotations.* St. Louis, MO: Amerisearch, Inc.

Fincher, J. (1982). *The human brain: Mystery of matter and mind.* Washington, DC: U.S. News Books.

Forbes, K., & Zampelli, E. (2013). The impacts of religion, political ideology, and social capital on religious and secular giving: evidence from the 2006 Social Capital Community Survey. *Applied Economics, 45*: 2481-2490.

Forget, F. (2013). On the probability of habitable planets. *International Journal of Astrobiology, 12:* 177-185.

Flew, A., & Varghese, R. A. (2007). *There is a God.* New York: HarperCollins.

Folger, T. (2008). Science's alternative to an intelligent creator: The multiverse theory. *Discover Magazine,* December 10th. http://discovermagazine.com/2008/dec/10-sciences-alternative-to-an-intelligent-creator.

Fowler, D., Coyle, M., Skiba, U., Sutton, M., Cape, J., Reis, S., Sheppard, L., Jenkins, A., Grizzetti, B., Galloway, J. & Vitousek, P. (2013). The global nitrogen cycle in the twenty-first century. *Philosophical Transactions of the Royal Society B: Biological Sciences, 368:*.20130164.

Frankenberry, N. (2008). *The faith of scientists: In their own words.* Princeton, NJ: Princeton University Press.

Freedom House (2018). *About Freedom in the World: An annual study of political rights and civil liberties.* https://freedomhouse.org/report-types/freedom-world

Fragile Families Research Brief (2005). Religion and marriage in urban America. Bendheim-Thoman Center for Research on Child Wellbeing, Princeton University Social Indicators Survey Center, Columbia University.

Frankl, V. (1985). *Man's Search for Meaning.* New York: Pocket Books.

Freeland, S. (2008). Could an intelligent alien predict earth's biochemistry? In Barrow et al, (eds.), *Fitness of the cosmos for life: Biochemistry and fine-tuning.* Pp. 280-315. Cambridge: Cambridge University Press.

Fudge, E. (2012). *The fire that consumes: A Biblical and historical study of the doctrine of final punishment.* Cambridge: The Lutteworth Press.

Gaither, C. & Cavazos-Gaither, A. (2001). *Naturally speaking: a dictionary of quotations on biology, botany, nature and zoology.* London: Institute of Physics Publishing.

Galison, P., Holton, G. & Schweber, S. (2008). *Einstein for the 21st century: His legacy in science, art, and modern culture.* Princeton University Press.

Garay, A. (1993). Theoretical and experimental studies of the possibility of chirality dependent time direction in molecules. In *Chemical Evolution: Origin of Life* (Ed. Ponnamperuma, C. & Chela-Flores, J.), pp. 165–179. Hampton, VA: Deepak Publishing.

Gauger, A., & Axe, D. (2011). The evolutionary accessibility of new enzymes functions: A case study from the biotin pathway. *Bio-Complexity, 2011: 1-17.*

Gervais, W. (2014). Everything is permitted? People intuitively judge immorality as representative of atheists. *PloS one, 9*(4): 1-9. e92302.

Giedd, J. (2004). Structural Magnetic Resonance Imaging of the Adolescent Brain. *Annals of the New York Academy of Science,* 1021:77-85.

Geisler, N. & Corduan, W. (2002). *Philosophy of religion* (2nd ed.). Eugene, OR: Wipf & Stock.

Gefter, A. (2008). Why it's not as simple as God vs the multiverse. *New Scientist, 2685*(04).

Gillen, A. (2001), *Body by Design.* Green Forest, AR: Master Books.

Gitt, W. (1996). Information, science, and biology. *CENTech Journal,* 10:181-187.

Gitt, W. (2006). *In the beginning was information: A scientist explains the incredible design in nature.* Portland, OR: New Leaf Publishing Group.

Gitt, W., Compton, B. & Fernandez, J., (2011). *Without Excuse.* Atlanta, GA: Creation Book Publishers.

Glasco, D. (2016). Beyond the DNA-protein paradox: a "clutch" of other chicken-egg paradoxes in cell and molecular biology. *Answers Research Journal, 9:* 209-227.

Goldberg, E. (2001). *The executive brain: Frontal lobes and the civilized mind.* New York: Oxford University Press.

Gonen, N., Futtner, C., Wood, S., Garcia-Moreno, S., Salamone, I., Samson, S., Sekido, R., Poulat, F., Maatouk, D., & Lovell-Badge, R. (2018). Sex reversal following deletion of a single distal enhancer of Sox9. *Science,* p.eaas9408.

Gonzales, G., Brownlee, D. & Ward, P. (2001). Refugees for life in a hostile universe. *Scientific American,* 285: 60-67.

Gonzalez, G., & Richards, J. (2004). *The privileged planet: how our place in the cosmos is designed for discovery.* New York: Regnery Publishing.

Gonzalo, J. (2008). *The intelligible universe: an overview of the last thirteen billion years.* Hackensack, NJ: World Scientific.

Gorelick, R., & Heng, H. (2011). Sex reduces genetic variation: a multidisciplinary review. *Evolution: International Journal of Organic Evolution,* 65:1088-1098.

Goswami, A. (2014). *Creative evolution: A physicist's resolution between darwinism and intelligent design.* Wheaton, IL: Quest Books.

Gottlieb, G. (2007). Probabilistic epigenesis. *Developmental Science,* 10: 1–11.

Gould, S. (1977). Evolution's erratic pace. *Natural History, 86:* 12-16.

Gould, S. (1992). Impeaching a self-appointed judge. *Scientific American* 267:118-121.

Gould, S. (1997). Foreword: The positive power of skepticism," *Why people believe weird things,* Michael Shermer. New York: W.H. Freeman.

Graebner, N. (1976). Christianity and democracy: Tocqueville's views of religion in America. *The Journal of Religion, 56:* 263-273.

Grant, E. (2004). *Science and religion, 400 BC to AD 1550: From Aristotle to Copernicus.* Westport, CT: Greenwood.

Grassé, P. (1977). *Evolution of living organisms: evidence for a new theory of transformation.* New York: Academic Press.

Graur, D. (2013). Update version of the SMBE/SESBE Lecture on ENCODE & junk DNA. https://www.slideshare.net/dangraur1953/update-version-of-the-smbesesbe-lecture-on-encode-junk-dna-graur-december-2013.

Greene, B. (2004). *The fabric of the cosmos*. New York: Vintage.

Greenstein, G. (1988). *The symbiotic universe: Life and mind in the cosmos*. New York: William Morrow.

Grelle, B. (2016). *Antonio Gramsci and the Question of Religion: Ideology, Ethics, and Hegemony*. New York: Taylor & Francis.

Greyson, B. (2010). Seeing dead people not known to have died: "Peak in Darien" experiences. *Anthropology and Humanism, 35:*159-171.

Gribbin, J. (2018). Alone in the Milky Way. *Scientific American,* 319: 94-99.

Gribbin, J. & Rees, M. (1989). *Cosmic coincidences: Dark matter, mankind, and anthropic cosmology*. New York: Bantam Books.

Grim, B., & Grim, M. (2016). The socio-economic contribution of religion to American society: An empirical analysis. *Interdisciplinary Journal of Research on Religion,* 12:2-31.

Grizzetti, B., Galloway, J. & Vitousek, P. (2013). The global nitrogen cycle in the twenty-first century. *Philosophical Transactions of the Royal Society B: Biological Sciences, 368:*.20130164.

Gross, N., & Simmons, S. (2009). The religiosity of American college and university professors. *Sociology of Religion, 70:*101-129.

Grossman, L. (2011). Water's quantum weirdness makes life possible. *New Scientist,* October 25th. 14.

Grossman, L. (2012). Death of the eternal cosmos, *New Scientist,* 213:6–7.

Gu, L., Baldocchi, D., Wofsy, S., Munger, J., Michalsky, J., Urbanski, S., & Boden, T. (2003). Response of a deciduous forest to the Mount Pinatubo eruption: Enhanced photosynthesis. *Science,* 299: 2035-2038.

Guiso, L., Sapienza, P., & Zingales, L. (2003). People's opium? Religion and economic attitudes. *Journal of monetary economics, 50*: 225-282.

Gupta, N., Agogino, A., & Tumer, K. (2006). Efficient agent-based models for non-genomic evolution. In *Proceedings of the fifth international joint conference on Autonomous agents and multiagent systems* (pp. 58-64). ACM.

Gust, D., Moore, T., & Moore, A. (2009). Solar fuels via artificial photosynthesis. *Accounts of chemical research, 42:*1890-1898.

Habermas, J. (2006). Time of transitions, C. Cronin & M.Pensky (trans.). Cambridge: Polity.

Hall. D. (2006). Religious attendance: More cost effective than lipitor? *Journal of the American Board of Family Practice,* 19:431-433.

Hall, S. (2012). Hidden treasures in junk DNA. *Scientific American,* October 1. https://www.scientificamerican.com/article/hidden-treasures-in-junk-dna/

Hamilton, W. (1999). *Narrow roads to gene land: evolution of sex,* Vol. 2. Oxford: Oxford University Press.

Harris, S. (2005). *The end of faith: Religion, terror, and the future of reason*. New York: WW Norton & Company.

Harrison, P. (2012). Christianity and the rise of western science. *ABC Religion and Ethics*.
http://www.abc.net.au/religion/articles/2012/05/08/3498202.htm

Hartsfield, T. (2016). String theory has failed as a scientific theory, Real Clear Science, January 8th.
http://www.realclearscience.com/blog/2016/01/string_theory_has_failed_as_a_scientific_theory.html

Hass, B. (2018). "We are scared, but we have Jesus": China and its war on Christianity. *The Guardian*, September 28th.
https://www.theguardian.com/world/2018/sep/28/we-are-scared-but-we-have-jesus-china-and-its-war-on-christianity

Haught, J. (2008a). Is fine-tuning remarkable? In J Barrow et al, *Fitness of the cosmos for life: Biochemistry and fine-tuning.* pp. 31-48, Cambridge: Cambridge University Press.

Haught, J. (2008b). *God after Darwin: A theology of evolution.* Boulder, CO: Westview Press.

Haught, J. (2015). Life, design, and drama: A theological response to evolutionary naturalism. *Journal of Religion and Society*. Supplement 11: 128-137.

Harrub, B. (2005). The unevolvable circulatory system. *Reason & Revelation*. 25:81-87.

Hawking, S. (n.d.). The beginning of time. The Stephen Hawking website.
http://www.hawking.org.uk/the-beginning-of-time.html.

Hawking, S. (1988). *A brief history of time.* New York: Bantam Books.

Hawking, S. (2001). *The universe in a nutshell.* New York: Bantam books.

Hawking, S. (2018). *Brief answers to the big questions.* New York: Bantam Books.

Hawking, S. & Mlodinow, L. (2010). The *Grand Design.* New York: Bantam Books.

Hazard, J., Butler, W. & Maggs, P. (1977). *The Soviet legal system.* Dobbs Ferry, NY: Oceana.

Hazen, R. & Trefil, J. (2009). *Science matters: Achieving scientific literacy.* New York: Anchor.

Heeren, F. (2000). *Show Me God: What the Message from Space Is Telling Us About God.* Miamitown, Oh: Day Star Publications.

Heile, F. (2016). Is it theoretically possible to build a collider that can test the predictions of string theory? https://www.quora.com/Is-it-theoretically-possible-to-build-a-collider-that-can-test-the-predictions-of-string-theory.

Hamilton, A. (1788). *The Judiciary Department. Federalist Papers No 78.*
http://www.constitution.org/fed/federa78.htm.

Hannay, T. (2019). Life versus entropy. *Science*, 565:427-428.

Hauser, M., Yang, C., Berwick, R., Tattersall, I., Ryan, M., Watumull, J., Chomsky, N. & Lewontin, R. (2014). The mystery of language evolution. *Frontiers in Psychology*, 5:1-12.

Hewitt, D. (2016). Senior Chinese religious advisor calls for promotion of atheism in society. International Business Review, August 27th.

http://www.ibtimes.com/senior-chinese-religious-advisor-calls-promotion-atheism-society-2363850

Hewlett, M. (2008). Molecular biology and religion. In P. Clayton & Z. Simpson (eds.), *The Oxford handbook on religion and science*, pp.172-186. Oxford: Oxford University Press.

Heyer, W. (2011). Public school LGBT programs don't just trample parental rights. They also put kids at risk. The Witherspoon Institute. http://www.thepublicdiscourse.com/2015/06/15118/

Hick, J. (1963). *Philosophy of religion.* Englewood Cliffs, NJ: Prentice-Hall.

Hick, J. (1977). *Evil and the God of love,* 1st ed. New York: Harper & Row.

Hick, J. (1998). The theological challenge of religious pluralism. In: Introduction to Christian Theology: Contemporary North American Perspectives. In R. Badham, ed.), pp. 24-36. Louisville, KY: Westminster John Knox Press,

Hick, J. (2007) *Evil and the God of Love,* 2d. ed. New York: Palgrave Macmillan.

Hiller, J. (2004). Speculations on the links between feelings, emotions and sexual behaviour: Are vasopressin and oxytocin involved? *Sexual and Relationship Therapy,* 19:1468-1479.

Hitchens, C. (2003). Mommie dearest: The pope beatifies Mother Teresa, a fanatic, a fundamentalist, and a fraud. *Slate,* October 20th.

Hitchens, C. (2007). *God is not great:* London: Atlantic Books.

Hitchens, P. (2010). *The rage against God.* London: Bloomsbury Publishing.

Hitchens, P. (2012). What's socialist about state ownership? Beats me. *Daily Mail,* October 11th.

Hitler, A. (2007). *Hitler's table talk 1941-1944: Secret conversations,* Roper, T. ed. New York: Enigma Books.

Hjorth-Jensen, M. (2011). The carbon challenge. *Physics, 4,* 38

Hoffman, N. (2001). The Moon and plate tectonics: Why we are alone. Space Daily. http://www.spacedaily.com/news/life-01x1.html.

Holder, R. (2013). Lemaître and Hoyle: Contrasting characters in science and religion. In Holder, R. & Mitton, S. (Eds.), *Georges Lemaître: Life, science and legacy,* pp. 39-54. Heidelberg: Springer Science & Business Media.

Holder, M., Coleman, B., & Wallace, J. (2010). Spirituality, religiousness, and happiness in children aged 8–12 years. *Journal of happiness studies, 11*:131-150.

Holt, J. (1997). Science resurrects God. *Wall Street Journal,* December 24. https://www.wsj.com/articles/SB882911317496560000

Holt, J. (2012). *Why does the world exist? An existential detective story.* New York: WW Norton

Holt, J. (2018). *When Einstein walked with Gödel: Excursions to the edge of thought.* New York: Farrar, Straus and Giroux.

Horgan, J (2012). Stanley Miller and the quest to understand life's beginning. *Scientific American,* July 29th. https://blogs.scientificamerican.com/cross-check/stanley-miller-and-the-quest-to-understand-lifes-beginning/

Horgan, J. (2014). The philosophy of guessing has harmed physics, expert says. *Scientific American,* August 21st.

https://blogs.scientificamerican.com/cross-check/the-philosophy-of-guessing-has-harmed-physics-expert-says/

Horgan, J. (2017). Why string theory is still not even wrong. *Scientific American*, April 27th. https://blogs.scientificamerican.com/cross-check/why-string-theory-is-still-not-even-wrong/

Hosking, G. (1985). *The first socialist society: A history of the Soviet Union from within.* Cambridge: Harvard University Press.

Howard, D. (2005). Albert Einstein as a philosopher of science. *Physics Today*, Dec.: 34-40.

Hoyle, F. (1982). The universe: Past and present reflections. *Annual Review of Astronomy and Astrophysics, 20:* 1-36.

Hoyle, F. (1999). *Mathematics of evolution.* Memphis, TN: Acorn Enterprises.

Hoyle, F., & Wickramasinghe, C. (1981). *Evolution from space.* London: JM Dent.

Hughes, K., Moore, R., Morris, P., & Corr, P. (2012). Throwing light on the dark side of personality: Reinforcement sensitivity theory and primary/secondary psychopathy in a student population. *Personality and Individual Differences*, 52:532-536.

Hulme, C., & Salter, M. (2001). The Nazi's persecution of religion as a war crime: The Oss's response within the Nuremberg Trials process. *Rutgers Journal of Law & Religion, 3*:1-27.

Hummer, R., Rogers, R. Nam, C. & Ellison, C. (1999). Religious Involvement and U.S. Adult Mortality. *Demography* 36:273–285.

Hutchinson, I. (2007). Warfare and wedlock: redeeming the faith-science relationship. *Perspectives on Science and Christian Faith, 59:* 91-101.

IBM (1999). IBM announces $100 million research initiative to build world's fastest supercomputer. Press Release, December 6th. https://www-03.ibm.com/press/us/en/pressrelease/1950.wss

International Christian Concern (2017). *2016 hall of shame.* Washington, DC: ICC.

Isaacson, W. (2007). *Einstein: His life and universe.* New York: Simon and Schuster.

Jal. M. (2010), Interpretation as phantasmagria: Variations on a theme on Marx's Theses on Feuerbach. *Critique*, 38: 117-142.

Jammer, M (1999). *Einstein and religion: Physics and theology.* Princeton, NJ: Princeton University Press.

Jastrow, R. (1978). *God and the astronomers (1978),* W. W. Norton

Jastrow, R. (1981). *The enchanted loom: Mind in the universe.* New York: Simon & Schuster.

Jastrow, R. (1992). *God and the Astronomers.* New York: WW Norton.

Jay, J. (1784). letter to Peter Augustus Jay, April 9. https://selfeducatedamerican.com/2011/11/19/john-jay-on-the-holy-bible-the-best-of-all-books/

Jeans, J. (1930). *The mysterious universe.* Cambridge: Cambridge University Press.

Jellinek, A. & Jackson, M. (2015). Connections between the bulk composition, geodynamics and habitability of Earth. *Nature Geoscience, 8*(8), 587.

Jennings, B. (2015). *In defense of scientism: An insider's view of science.* Vancouver, BC: Byron K. Jennings.

Jenkins, A., & Perez, G. (2010). Looking for life in the multiverse. *Scientific American, 302*:42-51.

Johnson, B. & Jang, S. (2011). Crime and religion: Assessing the role of the faith factor. In, Rosenfeld, R., Quinet, K., & Carcia, C. (eds.) *Contemporary issues in criminological theory and research: The role of social institutions*, pp. 117-149. Belmont, CA: Wadsworth.

Johnson, D. (2009). *Probability's Nature and Nature's Probability-Lite: A Sequel for Non-Scientists and a Clarion Call to Scientific Integrity.* Kingston, TN: Big Mac Publishers.

Johnson, J. & Potter, J. (2005). The argument from language and the existence of God. *The Journal of religion, 85*: 83-93.

Johnson, P. (1999). The Church of Darwin. *Wall Street Journal*, August 16th. https://www.wsj.com/articles/SB934759227734378961

Jordan, J. (2012). The topography of divine love. *Faith and Philosophy, 29:* 53-69.

Kainz, H. P. (2010). *The existence of God and the faith-instinct.* Selinsgove, PA: Susquehanna University Press.

Kant, I. (1999). *Critique of pure reason.* Cambridge. Cambridge University Press.

Karplus, M. (1997). The Levinthal paradox: yesterday and today. *Folding and Design, 2:*69-75.

Kendall, T. & Tamura, R. (2010). Unmarried fertility, crime, and social stigma. *The Journal of Law and Economics, 53*(1), 185-221.

Kennedy, D. (1907). St. Albertus Magnus. In *The Catholic Encyclopedia.* New York: Robert Appleton Company. New Advent: http://www.newadvent.org/cathen/01264a.htm. p. 265.

Kenny, M. (2008). Atheists, enjoy life? *The Guardian*, October 24th.

Kenyon, D. (2002). *Unlocking the mystery of life*: Script draft of video. http://www.divinerevelations.info/documents/intelligent_design/unlockingthemysteryoflifescript.pdf.

Keyser, C. (1915). *The new infinite and the old theology.* Yale University Press.

King, V., Ledwell, M., & Pearce-Morris, J. (2013). Religion and ties between adult children and their parents. *The Journals of Gerontology Series B: Psychological Sciences and Social Sciences, 68*: 825–836.

Kipnis, J. (2018), Immune system: The "seventh sense." *Journal of Experimental Medicine*, 215:397-398.

Klyce, B. (nd). The RNA world and other origin-of-life theories. https://www.panspermia.org/rnaworld.htm.

Koenig, H., Pearce, M., Nelson, B., & Erkanli, A. (2016). Effects on daily spiritual experiences of religious versus conventional cognitive behavioral therapy for depression. *Journal of Religion and Health, 55*:1763-1777.

Kohler T; Pratt J; Debarbieux B; Balsiger J; Rudaz G; Maselli D; (eds) (2012). *Sustainable mountain development, green economy and institutions. From Rio 1992 to Rio 2012 and beyond.* Swiss Agency for Development and

Cooperation (SDC); Centre for Development and Environment (CDE), University of Bern

Kolakowski, L. (1981). *Main currents of Marxism: Volume III, the breakdown.* Oxford: Oxford University Press.

Kolb, E. (2018). *The early universe.* Boca Raton, FL: CRC Press.

Kolb, B., Mychasiuk, R., & Gibb, R. (2014). Brain development, experience, and behavior. *Pediatric Blood & Cancer, 61*:1720-1723.

Koonin, E. (2000). How many genes can make a cell: the minimal-gene-set concept. *Annual review of genomics and human genetics, 1:* 99-116.

Koonin, E. (2007). The cosmological model of eternal inflation and the transition from chance to biological evolution in the history of life. *Biology Direct, 2*: 1-21.

Kopparapu, R., Ramirez, R., SchottelKotte, J., Kasting, J., Domagal-Goldman, S., & Eymet, V. (2014). Habitable zones around main-sequence stars: dependence on planetary mass. *The Astrophysical Journal Letters, 787*(2), L29.

Korthof, G. (2006). Fred Hoyle's The Intelligent Universe: A summary & review. http://wasdarwinwrong.com/kortho47.htm.

Kraatz, B., Luo, Z., Meng, J. and Ni, X. (2013). The placental mammal ancestor and the post–K-Pg radiation of placentals. *Science, 339:* 662-667.

Krebs, R. (2006). *The history and use of our earth's chemical elements: a reference guide.* Westport: CT: Greenwood.

Kristeva, J. (1989). *Black sun: Depression and melancholia.* Columbia University Press.

Kruk, E. (2012). Arguments for an equal parental responsibility presumption in contested child custody. *The American Journal of Family Therapy, 40*:33-55.

Ktorupić, D., Gračanin, A., & Corr, P. (2016). The evolution of the Behavioural Approach System (BAS): Cooperative and competitive resource acquisition strategies. *Personality and Individual Differences, 94*:223-227.

Kuhn, J. (2012). Dissecting Darwinism. *Proceedings Baylor University Medical Center, 25*:41-47.

Kurland, C. (2010). The RNA dreamtime. *Bioessays, 32*:866-871.

Labin, A. & Ribak, E. (2010). Retinal glial cells enhance human vision acuity. *Physical Review Letters, 104*: 158102.

Lampert, L. (2001). *Nietzsche's task: an interpretation of beyond good and evil.* New Haven, CT: Yale University Press.

Lane, A. (1981). Did Calvin believe in free-will? *Vox Evangelica, 12*, 72-90.

Lane, N., Allen, J., & Martin, W. (2010). How did LUCA make a living? Chemiosmosis in the origin of life. *BioEssays, 32*: 271-280.

Langston, J., Powers, H., & Facciani, M. (2019). Toward faith: A qualitative study of how atheists convert to Christianity. *Journal of Religion & Society, 21*:1-23.

Laughlin, R. (2005). *A different universe: Reinventing physics from the bottom down.* New York: Basic Books.

Leiter, B. (2014). *Why tolerate religion?* Princeton, NJ: Princeton University Press.

Lemley, B. (2000). Why is there life? *Discover, 21:* 64-69.

Lennox, J. (2009). *God's Undertaker: Has Science Buried God?* Oxford: Lion.

Lennox, J. (2011). *God and Stephen Hawking: Whose Design is it Anyway?* Oxford: Lion Books.

Lennox, J. (2012). Not the God of the gaps, but the whole show. The Cristian Post, August 20th. https://www.christianpost.com/news/the-god-particle-not-the-god-of-the-gaps-but-the-whole-show.html

Leslie, J. (1989). *Universes*. London: Routledge.

Levinton, J. (1992). The Big Bang of animal evolution. *Scientific American*, November.

Lewis, B. (1993). *Islam and the West*. New York: Oxford University Press.

Lewis, C. (1986). *The grand miracle: And other selected essays on theology and ethics from God in the dock*. New York: Ballantine Books.

Lewis, C. (2001). *Mere Christianity*. New York: Harper Collins.

Lewis, G., & Barnes, L. (2016). *A fortunate universe: Life in a finely tuned cosmos*. Cambridge: Cambridge University Press.

Lewontin, R. (1972). Testing the theory of natural selection. *Nature, 236*:181-182.

Lewontin, R. (1997). Billions and billions of demons. *New York Review of Books*, January 9th.

Lewontin, R. (2014). The mystery of language evolution. *Frontiers in Psychology, 5*:1-12.

Libet, B. (1999). Do we have free will? *Journal of Consciousness Studies, 6*: 47-57.

Libet, B., Wright Jr, E. W., & Gleason, C. A. (1983). Preparation-or intention-to-act, in relation to pre-event potentials recorded at the vertex. *Electroencephalography and clinical Neurophysiology, 56*:367-372.

Lightman, A. (2011). The accidental universe: Science's crisis of faith. *Harper's Magazine*, December.

Lim, R. (2017). *Self and the Phenomenon of Life: A Biologist Examines Life from Molecules to Humanity*. Hackensack, NJ: World Scientific.

Lipka, M. (2016). 10 facts about atheists. Washington, DC.: Pew Research Center.

Lipton, P. (2000). Inference to the best explanation. In W. Newton-Smith (ed.), *A companion to the philosophy of science*. pp. 184.193. Hoboken, NJ: Blackwell.

Livio, M., & Rees, M. J. (2005). Anthropic reasoning. *Science, 309:* 1022-1023.

Loeb, A. (2010). *How did the first stars and galaxies form?* Princeton, NJ: Princeton University Press.

Lohr, S. (1999)) I.B.M. plans a supercomputer that works at the speed of life. *New York Times*, December 6th. https://www.nytimes.com/1999/12/06/business/ibm-plans-a-supercomputer-that-works-at-the-speed-of-life.html

Long, K. (nd). What 1000s of near death experiences can teach us about dying. *Goop Wellness*. https://goop.com/wellness/spirituality/1000s-near-death-experiences-can-teach-us-dying/.

Lovelock, J. (1995). *The ages of Gaia: A biography of our living earth*. Oxford: Oxford University Press.

Maddox, J. (1989). Down with the big bang. *Nature, 34:* 425.

Madison, J. (1785). *Memorial and Remonstrance* (1785). Bill of Rights Institute. ttps://billofrightsinstitute.org/founding-documents/primary-source-documents/memorial-and-remonstrance/

Malachi, M. (2008). *Keys of this Blood: Pope John Paul II versus Russia and the West for control of the New World Order.* New York: Simon and Schuster.

Malo, A. (2017). Nihilism and freedom in the legend of the Grand Inquisitor. *Church, Communication and Culture, 2*:259-271.

Margenau, H. & Varghese, R. (1997). *Cosmos, Bios, Theos: Scientists Reflect on Science, God, and the Origins of the Universe, Life, and Homo sapiens.* 4th ed. Chicago and La Salle, Illinois: Open Court Publishing Company.

Markham, I. (2010). *Against atheism: Why Dawkins, Hitchens, and Harris are fundamentally wrong.* New York: Wiley and Sons.

Marks, J. (2012). *Ethics without morals: in defence of amorality.* New York: Routledge.

Marin, I., & Kipnis, J. (2013). Learning and memory... and the immune system. *Learning & memory, 20:* 601-606.

Marsh, J. (2012). *The Liberal Delusion: The Roots of Our Current Moral Crisis.* Bury St. Edmunds, England: Arena books.

Maurin, A. (2013). Infinite regress arguments. *In Johanssonian investigations,* Svennerlind, C. Almäng, J. & Ingthorsson R. (eds.) pp.421-438. Heusenstamm, Germany: Ontos Verlag.

Maxmen, A. (2011). Evolution: A can of worms. *Nature News, 470*:161-162.

Mayer, A. (1993). Universal versus Islamic human rights: A clash of cultures or clash with a construct. *Michigan Journal of International Law,* 15: 307-429.

Mayford, M., Siegelbaum, S., & Kandel, E. (2012). Synapses and memory storage. *Cold Spring Harbor perspectives in biology, 4*(6), a005751.

Mayr, E. (2004). *What makes biology unique?* New York: Cambridge University Press.

McCombs, C. (2004). Evolution hopes you don't know chemistry: The problem of control. *Acts & Facts.* 33 (8).

McGilchrist, I. (2009). *The master and his emissary: The divided brain and the making of the western world.* Yale University Press

McGrath, A. (2010). *Mere Theology.* London: SPCK Publishers

McIntosh, A. (2009). Information and entropy–top-down or bottom-up development in living systems? *International Journal of Design & Nature and Ecodynamics, 4:* 351-385.

McKay, B., & Zink, R. (2015). Sisyphean evolution in Darwin's finches. *Biological Reviews, 90:* 689-698.

McLean, E. (2017). Reasons to panic about the hierarchy problem. https://massgap.wordpress.com/2017/03/26/reasons-to-panic-about-the-hierarchy-problem/

McLeish, T., Bower, R.., Tanner, B., Smithson, H., Panti, C., Lewis, N., & Gasper, G. (2014). A medieval multiverse. *Nature,* 507: 161-163.

McLendon, K. (2017). Stephen Hawking, Michio Kaku, and other scientists on God: The existence and mind of God. *Inquisitr.* August 17th.

https://www.inquisitr.com/opinion/4441013/stephen-hawking-michio-kaku-and-other-scientists-on-god-the-existence-and-mind-of-god/

McNichol, J. (2008). Primordial soup, fool's gold, and spontaneous generation. *Biochemistry and Molecular Biology Education, 36*(4), 255-261.

Meirmans, S., & Strand, R. (2010). Why are there so many theories for sex, and what do we do with them? *Journal of Heredity, 101*(suppl_1), S3-S12.

Melchior, J. (2014). China still persecuting Christians. *National Review*, February 19th. https://www.nationalreview.com/2014/02/china-still-persecuting-christians-jillian-kay-melchior/

Meyer, S. (1999). The return of the God hypothesis. *Journal of Interdisciplinary Studies*, 11:1-31.

Meyer, S. (2003). DNA and the origin of life: Information, specification, and explanation. In J. Campbrell & S. Meyer, *Darwinism, Design and Public Education*, pp. 223-285. East Lansing, MI: Michigan State University Press.

Meyer, S. (2009). *Signature in the Cell: DNA and the Evidence for Intelligent Design*. Grand Rapids, MI: Zondervan.

Miller, K. (1999). *Finding Darwin's God*. New York: Harper-Collins.

Millikan, R. (1927). *Evolution in science and religion*. New Haven, CT: Yale University Press.

Mills, G. & Kenyon, D. (1996). The RNA world: A critique. *Origins & Design*, 17:9-16.

Milner, G. (2017). Nobody knows what lies beneath New York City. *Bloomberg Businessweek*, August 10th. https://www.bloomberg.com/news/features/2017-08-10/nobody-knows-what-lies-beneath-new-york-city.

Mitchell, K. (2007). The genetics of brain wiring: From molecule to mind. *PLoS Biology*, 4:690-692.

Mithani, A., & Vilenkin, A. (2012). Did the universe have a beginning? *arXiv:1204.4658*.

Montagu, A. (1978). Touching: The human significance of the skin. New York: Harper & Row.

Montefiore, H. (2016). *Reclaiming the high ground: a Christian response to secularism*. New York, St. Martin's Press.

Monton, B. (2009). *Seeking God in science: An atheist defends intelligent design*. Peterborough, Ontario: Broadview Press.

Mua, K. (2016). Religion as a Tool for Economic/Political Transformation. *Political Transformation, Working Paper, 1-22*.

Müller, G. (2017). Why an extended evolutionary synthesis is necessary. *Interface focus, 7*, 20170015.

Muravchik, J. (2003). *Heaven on Earth: The rise and fall of socialism*. New York: Encounter Books.

Mayr, E. (2001). *What evolution is*. New York: Basic Books.

Nagel, T. (1997). *The last word*. New York: Oxford University Press.

Nagel, T. (2012). *Mind and cosmos: why the materialist neo-Darwinian conception of nature is almost certainly false*. New York: Oxford University Press.

Nahmias, E. (2015). Why we have free will. *Scientific American*, January: 77-79.

National Aeronautics and Space Administration (nd). *Tests of Big Bang: The CMB*. https://wmap.gsfc.nasa.gov/universe/bb_tests_cmb.html.

National Aeronautics and Space Administration (nd). Tests of Big Bang: The light elements. https://map.gsfc.nasa.gov/

National Aeronautics and Space Administration (2016). The Magnetosphere: Our shield in space. NASA History Office. https://history.nasa.gov/EP-177/ch3-4.html.

National Aeronautics and Space Administration (2018). Dark energy, dark matter. https://science.nasa.gov/astrophysics/focus-areas/what-is-dark-energy.

National Aeronautics and Space Administration (2017). The Sun, August 3rd https://www.nasa.gov/sun.

National Aeronautics and Space Administration (2019). Wilkinson Microwave Anisotropy Probe https://map.gsfc.nasa.gov/universe/WMAP_Universe.pdf

National Institute of Heath (2012). ENCODE data describes function of human genome. National Human Genome Research Institute https://www.genome.gov/27549810/2012-release-encode-data-describes-function-of-human-genome/

Neveu, M., Kim, H., & Benner, S. (2013). The "strong" RNA world hypothesis: Fifty years old. *Astrobiology, 13:* 391-403.

Newport, F. (2010). Americans' church attendance inches up in 2010. Pew Report, http://www.gallup.com/poll/141044/americans-church-attendance-inches-2010.aspx

Newton, I. (1846). *Newton's Principia: The Mathematical Principles of Natural Philosophy/by Sir Isaac Newton*. (N. Chittenden, Ed.). New York: Daniel Adee.

Nicholl, D. (2008). An introduction to genetic engineering (3rd. edition). Cambridge: Cambridge University Press.

Nietzsche, F. (1997) *Daybreak: Thoughts on the Prejudices of Morality*, R.J. Hollingdale (trans.), Cambridge: Cambridge University Press.

Nietzsche, F. & Hollingdale, R. (1990). Expeditions of an untimely man. In *Twilight of the Idols; and, The Anti-Christ* (Penguin Classics, London: Penguin Books.

Nitardy, C. (2012). *Stumbling blocks of evolutions*. Maitland, FL: Xulaon.

The Nobel Prize (nd). All Nobel Prizes. https://www.nobelprize.org/prizes/lists/all-nobel-prizes/

Nuland, S. (1997). *The wisdom of the body*. New York: Alfred A. Knopf.

Olasky, M. & Perry, J. (2005). *Monkey business: The true story of the Scopes Trial*. Nashville, TN: Broadman & Holman.

O'leary, M., Bloch, J., Flynn, J., Gaudin, T., Giallombardo, A., Giannini, N., Goldberg, S., Kraatz, B., Luo, Z., Meng, J. and Ni, X. (2013). The placental mammal ancestor and the post–K-Pg radiation of placentals. *Science, 339:* 662-667.

Olsen, B. (2013). *Future Esoteric: The Unseen Realms*. San Francisco: CCC Publishing.

Orgel, L. (2008). The implausibility of metabolic cycles on the prebiotic Earth. *PLoS biology, 6*(1), e18.

Osgood, D. & Chambers, J. (2003). Community correlates of rural youth violence. *Juvenile Justice Bulletin*, May. Washington, DC: U.S. Department of Justice.

Overman, D. (2008). *A case for the existence of God*. Lanham, MD: Rowman & Littlefield.

Overy, R. (2004). The dictators: Hitler's Germany and Stalin's Russia. New York: W. W. *Norton*.

Page, D. (2007). Predictions and test of multiverse theories. in B. Carr (ed.), *Universe or multiverse*, pp. 411-430. Cambridge: Cambridge University Press.

Page, D. (2014). Cosmological ontology and epistemology. Paper presented for the proceedings of the Philosophy of Cosmology UK/US Conference, Tenerife, Spain.

Pagels, H. (1985). A cozy cosmology. *The Sciences* 25:34-38.

Parkin, J. & Cohen, B. (2001). An overview of the immune system. *The Lancet*, *357*:1777-1789.

Pascal, B. (1958) Pascal's Pensees. (Introduction by T. S. Elliot). New York: E. P. Dutton.

Patterson, C., Williams, D., & Humphries, C. (1993). Congruence between molecular and morphological phylogenies. *Annual Review of Ecology and Systematics*, *24*: 153-188.

Pearson, C. (2009). *The gospel of inclusion: Reaching beyond religious fundamentalism to the true love of God and self*. New York: Simon and Schuster.

Penn, A. (2001). Early brain wiring: Activity-dependent processes. *Schizophrenia Bulletin*, 27:337-348.

Penrose, R. (2010). Scientist debunks Hawking's 'no God needed' theory. *Independent Catholic* September 29. http://www.indcatholicnews.com/news.php?viewStory=16815

Penrose, R. (2016). *The emperor's new mind: Concerning computers, minds, and the laws of physics*. New York Oxford University Pres.

Percharde, M., Lin, C., Yin, Y., Guan, J., Peixoto, G., Bulut-Karslioglu, A., Biechele, S., Huang, B., Shen, X. & Ramalho-Santos, M. (2018). A LINE1-Nucleolin partnership regulates early development and ESC identity. *Cell*, 174: 391–405.

Pereira, C. & Shashi Kiran Reddy, J., (2016). Near-death cases desegregating non-locality/disembodiment via quantum mediated consciousness: An extended version of the cell-soul pathway. *Journal of Consciousness Exploration & Research*, 7: 951-968.

Perlmutter, S., Aldering, G., Goldhaber, G., Knop, R.., Nugent, P., Castro, P.., Deustua, S., Fabbro, S., Goobar, A., Groom, D., & Hook, I. (1999). Measurements of Ω and Λ from 42 high-redshift supernovae. *The Astrophysical Journal*, *517*:.565-586.

Persaud, C. (2007). *Evolution: Beyond the Realm of Real Science*. Maitland, FL: Xulon Press.

Peskin, M. & Schroeder, D. (2018). *An introduction to quantum field theory*. Boca Raton, FL. CRC Press.

Peterson, K., Dietrich, M., & McPeek, M. (2009). MicroRNAs and metazoan macroevolution: insights into canalization, complexity, and the Cambrian explosion. *Bioessays, 31*:736-747.

Petherick, A. (2010). Development: mother's milk: a rich opportunity. *Nature, 468:* S5-S7.

Petroski, H. (1996). *Invention by design: How engineers get from thought to thing.* Cambridge, MA: Harvard University Press.

Pew Forum on Religion & Public Life (2009). Eastern, New Age beliefs widespread: Many Americans mix multiple faiths. Washington, DC: Pew Research Center.

Pew Research Center (2016). Religious landscape study. http://www.pewforum.org/religious-landscape-study.

Phillips, T. (2014). China on course to become 'world's most Christian nation' within 15 years. *The Telegraph*, April 19th.

Physics.org (2017). What causes the magnetic shield? http://www.physics.org/article-questions.asp?id=64

Physics.Org (2018). Understanding how to control 'jumping' genes. June 22nd. https://phys.org/news/2018-06-genes.html

Pipes, R. (2011). *Russia under the Bolshevik regime.* New York: Vintage.

Planinić, J. (2010). The design argument – Anthropic principle. *Journal of Philosophy and Religious studies*, 65: 47-54

Plank, M. (1949). *Scientific Autobiography and Other Papers* (trans. F. Gaynor). New York: Philosophical Library

Plaxco, K., & Gross, M. (2006). *Astrobiology: a brief introduction.* Baltimore, MD: Johns Hopkins University Press.

Polkinghorn, J. (2001). Kenotic creation and Divine action, in Polkinghorne, J. (ed.). *The work of love: Creation as kenosis*, pp. 90-106.Grand Rapids, MI: Eerdmans.

Polkinghorne, J. (2007). *One World: The interaction of science and theology.* West Conshohocken, PA: Templeton Press.

Ponomarenko, E., Poverennaya, E., Ilgisonis, E., Pyatnitskiy, M., Kopylov, A., Zgoda, V., Lisitsa, A. & Archakov, A. (2016). The size of the human proteome: the width and depth. *International Journal of Analytical Chemistry, 2016:1-6.*

Powell, B. (2005). "Explore as much as we can:" Nobel Prize winner Charles Townes on evolution, intelligent design, and the meaning of life. UCBerkley News, June17th.
https://www.berkeley.edu/news/media/releases/2005/06/17_townes.shtml

Prager, D. (2019). Why so many mass shootings? Ask the right questions and you might find out. Real Clear Politics, June 4th.
https://www.realclearpolitics.com/articles/2019/06/04/why_so_many_mass_s hootings_ask_the_right_questions_and_you_might_find_out_140486.html

Pross, A. (2012). What is Life? How chemistry becomes biology. Oxford: Oxford University Press.

Radford, T. (2010). The Grand Design: New answers to the ultimate questions of life by Stephen Hawking and Leonard Mlodinow, September 17th.
https://www.theguardian.com/books/2010/sep/18/questions-life-cosmology-stephen-hawking

Rands, C., Meader, S., Ponting, C., & Lunter, G. (2014). 8.2% of the human genome is constrained: variation in rates of turnover across functional element classes in the human lineage. *PLoS Genetics, 10*:e1004525.

Rasic, D., Belik, S., Elias, B., Katz, L., Enns, M., Sareen, J., & Team, Swampy Cree Suicide Prevention (2009). Spirituality, religion and suicidal behavior in a nationally representative sample. *Journal of Affective Disorders, 114*: 32-40.

Raymo, C. (2002). Intelligent design happens naturally. Boston Globe, May 14th. http://www.boston.com/dailyglobe2/134/science/Intelligent_design_happens_naturallyP.shtml

Reeves, C. Comment made with signature to the *Dissent from Darwinism Statement*. https://dissentfromdarwin.org/2008/09/22/professor_colin_reeves_coventr/

Reeves, M., Gauger, A., & Axe, D. (2014). Enzyme families--shared evolutionary history or shared design? A study of the GABA-Aminotransferase family. *BIO-Complexity*, 2014:1-16.

Regalado, A. (2013). The brain is not computable. *MIT Technology Review*, February 18th.

Rensselaer Polytechnic Institute (2011). Setting the stage for life: Scientists make key discovery about the atmosphere of early Earth. *ScienceDaily*. November 13th. www.sciencedaily.com/releases/2011/11/111130141855.htm.

Renthal W. & Nestler, E. (2009). Chromatin regulation in drug addiction and depression. *Dialogues in Clinical Neuroscience*, 11:257-68.

Richerson, P. & R. Boyd (2010). *Evolution since Darwin: The first 150 years*. In M. Bell, D. Futuyma, W. Eanes & J. Levinton, (eds.) Sunderland, MA: Sinauer, pp. 561-588.

Ridley, M. (2001). *The cooperative gene: How Mendel's demon explains the evolution of complex beings*. New York: Simon and Schuster.

Ring, K. (1984). *Heading toward omega: In search of the meaning of the near-death experience*. New York, NY: Coward, McCann & Geohegan.

Robertson, M., & Joyce, G. (2012). The origins of the RNA world. *Cold Spring Harbor perspectives in biology, 4*(5), a003608.

Rokas, A., & Carroll, S. (2006). Bushes in the tree of life. *PLoS biology, 4*:1899-1904.

Ross, H. (1993). The Creator and the Cosmos: How the greatest scientific discoveries of the century reveal God. *Colorado Springs, CO: NavPress*.

Ross, H. (1994). Astronomical evidences for a personal, transcendent God. In J. Moreland (ed.) *The Creation Hypothesis: Scientific Evidence for an Intelligent Designer*, pp. 141-172. Downers Grove, IL: InterVarsity Press.

Ross, H. (2016). *Improbable planet: How Earth became humanity's home*. Grand rapids, MI: Baker Books.

Rosing, M. (1999). 13C-depleted carbon microparticles in> 3700-Ma sea-floor sedimentary rocks from West Greenland. *Science, 283*: 674-676.

Russell, R. (2008). Quantum physics and the theology of non-interventionist objective divine action. In P. Clayton & Z. Simpson (eds.), *The Oxford handbook of religion and science*, pp. 579-595. Oxford: Oxford University Press.

Rux, D. & Wellik, D. (2017). Hox genes in the adult skeleton: Novel functions beyond embryonic development. *Developmental Dynamics, 246:* 310-317.

Salas-Wright, C. Vaughn, M. & Maynard, B. (2014). Buffering effects of religiosity on crime: Testing the invariance hypothesis across gender and developmental period. *Criminal Justice and Behavior, 41:* 673-691.

Sample, I. (2018). Stephen Hawking's final theory sheds light on the multiverse. *The Guardian*, May 2nd. https://www.theguardian.com/science/2018/may/02/stephen-hawkings-final-theory-sheds-light-on-the-multiverse

Sampson, R. (2005). *The Heart of Wisdom Teaching Approach: Bible Based Homeschooling.* Merritt Island, FL: Heart of Wisdom Publishing.

Sandage, A. (1985). A scientist reflects on religious belief. *Truth: An International, Inter-disciplinary Journal of Christian Thought, 1:*53-54.

Sanford, J., Brewer, W., Smith, F., & Baumgardner, J. (2015). The waiting time problem in a model hominin population. *Theoretical Biology and Medical Modelling, 12: 1-28.*

Sayers, D. (1994). *Creed or chaos?* New York: Harcourt Brace.

Schaefer, H. (2003). *Science and Christianity: Conflict or coherence?* Watkinsville, GA: The Apollos Trust.

Schafer, L. (2006). Quantum reality and the consciousness of the universe. *Zygon*, 41:505-532.

Schratz, P. (2019). Astronauts wore faith in God on their spacesuits, *Catholic News Service*, July 19th.

Schroeder, G. (1997). *The science of God.* New York: Broadway Books.

Scharf, C. (2014). *The Copernicus Complex: Our Cosmic Significance in a Universe of Planets and Probabilities.* New York: Scientific American/Farrar, Straus and Giroux.

Schmacher, R. (2012). The myth that religion is the #1 cause of war. The Christian Apologetics & Research Ministry. https://carm.org/religion-cause-war

Schmidt, A. (2001). *Under the influence: How Christianity transformed civilization.* Grand Rapids, MI: Zondervan.

Schmidt, A. (2004). H*ow Christianity changed the world.* Grand Rapids, MI: Zondervan.

Schulzke, M. (2013). The politics of new atheism. *Politics and Religion, 6:*778-799.

Scornavacchi, M. (2015). *Superintelligence, Humans, and War.* Joint Forces Staff College Joint Advanced Warfighting School Staff College, Norfolk, VA.

Scott, M. (2010). Suffering and soul-making: Rethinking John Hick's theodicy. *The Journal of Religion*, 90:313-334.

Seckbach, J., & Gordon, R. (2009). *Divine action and natural selection: science, faith and evolution.* Hackensack, NJ: World Scientific.

Shackelford, K. (2017). Undeniable: The survey of hostility to religion in America. First Liberty Institute, Plano: TX.

Shalev, B. (2003). *Religion of Nobel Prize winners. 100 years of Nobel prizes.* New Delhi: Atlantic Publishers & Distributors.

Shapiro, R. (2007). A simpler origin of life. *Scientific American*, 296:46-53.

Sharma, S., & Ghoshal, S. (2015). Hydrogen the future transportation fuel: from production to applications. *Renewable and Sustainable Energy Reviews*, 43: 1151-1158

Shaviv, G. (2015). Who discovered the Hoyle Level? *Acta Polytechnica CTU Proceedings, 2:* 311-320.

Sheen, F. (1948). *Communism and the Conscience of the West.* Indianapolis and NY: Bobbs-Merrill.

Sherwood, M. (2007). *After Abolition: Britain and the Slave Trade Since 1807.* London: I.B. Tauris & Co.

Sherrington, C. (1951). Man: On his nature. Oxford: Oxford University Press.

Shi, S., Cheng, T., Jan, L. & Jan, Y. (2004). The immunoglobin family member dendrite arborization and synapse maturation 1 (Dasm1) controls excitatory synapse maturation. *Proceedings of the National Academy of Sciences*, 101:13246-13351.

Shostak, S. (2011). Who or what built the universe? *HuffPost*, May 25th. https://www.huffingtonpost.com/seth-shostak/who-or-what-built-the-uni_b_706185.html

Shrader, D. (2005). Thou shalt not: Supreme Court rulings concerning public displays of religious symbols. In *Proceedings of the Fourth Annual Hawaii International Conference on Arts and Humanities*, pp. 5564-5608.

Silva, I. (2015). A cause among causes? God acting in the natural world. *European Journal for Philosophy of Religion* 7: 99-114.

Silver, C. (2013). Atheism, agnosticism, and nonbelief: A qualitative and quantitative study of type and narrative. Doctoral dissertation, The University of Tennessee at Chattanooga

Simon, S. (2004). Big Bang: The most important scientific discovery of all time and why you need to know about it. New York: Harper-Perennial.

Slick, M. (2013). A look at two common atheist arguments. Christian Apologetics & Research Ministry, July 2nd. https://carm.org/a-look-at-two-common-atheist-arguments.

Smith, G. (2017). Does faith make you healthy and happy? The case of evangelicals in the UK. *Journal of Religion and Society*, 19:1-15.

Smith, W. (1981), *Therefore Stand.* New Canaan, CT: Keats Publishing.

Solzhenitsyn, A. (2006). Templeton Lecture, May 10, 1983," in *The Solzhenitsyn reader: New and essential writings, 1947-2005*, eds. E. Ericson, Jr. and D. Mahoney. Wilmington, DE: Intercollegiate Studies Institute p.

Sompayrac, L. (2015). *How the immune system works.* New York: John Wiley & Sons.

Sørensen, T., Danbolt, L., Lien, L., Koenig, H., & Holmen, J. (2011). The relationship between religious attendance and blood pressure: the HUNT study, Norway. *The International Journal of Psychiatry in Medicine, 42:* 13-28.

Sowell, E., Thompson, P. & Toga, A. (2004). Mapping Changes in the Human Cortex throughout the Span of Life. Neuroscientist, 10:372-392.

Spoel, D., Xu, Y. & Garcia, A., (2016). Water determines the structure and dynamics of proteins. *Chemical Reviews, 116:* 7673-7697.

Spradley, J. (2010). Ten lunar legacies: Importance of the Moon for life on Earth. *Perspectives on Science & Christian Faith, 62*:267-275.

Stark, R (2003). *For the Glory of God.* Princeton, NJ: Princeton University Press.

Stark, R. (2012). *America's Blessings: How Religion Benefits Everyone, Including Atheists.* West Conshohocken, PA: Templeton Foundation Press.

Stenger, V. (2009). *The new atheism: Taking a stand for science and reason.* Amherst, NY: Prometheus Books.

Stephan, A. (2000). Religion, democracy, and the "Twin Tolerations." *Journal of Democracy,* 11:35-57.

Stevens, J. (1985). Reverse engineering the brain. *Byte, 10:* 287-299.

Stiles, J., & Jernigan, T. (2010). The basics of brain development. *Neuropsychology Review, 20:* 327-348.

Straumann, B. (2008). The peace of Westphalia as a secular constitution. *Constellations, 15*: 173-188.

Strauss, M. (2017). The God Particle...and God. https://www.michaelgstrauss.com/2017/01/the-god-particleand-god.html.

Strawbridge, W., Cohen, R., Shema, S., & Kaplan, G. (1997). Frequent attendance at religious services and mortality over 28 years. *American Journal of Public Health, 87:* 957-961.

Substance Abuse and Mental Health Services Administration (2016). 2015 National survey on drug use and health. Rockville, MD: SAMHSA.

Sullivan, A. (2018). America's New Religions. New York Intelligencer, December 7th. http://nymag.com/intelligencer/2018/12/andrew-sullivan-americas-new-religions.html

Susskind, L. (2005). *The Cosmic Landscape: String Theory and the Illusion of Intelligent Design,* New York: Little, Brown and Company.

Świeżyński, A. (2016). Where/when/how did life begin? A philosophical key for systematizing theories on the origin of life. *International Journal of Astrobiology, 15:* 291-299.

Swinburne, R. (2010). *Is there a God?* Oxford: Oxford University Press.

Swindell, R. (2003). Shining light on the evolution of photosynthesis. *Journal of Creation,* 17:74-84.

Tagliaferre, L. (2013). Theofatalism: Theology for agnostics and atheists. Bloomington, IN: iUniverse Press.

Tarter, J., Backus, P., Mancinelli, R., Aurnou, J., Backman, D.., Basri, G., Boss, A., Clarke, A., Deming, D., Doyle, L., & Feigelson, E. (2007). A reappraisal of the habitability of planets around M dwarf stars. *Astrobiology, 7:* pp.30-65

Tatara, C. (2013). Hitler, Himmler, and Christianity in the Early Third Reich. *Constructing the Past, 14*: 39-44.

Tayler, J. (2015). The right hides behind a fictional Bible: Memo to Ted Cruz and Donald Trump --your favorite book is made up. *Salon,* September 20th. https://www.salon.com/2015/09/20/the_right_hides_behind_a_fictional_bible_memo_to_ted_cruz_and_donald_trump_your_favorite_book_is_made_up/

Taylor, B. (2015). Muslim writer explains why Islam and democracy are incompatible. Communities Digital News, August 26th.

https://www.commdiginews.com/world-news/muslim-writer-explains-why-islam-and-democracy-are-incompatible-47245/

Taylor, D. (2012). My godless city is full of Christian charity. *Independent*, December 16th .https://www.independent.co.uk/voices/comment/my-godless-city-is-full-of-christian-charity-8420317.html

Taylor, S. (1998). On the difficulties of making earth-like planets. *Meteoritics and Planetary Science, 34: 317-329.*

Tegmark, M. (2009). The multiverse hierarchy. *arXiv preprint arXiv:0905.1283.*

Tegmark, M. (2014). Is the Universe made of math? *Scientific American,* December. https://www.scientificamerican.com/article/is-the-universe-made-of-math-excerpt/

Tipler, F. J. (1988). The anthropic principle: a primer for philosophers. In *PSA: Proceedings of the Biennial Meeting of the Philosophy of Science Association* Vol. 1988: 27-48).

Tipler, F. (1994). *The physics of immortality: Modern cosmology, God, and the resurrection of the dead.* New York: Anchor.

Tipler, F. (2003). Intelligent life in cosmology. *International Journal of Astrobiology*, 2: 141-148.

Todd, S. (1999). A view from Kansas on that evolution debate. *Nature, 401*: 423-423.

Tertullian, Q. (1842). *Tertullian: V. 1. Apologetic and Practical Treatises.* London: John Henry Parker. p. 349

Trefil, J., & Hazen, R. (2007). *The sciences: An integrated approach.* New York, Wiley.

Trevors, J. & Abel, D. (2004). Chance and necessity do not explain the origin of life. *Cell Biology International, 28:* 729-739.

United Nations Office on Drugs and Crime. (2014). Some 437,000 people murdered worldwide in 2012, according to new UNODC study. http://www.unodc.org/unodc/en/press/releases/2014/April/some-437000-people-murdered-

University of California, Davis (nd). The electromagnetic spectrum. Online educational course. http://earthguide.ucsd.edu/virtualmuseum/ita/07_1.shtml

University of Warwick (2003). Psychology researcher says spiritual meaning of Christmas brings more happiness than materialism. News and Events. December 8th. https://warwick.ac.uk/newsandevents/pressreleases/ne100000008548/.

U.S. Department of Health and Human Services (2009). Understanding the effects of maltreatment on brain development. https://www.childwelfare.gov/pubs/issue_briefs/braindevelopment/brain_development.pdf.

U.S. Department of Health and Human Services (2011). *Births: Preliminary data for 2010.* Washington, DC: U.S. Government Printing Office.

U.S. Department of Health & Human Services (2014). AIDS.gov. https://www.aids.gov/hiv-aids-basics/hiv-aids-101/statistics/

Valencia, D., O'Connell, R., & Sasselov, D. (2007). Inevitability of plate tectonics on super-earths. *The Astrophysical Journal Letters, 670*:45-48.

Varghese, R. (2013). *The Missing Link: A Symposium on Darwin's Framework for a Creation-evolution Solution.* Lanham, MD: Rowman & Littlefield.

Vasas, V., Szathmáry, E., & Santos, M. (2010). Lack of evolvability in self-sustaining autocatalytic networks constraints metabolism-first scenarios for the origin of life. *Proceedings of the National Academy of Sciences, 107:* 1470-1475.

Vieru, T. (2011). Moons like our own are extremely rare in the universe. Softpedia News, June 29th. https://news.softpedia.com/news/Moons-Like-Our-Own-Are-Extremely-Rare-in-the-Universe-214242.shtml

Vohs, K. & Schooler, J. (2008). The value of believing in free will: Encouraging a belief in determinism increases cheating. *Psychological Science, 19*: 49-54.

Wald, G. (1954). The origin of life," *Scientific American,* 191: 45–53.

Wald, G. (1984). Life and mind in the universe. *International Journal of Quantum Chemistry, 26*: 1-15.

Wallace, P. (2016). *Stars beneath us: Finding God in the evolving cosmos.* Minneapolis, MN: Fortress Press.

Walker, S., & Davies, P. (2013). The algorithmic origins of life. *Journal of the Royal Society Interface, 10:* 20120869.

Walker, S., & Davies, P. (2016). The "hard problem" of life. *arXiv preprint arXiv:1606.07184.*

Walsh, A. (1998). Religion and hypertension: Testing alternative explanations among immigrants. *Behavioral Medicine,* 24:122-130.

Walsh, A. (2009). *Biology and Criminology: The biosocial synthesis.* New York: Routledge.

Walsh, A. (2016). *Love: The biology behind the heart.* New Brunswick: NJ: Transaction Press.

Walsh, A. (2017). Taboo issues in social science: Questioning conventional wisdom. Wilmington, DE: Vernon Press.

Walsh, A. (2018). *The Gavel and Sickle: The Supreme Court, Cultural Marxism, and the Assault on Christianity.* Wilmington, DE: Vernon Press.

Walsh, A. (2019). *Reinforcement Sensitivity Theory: A Unifying Framework for Biosocial Criminology.* New York: Routledge.

Walsh, J. (2013). *Old time makers of medicine.* New York: Simon and Schuster.

Ward, P. & Brownlee, D. (2000). *Rare Earth.* New York: Copernicus Books.

Washington, G. (1796). Washington's farewell address, 1796. The Avalon Project, http://avalon.law.yale.edu/18th_century/washing.asp.

Weitnauer, C. (2013). The irony of atheism. In T. Gilson & C. Weitnauer (eds.) *True reason,* pp. 25-36. Grand Rapids, MI: Kregel.

Wei-Haas, M. (2018). Volcanoes, explained. *National Geographic,* January 15th. https://www.nationalgeographic.com/environment/natural-disasters/volcanoes/

Weitoft, G., Hjern, A., Haglund, B. & Rosén, M. (2003). Mortality, severe morbidity, and injury in children living with single parents in Sweden: a population-based study. *The Lancet, 361:* 289-295.

Wells, J. (2017). *Zombie science: More icons of evolution.* Seattle, WA: Discovery Institute.

Whitfield, J. (2008). Biological theory: Postmodern evolution? *Nature News, 455:* 281-284.

Weinberger, D., Elvevag, B., & Giedd, J. (2005) *The Adolescent Brain: A Work in Progress.* Washington, DC: The National Campaign to Prevent Teen Pregnancy.

Weinhold, B. (2006). Epigenetics: The science of change. *Environmental Health Perspectives,* 114:161-167.

Wigner, E. (1990). The unreasonable effectiveness of mathematics in the natural sciences. *Mathematics and Science* 13:1-14

Wigner, E. (2013). *The collected works of Eugene Paul Wigner: Historical, philosophical, and socio-political papers. Historical and Biographical Reflections and Syntheses.* Berlin: Springer-Verlag.

Wiker, B. (2005). How the world's most notorious atheist changed his mind. *Strange Notions.* http://www.strangenotions.com/flew/

Wilcox, B. (2004). *Soft patriarchs, new men: How Christianity shapes fathers and husbands.* Chicago: University of Chicago Press.

Wilkins, M., & Moreland, J. (2010). *Jesus Under Fire: Modern Scholarship Reinvents the Historical Jesus.* Grand Rapids, MI: Zondervan. "

Wilkins, A., & Holliday, R. (2009). The evolution of meiosis from mitosis. *Genetics, 181:* 3-12.

Will, G. (2008). Bleeding hearts but tight fists. The Washington Post, March 27th http://www.washingtonpost.com/wp-dyn/content/article/2008/03/26/AR2008032602916.html

Williams, G. (1992). Natural selection: Domains, levels and challenges. New York: Oxford University Press.

Wilson, E. (1993). *The moral sense.* New York: Free Press.

Winthrop, R. (1852). Addresses and Speeches on Various Occasions, Boston: Little, Brown.

Woit, P. (2006). *Not even wrong: The failure of string theory and the search for unity in physical law.* New York: Basic Books.

Wolfenden, R. (2008). Without enzymes, biological reaction essential to life takes 2.3 billion years: UNC study. UNC School of Medicine, December. https://www.med.unc.edu/biochem/news/without-enzyme-biological-reaction-essential-to-life-takes-2-3-billion-years-unc-study/

Wood, R., Goesling, B., & Avellar, S. (2007). The effects of marriage on health: a synthesis of recent research evidence. *Washington DC: Mathematical Policy Research.*

Wolchover, N. (2013). Is nature unnatural? *Quanta,* 24 May.

Woollett, K., & Maguire, E. (2011). Acquiring "the Knowledge" of London's layout drives structural brain changes. *Current biology, 21:* 2109-2114.

Wu, J., Desch, S., Schaefer, L., Elkins-Tanton, L., Pahlevan, K., & Buseck, P. (2018). Origin of Earth's water: Chondritic inheritance plus nebular ingassing and storage of hydrogen in The core. *Journal of Geophysical Research: Planets, 123*: 2691-2712.

Wuethrich, B. (1998). Putting theory to the test. *Science.* 281:1980-1982.

Xu, Q. (2014). *The evolutionary feminism of Zhang Kangkang and the developingdialogue between Darwinism and gender studies*. PhD dissertation, University of Helsinki, Finland.

Yahya, H. (1999). *The Creation of the Universe*. Istanbul: Global Yayincilik.

Yale Law School (nd). Northwest Ordinance; July 13, 1787, Avalon Project.. http://avalon.law.yale.edu/18th_century/nworder.asp

Yockey, H. (2005). *Information theory, evolution, and the origin of life*. Cambridge: Cambridge University Press.

Zacharias, R. (2008). *End of reason: A response to the New Atheists*. New York: Harper Collins.

Zalasiewicz, J. & Williams, M. (2014). Weird wet worlds: Why Earth is lucky to have oceans. *New Scientist*, October 29th. https://www.newscientist.com/article/mg22429930-600-weird-wet-worlds-why-earth-is-lucky-to-have-oceans/

Zauzmer, J. (2016). In space, John Glenn saw the face of God: 'It just strengthens my faith." *The Washington Post*, December 11th.

Ziegler Hemingway, M. (2008). Look who's irrational now. *Wall Street Journal*, Sept. 19th.

Zimmermann, A. (2013). A law above the law: Christian roots of the English common law. *Glocal Conversations, 1*: 85-98.

Figures

Figure 1.1. "Distribution of Atheists, agnostics, and Freethinkers in Nobel Prizes between 1901-2000" by Jobas is licensed under CC BY-SA 4.0 (https://commons.wikimedia.org/wiki/File:Distribution_of_Atheists,_agnostics,_and_Freethinkers_in_Nobel_Prizes_between_1901-2000.png).

Figure 4.1. Reprinted from "CMB Timeline300 no WMAP" by NASA/WMAP Science Team, ca. 2006 (https://commons.wikimedia.org/wiki/File:CMB_Timeline300_no_WMAP.jpg). NASA image; Government agency. Public domain.

Figure 5.1. Reprinted from "Particle overview" by Headbomb is licensed under CC BY-SA 3.0 (https://commons.wikimedia.org/wiki/File:Particle_overview.svg).

Figure 5.2. Reprinted from "Triple-Alpha Process" by Borb is licensed CC BY-SA 3.0 (https://commons.wikimedia.org/wiki/File:Triple-Alpha_Process.svg).

Figure 6.1. Reprinted from "End of universe" by NASA, 2006 (https://commons.wikimedia.org/wiki/File:End_of_universe.jpg). NASA image; Government agency. Public domain.

Figure 8.1. Reprinted from "Tectonic plate boundaries" by Jose F. Vigil. United States Geological Survey, 1997 (https://commons.wikimedia.org/wiki/File:Tectonic_plate_boundaries.png). Public domain in the USA.

Figure 9.1. Reprinted from "Nitrogen Cycle" by U.S. Environmental Protection Agency, 2003 (https://commons.wikimedia.org/wiki/File:Nitrogen_Cycle.jpg). Public domain.

References

Figure 12.1. Source: U.S. Department of Energy Human Genome project. Public domain.

Figure 12.2. Reprinted from "225 Peptide Bond-01" by OpenStax College is licensed under CC BY 3.0, 2013
(https://commons.wikimedia.org/wiki/File:225_Peptide_Bond-01.jpg).

Figure 13.1. Reprinted from "Chirality with hands" by NASA, ca. 2008
(https://commons.wikimedia.org/wiki/File:Chirality_with_hands.svg).
NASA image; Government agency. Public domain.

Figure 14.1. Reprinted from "Cell membrane detailed diagram en" by Mariana Ruiz Villarreal is in the public domain, 2007
(https://commons.wikimedia.org/wiki/File:Cell_membrane_detailed_diagram_en.svg).

Figure 15.1. Reprinted from "Epigenetic mechanisms" by National Institutes of Health, 2005
(https://commons.wikimedia.org/wiki/File:Epigenetic_mechanisms.jpg).
Public domain in the USA.

Figure 18.1. Reprinted from "Three cell growth types" by JWSchmidt is licensed under CC BY-SA 3.0, 2004
(https://en.wikipedia.org/wiki/File:Three_cell_growth_types.png).

Figure 19.1. Published by the Institute for Quality and Efficiency in Health Care (Germany).

Figure 20.1. Reprinted from "Brain side view" by the National Institute on Aging, 2016
(https://www.flickr.com/photos/nihgov/24414866102).Government Document. Public domain.

Figure 20.2. Reprinted from "Complete neuron cell diagram en" by Mariana Ruiz Villarreal is in the public domain, 2007
(https://commons.wikimedia.org/wiki/File:Complete_neuron_cell_diagram_en.svg).

Figure 20.3. Source: National Institute of Health.

Index

A

abduction 26-28
Abel, David 144
abiogenesis 129-131
Acton, John 279
Aczel, Amir 166
ad infinitum 30
Adamantius, Origen 246
adenine (A) 118-120, 141
adenosine triphosphate (ATP) 82, 125, 212
Advances in Biological Chemistry 147
aerosols 84
agape 200, 228
agency 35, 232
agreeableness 263
Aldrin, Buzz 269
Alexander, Eban 221
alleles 119, 158, 165
Alliance Defending Religious Freedom 2
America's Blessings: How Religion Benefits Everyone, Including Atheists 258
American Atheists 4, 20
American Family Association 2
American Grace 268
American Humanist Association 279
American Journal of Psychiatry 262
American Sexual Health Association 283
amino acids 121-122
    to proteins 132-135
Ammar, Salem Ben 276
amygdale 211, 215
Anfinsen, Christian 8
anisotropy 58
annihilation 245, 247
Annual Review of Astronomy and Astrophysics 53
Anthropic principle 23-25, 113
apobetics 149
Appolloni, Simon 34
Aquinas, St. Thomas 241
Aristotle 33, 110
aseity 30
Association of American Physicians and Surgeons 192
astronomical units (AUs) 68
atheism
    and China 256-257
    and Darwin 155
    and free will 231
    and Marxism 257
    and morality 251-254, 285
        fundamentalist 4
    Flew on 127
    Gettysburg Address of 36
    militant 4
    Nagel on 6
    new 2-4
    of materialism/naturalism 20
    rationality of 8-9
    rock of 241
    soft 253
    versus Christianity 252, 261-263
atomic theory 17, 104, 155, 231
Australopithecus 173
autonomic nervous system (ANS) 215, 235

Aviezer, Nathan 179
axon 210, 212-213, 217
Ayala, Francisco 95

B

Bacon, Francis 9
Bacon, Roger 17
Barnes, Luke 24
Bates, Elizabeth 226
Batygin, Konstantin 73
Bauchau, Vincent 166
Bechly, Gunter 182
behavioral approach system (BAS) 214
behavioral inhibition system (BIS) 214
Behe, Michael
    on blood clotting 205
    on E.coli 170
    on ID 178
    on immune system 204
Benner, Steven 136
Benzmüller, Christoph 110
Bernhardt, Harold 142
beryllium 52
beta decay 49
Bible
    and Luther 276
    in school 285, 287, 301
    on human life 192
Big Bang theory 17, 36, 38
big whack 72, 81
Biochemical Predestination 131-132
Biochemistry and Molecular Biology Education 143
BIO-Complexity 170
Biological Reviews 171
BioLogos Foundation 184
biota 93, 172
Birney, Ewan 158

blastocyst 190-191
blood 203-206
    clotting 214, 216
blood-brain barrier 203
Bloomberg Businessweek 204
B-lymphocytes (B-cells) 202
Boettner, Loraine 238
Bohr, Niels 113, 237
Bonaparte, Napoleon 13
Bonhoeffer, Dietrich 177
Boone, Pat 284
Born Atheist 4
Boscovich, Roger 17
bosons 49, 61
Bragg, William 7
brain 209-213
    and love 228-229
    and mind 219-223
    neurotransmitters 214-215
    social 216
    structures 216-217
Bray, Gerald 246
Brown, Arthur 78
Bruno, Giordano 113
Bruteau, Beatrice 7
Bryson, Bill 124
Buber, Martin 6
Burger, Warren 288

C

Calvin, John 238
Calvin, Melvin 17
Calvinism 238, 277
Cambrian explosion 171-173
Campbell, David 268
Canadian Community Health Survey 262
carbon 50-53
carbon dioxide 82-84
carbon 12 50-52
carbonates 133

Index

cardivoscular system 204, 206
Carey, Nessa 159
Carnoy, Jean-Baptiste 17
Carr, Bernard 97, 114
Carroll, Lewis 194
Carter, Brandon 23
Catholic Church
    and Galileo 16
    and health care 268
    and TE 184
    and University of Bologna 18
Cell Biology International 130
cell-soul pathway 225
cerebellum 212, 216
cerebral cortex 195, 211, 216
Chain, Ernst 149
Chalmers, David 219
Chan, Jun Yaun 183
chance and necessity 131-132, 174
chaperones 123
chemical reactivity 135
Chesterton, G.K. 10, 280
Child Trends 256
China 256
Chinese Cultural Revolution 256
chiral 134-135
chlorophyll 84, 85, 168
chloroplasts 84
Choroser, Michael 182
Christ, Jesus 238, 243, 246
Christian Legal Society (CLS) 290
Christian Legal Society v. Martinez 290
Christian Science Church 242
Christianity
    and antisocial behavior 263
    and family 265
    and freedom 279-281
    and giving 267
    and happiness 261
    and health 266
    and higher education 29
    and human rights 273-276
    and modern assault on 283-286
    and morality 251-254
    and Nazi Germany 256-257
    and science 15-18, 21
    and societal benefits 258-259
Circumstellar Habitable Zone (CHZ) 68
Cleaver, Gerald 114
Cliff, Harry 62
Clinton, Bill 117
codons 121
Coke, Edward 279
Collins, Francis
    and BioLogos Foundation 184
    and Human Genome Project 117
    and TE 186
    on DNA 126
    on God and science 18
Collins, Robin 46, 113
colostrum 201
communism 255, 257
compartmentalization 145
compatibilism 236-267
complementarity 237
Compton, Robert 118
Congressional Prayer Caucus Foundation 291
Congressional Record 285
conscience 252
consciousness 219-222, 225
consciousness 219-223
conservatives 9-10, 270
Constantine 275
Constantinople 276
convection 81
Copernican principle 23, 25, 94
Copernicus, Nicolaus 16
Corey, Michael 24
corotation circle 67
corpus callosum 211

Cory, Michael 51, 88
cosmic microwave background (CMB)
   radiation 39, 58
cosmological constant 34, 46, 59-60
cosmology 8, 28, 31, 115
cosyntics 148
Couenhoven, Jesse 237
Coyne, Jerry 199
Craig, William Lane 30, 36-37, 61, 257
creatio ex nihilo 33, 38
Cremonini, Cesare 16
Crick, Francis 20, 130
critical value (pcrit) 58-59
crusades 278
Cuomo, Eddie 225
cytokines 202
cytoplasm 120, 125
cytosine (C) 118, 141, 162
cytoskeleton 125

D

D (dextro) 134
daily spiritual experiences (DSE) 267
dark energy 34, 46-47
Darwin, Charles 155, 275
Darwin's Black Box 205
Darwin's finches 171
Darwinism 153-154, 182-183
Davies, Paul
   on Big Bang 35, 58
   on electromagnetism, 47, 49
   on Newton 18
   on OoL 130
Dawkins, Richard 3
   and The God Delusion 181
   on child abuse 265
   on eyes 207
   on Grand Design 113, 13
   on religion 26
   on religious wars 277
   on Replicator 131
de Duve, Christian 131
De Genesi ad Literam 156
De Luce 35
de Tocqueville, Alexis 280
decay 49
Declaration of Independence 273, 288
deduction 26-27
Dembski, William 42, 77, 160, 179-180
Demiurge 33
democracy 273-276
   and Christianity 279
Democracy in America 280
dendrites 210, 212-213, 217
Dennett, Daniel 3, 222
Descartes, Rene 219
design 179-180
Design-Centered Anthropic Principle (DCAP) 24
determinism 231
   and free will 236-237
   soft 232-233, 238
   strict 232
deuterium 41, 49
deuteron 41
dextro (D) 134
dike 241
Dingle, Herbert 109
diploids 193
Dirac, Paul 107
DNA 117
   and RNA 142, 145
   as a language 126
   as information 221
   helicase 119
   junk 158-159, 181

Index 329

methylation 161
methyltransferase 162
Dobzhansky, Theodosius 154
Donoghue, John 60
Donovan, William 256
dopamine 214
Doppler Effect 35
Dostoevsky, Fyodor 243
double helix ladder 118
Drummond, Henry 178
Durant, Ariel 281
Durant, Will 281
Dyson, Freeman 56

E

Eccles, John 22
Eddington, Arthur 21, 38, 108
efficient cause 25
Einstein, Albert
   on free will and determinism 237
   on math 109
   on science and God 15, 17, 26
Ekstrom, Sylvia 52
electromagnetism 47, 49
electro-volts 51
Elk Grove Unified School District et al v. Newdow et al. 183
elliptical galaxies 66
Ellis, George 109
emotion 15-16
Encyclopedia of DNA Elements (ENCODE) 158-159
Encyclopedia of Wars 277
endoplasmic reticulum 125
energy density (p) 58
Engel v. Vitale 286
English Petition of Rights 279
enzyme 168-170
Epicurean chaos 271
Epicurus 231

epigenetics 160-161
epinephrine 215
epistasis 161
equality 274
ergodic 99
eros 228
erythropoietin 206
establishment clause 183, 286-290
eternal damnation 245-247
eukaryotes 193, 196
evil 241-242
   and free will 243
   and salvation 245-247
   and soul-making 244-245
   moral 248-249
   natural 248-249
Evolution from Space 137
Evolution 153-154
   and language 227
   by natural selection 15
   common sense 157
   icons of 158-159
   of gaps 154, 181
   queen of problems 193-195
   theistic 177, 184-186
   versus ID 181-182
ex cathedra 98
exon 119

F

falsifiability 105
family 263, 265
Farrer. T.H. 156
Father Brown 10
fermions 61
Fernandez, Jorge 118
Feynman, Richard 111
fibrin 205-206
fight/flee/freeze system (FFFS) 215
Final Anthropic Principle (FAP) 25

Final Cause 25, 156
First Amendment 2-3, 193, 273, 280
First Amendment Defense Act 2
first cause 29-30, 37
First Liberty Institute 2
fitness 157, 223
fixation 157-158
flagellum 180
Flew, Antony 127
Fordyce, John 155
Forget, Francois 78
Four Horsemen 3
Frankl, Viktor 247
Free Exercise Clause 183, 286, 288
free will 231-233
    absolute 233
    and determinism 231-232, 237-238 and evil 243
    and free mind 233-236
    and omniscience 238-239
Freedom in the World Index 273
Freeland, Stephen 156
freethinkers 4, 8
Freud, Sigmund 266
frontal lobe 216
Fu, Bob 256
Fudge, Edward 247
functional magnetic resonance (fMRI) 220

G

gaia hypothesis 93
galactic habitable zone (GHZ) 66-67
galaxy 65-68
Galileo 16, 117
gametes 194, 196
gamma rays 47, 70
gamma-amino butyric acid 201

general relativity theory 34, 45, 101, 108
genes 118-119
    hox 191
    jumping 159
    junk 158, 202
    overlapping 166
Genesis 33, 81, 91, 195, 225
genetic mutations 157, 194
genome 117, 159-160
geocentric model 16
Geospiza 171
Gervias, Will 252
Gilbert, Scott 153
Gillen, Alan 207
Gitt, Werner 118, 149
Glashow, Sheldon 111
Glen, John 269
glial cells 210, 217
globular clusters 66
gluons 48, 61
God
    and anthropic principle 25
    and Darwinism 153-156
    and free will 283-239
    and human body 189-192, 210, 227
    and mathematics 107, 110-111
    and multiverse 113-115
    and natural evil 248-249
    and NDEs 214
    and ozone layer 78
    and problem of evil 241-245
    and science 17-19
    and TE 184-186
    and the soil 91
    as cause 29-33
    as creator of DNA 117-119, 126
    existence of 5-9, 28
    Flew on 127
    hypothesis 13-14, 27
    in government 280-281

Index

in schools 183, 285-287, 291
losing 97
love 228, 246-247
morality from 251-253
opposition to 37
particle 61-62
Russell on 1
God is not Great: Religion Poisons Everything 4, 263
Godel, Kurt 110
God-of-the-gaps 177-180
golgi 125
Gonzales, Guillermo 33, 77
Gorelick, Root 194
Goswani, Amit 185
Gould, Stephen Jay 171, 184, 223
Graebner, Norman 275
grand deisgn 115
Grand Inquisitor 243-244, 249
Grassé, Pierre-Paul 183
Graur, Dan 158
gravity 45-47
   and big crunch 59-60
   and matter & spacetime 102-103,108
   and Milky Way 65-66
   just right 78
   of moon 72
gray matter 212
Gray, Asa 155
Greenstein, George 35, 53
Greyson, Bruce 225
Grib, Andrei 257
Gribbin, John 50, 95
Griffiths, Robert 8
Grim, Brian 258
Grim, Melissa 258
Grosseteste, Robert 35
Grossman, Lisa 90
guanine (G) 118, 141
Gutenberg, Johannes 276

H

Habermas, Jurgen 274
Haidt, Jonathan 254
Haldane, J.B.S. 74, 132
Hall of Shame Report 2
Hall, Daniel 267
Hall, Stephen 158
Hamilton, Alexander 273, 279
Hamilton, W.D. 195
haploids 193
Harris, Sam 3
   on morality 252-253, 285
   on religion and violence 277
Harrub, Brad 206
Hartsfield, Tom 112
hate groups 2
Haught, John 94, 168, 249
Hawking, Stephen
   and eternity 37
   on Anthropic principle 24
   on free will 234
   on fundamental forces 45
   on God 29
   on imaginary time 108
   on odds of suitability for life 74
Hayes, Carlton 267
Heile, Frank 105
Heisenberg uncertainty principle 90
heliocentric model 16
heliosphere 79
helium 41
   and stars 65-66, 71
   and sun 69-70
   in triple-alpha process 52
helper T-cell 202
Henderson, Lawrence 88
Heng, Henry 194
Hera 65
Heracles 65
Hertog, Thomas 112

Hick, John 185, 245-246
Higgs boson 61-63
Higgs field 61
Higgs, Peter 61
Higher-order automated theorem provers 110
Hilbert, David 30
hippocampus 211, 215
histone 118
   acetylation 161
   acetyltransferaces (HATs) 162
   deacetylases (HDACs) 162
Hitchens, Christioher 3, 5, 254, 255, 263
Hitchens, Peter 255
Hitler, Adolf 257
homeobox 191
Homo erectus 173
Homo rudolfensis 173
Homo sapiens 173, 223
Hox genes 191
Hoyle state 51-52
Hoyle, Fred
   and Big Bang 38, 105
   on carbon 50, 53
   on Darwinism 153
   on ID 137
   on the beginning 36
Hsp90 (heat shock protein 90) 172
hubble volume 99
Hubble, Edwin 34
Human Genome Project 117, 160
Human Proteome Project (HPP) 144
human rights 273-274, 279
Hummer, Robert 266
Huxley, Thomas 16
hydrogen 41
   and plants 84
   and soil 91-92
   and the galaxy 65
   and the sun 69-71
   and water 87-90
   in the forces 48
hydrologic cycle 90
hydrothermal vents 143
hypothalamus 211, 214
hypotheses 27-28
   in science 103-104

I

illegitimacy 255, 264
imaginary time 108-110
immune system: 200-204
induction 26-27
Infeld, Leopold 237
information 147-149
   and mind 221
   and neurons 212-214
   problems 165-167
information first hypothesis 141, 149
Institute of Physics 79
intelligent design (ID) 178-182
International Christian Concern (ICC) 2
International Conference of the Origins of Life 149
International Journal of Quantum Chemistry 131
intron 119
internal space 101-102
Irenaeus 245, 247
isotropy 58
Irenaeus 245, 247
irregular galaxies 66, 68,
irreducible complexity 144, 180, 184, 186
Islam 19, 269, 275-278

J

Jastrow, Robert 38, 94

Jeans, James 21
Jesuits 16
Jenkins, Alejandro 59
Johnson, Philip 171, 183
Jones, John 183
Journal of Family Practice 266
Journal of High Energy Physics 57
Journal of Theoretical Biology 121
Jupiter 73-74

K

Kaku, Michio 114
kalam cosmological argument 36
Karamazov, Ivan 243
Kenny, Mary 261-262
Kenosis 185
Kenyon, Dean 131
Kepler, Johann 16
KKK 2, 290
Kitzmiller v. Dover Area School District 183
Klyce, Brig 169
Kondrashov, Alexey 194
Koonin, Eugene 137, 146
Korthof, Gert 138
Krauss, Lawrence 63
Krebs cycle 148
Krebs, Robert 46
Kristeva, Julia 262
Kuhn, Joseph 181, 210
Kurland, Charles 142

L

L (levo) 134
Lagrange, Joseph-Louis 13
language
  and mind 225-228
  of DNA 126, 149
Laplace, Pierre-Simon 13

Last Universal Common Ancestor (LUCA) 132
Laughlin, Greg 73
Laughlin, Robert 154
Lee v. Weisman 289
Leiter, Brian 3
Lemaitre, Georges 34
Lennox, John 14
  on information first hypothesis 149
  on fruit flies 170
  on law and agency 103
  on meaning of universe 63
Leonard Award 95
Leslie, John 24
Levo (L) 134
Liberals 10, 270
Levinton, Jeffrey 172, 195
Lewis, Bernard 278
Lewis, C.S. 244, 251-252
Lewontin, Richard 5, 14, 98, 154
Libet, Benjamin 234
light year 66-67, 106, 132
light elements 40-41, 77
Lightman, Alan 98
lightning 93, 132
limbic system 200, 211, 215
Lincoln, Abe 285
Linde, Andrei 24, 97
lithium 41, 70
Lipton, Peter 27
lithosphere 80-81
Loeb, Abraham 58
Long, Kenneth 224
love
  agape 200, 220
  as top-down causation 20, 220
  mother 228-229
  of God 246, 275
Lovelock, James 93
LUCA (Last Universal Common Ancestor) 132

Lucretius 231
luminosity 71-72
Luther, Martin 276
Lutheranism 277
lymphocytes 202-203
lysosomes 125

M

Ma, Zeyang 159
macroevolution
    and speciation 171-173
    and the problem of time 168
    defined 157
    in curriculum 182-183
    Laughlin on 154
magnetic field 78-79, 81
macrophages 124, 202-203
Maddox, John 37
Madison, James 286
Magnus, Albertus 18, 186
Malo, Antonio 244
Manson, Neil 114
Margulis, Lynn 145
Markham, Ian 4
Marks, Joel 253
Marx, Karl 254, 266, 291
Mars 73-74
mass (m) 62
    and forces 48-49
    and gravity 45-46, 59
    bare 62
    mountain 80
    of Jupiter 73
    of stars 71
    of Sun 70
    planetary 78
material 104, 148
material cause 19, 25
matter 21, 31
    and anti-matter 40
    and Big Bang 40

    atomic theory of 104
    density 57-59
    origin of 129-130
materialism 19
    and abiogenesis 129
    and Earth-Moon 73
    and mind 219, 222
    Bechly on 182
    methodological 20
    Politizer on 37
mathematics certainty of 26-27
    and imagination 109
    as ultimate reality 10
    in science 107-108
Mathematics of Evolution 153
Mercury 69
Mayr, Ernst 173
McCombs, Charles 136
McGilchrist, Iain 235
McIntosh, Andrew 144
McLean, Euan 62
McNichol, Jesse 143
meiosis 195-198
Meitner, Lise 63
Mendel, Gregor 17, 117
Mendeleev, Dmitri 104
messenger RNA (mRNA) 119
metabolism first hypothesis 144-146
metabolism 130
methane 93, 132
Meyer, Stephen 28, 166, 180
microevolution 154, 157, 195
Milky Way Galaxy 65-66
Miller, Kenneth 184
Miller, Stanley 132, 145
Millikan, Robert 222
Mind and Cosmos: Why the Materialist Neo-Darwinian conception of Nature is Almost Certainly False 181
mind/brain dualism 219

Index

miracle 249
mitochondria 125
mitosis 195-197
Mlodinow, Leonard 45
modal realism theory 101
model-dependent realism 101
molecular systematic 174
monomers 133-135
Montague, Ashley 200
Monton, Bradley 180
moon 72-73
morality
   and Christianity 251-253
   atheist 254
   decline in 264
   in form of love 267
   in schools 285
mentalism 21
Morrison, Philip 38
mountains 80-81
M-theory 101-102
   and math 107, 109, 111-113
   as a true theory 105-106
Mua, Kelly 258
Muhammad 269, 278
Muller, Gerd 168
multiverse 30
   and cosmological constant 60
   and M theory 101-102
   and panspermia 136-137
   critics 107-111
   losing God in 97-98
   models 99-101
   possibility of existence 113-115
Murphy's Law 56-57
Murray O'Hair, Madalyn 20
mutation 121, 170-172
   and evolution 194, 218
myelin 210
myelin sheath 212
myelination 217

N

Nagel, Thomas 6, 181
National Academy of Sciences (NAS) 171
National Aeronautics and Space Administration (NASA) 39
National Socialist Worker's Party (NAZI) 257
natural laws 19, 249
nebular cloud 69
natural selection 154-156
   evolution by 157-158
natural theology 7, 177
naturalism 19-21
Nazi Germany 256
near-death experiences (NDEs) 224
nebular cloud 69
necessity 165, 179
Nernst, Walter 38
neural plate 210
neutrons 40-41, 48-49
Neuronal network simulator software (NEST) 213
neurons 210, 212-214
New York Times 5, 39, 123
Newton, Sir Isaac 18
Nicolelis, Miguel 214
Nietzsche, Friedrich 253
nihilism 252, 262
Nilsson, N. Heribert 171
Ninety-Five Theses 276
NIODA (non-interventionist objective divine action) 185
nitrates 92-93
nitrogen 91-93
nucleotides 118, 120-121, 142, 166
norepinephrine 215, 233
nuclear pore complex 120
nuclear force 40, 48-49
nucleosynthesis 36, 41

nucleus 119, 125
nucleus accumbens 201, 214
Nuland, Sherwin 204
null hypothesis 27-28

O

oogonia 196
omega 59, 154
Oparin, Alexander 132
orbital eccentricity 68
orbital resonance 74
O-region (observable region of the universe) 137
Orgel, Leslie 146
origin of life (OoL) 15, 129-130
    Hoyle on 137-138
    premise of 144
    and information 147-148
Orion Arm 67
oxytocin 201
oxidation 93
ozone 78, 133

P

Page, Don 108, 249
Pagels, Heinz 25
Paleo, Bruno 110
panspermia 136-137
pantheism 178
Pascal, Blaise 262
Patriarch Kirill I 255
Patterson, Collin 171
Peace of Augsburg 277
Peace of Westphalia 277
Penrose, Roger
    on brain 209
    on mathematical truth 107
    on M-theory 111
    on phase-space 57, 99
Pensees 262

Penzias, Arno 39
Pereira, Contzen 225
Perez, Gilad 59
Perfect accident 143
Period of heavy bombardment 74
periodic table 50, 104
Perseus 67
Petroski, Henry 199
Pew Research Center 1, 9, 262, 269
phase space 56-57
phenotypes 157
phosphorus 83
photons 41
    and light 206, 242
    of the sun 70
    virtual 47
photosynthesis 70, 83-86
Pilate, Pontius 278
planck time 40, 42
Planinić, Josip 25
Plank, Max 19
plasmin 206
plate tectonics 80-81
platelets 205
Platonism 101
plurality of worlds 113
Politizer, Georges 37
Polkinghorne, John 115, 185, 223
polymerization 133-134
polymers 141, 133
polymorphisms 119, 161
Pontifical Council for Pastoral Assistance to Health Care Workers 268
positron 40
pragmatics 149
Pratt, Wallace 81
predictive accuracy 105
predictive scope 105
prefrontal cortex (PFC) 215-217
presynaptic knob 212-213
Primer 285

Index

primeval atom 34
primordial soup 132
principle of complementarity 237
principle of entropy 57
principle of mediocrity 23
principle of uncertainty 232
probability 42-43; 231
    boundary 42-43, 137
problem of evil 241-243
prokaryotes 193
Prost, Addy 145
protein folding 121-123
proteome 144
Protestant Reformation 276-277
protons 40-41, 48-49, 104
Putin, Vladimir 255
Putnam, Robert 268

Q

quantum entanglement 225
quantum mechanics 21, 84, 101, 185
quantum theory 61, 237
Quaternary Triplet Code 121
queen of problems 193, 219
Qur'an 278

R

R group 134
racemic 134
Radford, Tim 103
radiation 39-41
    deadly 66-68
    particle 78
    ultraviolet (UV) 134
rationalism 26
Raymo, Chet 199
reason 15, 18
Red Queen hypothesis 194-195
Reddy, Shashi 225

Rees, Martin 48, 50
Reeves, Colin 154
registered student organization (RSO) 290
Rehnquist, William 288
relational virtues 266
Religion and Natural Science 19
Religious Landscape Study 1
religious pluralism 277
Renaissance 276
replication 137
    DNA 157, 162
reproductive capacity 130
Republic 291
Republicans 9-10, 269-270
resonance 51-53
ribosomal RNA (rRNA) 121
ribosomes 121, 125
ribozyme replicase 142
Richards, Jay 33, 77
Ridley, Mark 197
RNA polymerase (RNAP) 119-120, 162
RNA- world hypothesis 141-143
rods 207
Ross, Hugh
    on exponential numbers 42
    on galaxies 66, 68
    on water 88
Rovelli, Carlo 112
Russell, Bertrand 1, 263
Russell, Colin 17
Russell, Robert 185
Russia 255-257

S

Sagan, Carl 5, 50, 77
Sagittarius 67
Salon 62
Sandage, Alan 208
Sanford, John 169

Santa Fe Independent School District v. Jane Doe 289
Sartre, J.P. 231
Saturn 73-74
Sayers, Dorothy 242
Schmidt, Alvin 267
Schopenhauer, Arthur 233
Schroeder, Gerald 193
Schulzke, Marcus 3-4
Schumacher, Robin 277
science 5, 7
    and atheism 8
    and Christianity 15-16, 21-22
    and mathematics 107, 111
    Christian origin of 17-19
    explanation in 26-28
    hypotheses in 103-105, 177-178
    settled 153
    supporting God 13-15
Science 149
Scientific American 101, 158
scientific method 15, 17, 112
Scott, Mark 247
secondary causes 115, 156
secularism 287
Seiberg, Nathan 74
semantics 149
serotonin 214-215
Serrate 159
sexual reproduction 190, 193-5
sexually transmitted disease (STD) 283-285
Shalev, Baruch 8
Shapiro, Robert 142
sharia law 274
Sherrington, Charles 191, 193
Shostak, Seth 103
Signature in the Cell 180
silicon 50
Silva, Ignacio 185
Silver, Christopher 263
simplicity 105

singularity 34
Skinner, B.F. 231
slavery 275
Slick, Matt 269
Smalley, Richard 114
Smith, Quentin 36
socialism 243, 254, 257, 274
soil 91-93
Solzhenitsyn, Aleksandr 255
Southern Poverty Law Center (SPLC) 2
SOX9 159
special providence 185
speciation 171
sperm 190, 196
Sperry, Roger 209
SRY gene (sex-determining region of the Y) 191
St. Augustine 156, 189, 237
stagnant lid 81
standard Model 61-62
Stark, Rodney 19, 258
Steinman, Gary 131
stellar nucleosynthesis 41, 53, 138
Stenger, Victor 269
Steno, Nicolas 17
Stephan, Alfred 276
Stevens, John 207
stomata 83
Stone v. Graham 287
Strauss, Michael 62
string theory 101-102, 111-112
strings 101
Strong Anthropic Principle (SAP) 24
strong force 48, 52
structures (primary, secondary, tertiary, quaternary) 122-123
subduction 80
subjectivism 15
Substance Abuse and Mental Health Services 283

Index

substantia nigra (SN) 214
sun 69-71, 79
Supernova Cosmology Project 59
supernovae explosions 49, 66, 79
superposition 84, 100
Supreme Court of the United
    States (SCOTUS) 285
survival of the fittest 153
Swinburne, Richard 113, 248
Swindell, Rick 85
synaptic gap 212-213
synaptogeneis 216-217, 228

T

Tayler, Jeffrey 62
Taylor, D.J. 268
Tegmark, Max 99, 108
Tegmark's multiverse levels 99-101, 112
teleologist 156
Ten Commandments 284, 287
tertiary stage 122-123
Tertullian, Quintus 18
thalamus 211
The Astonishing Hypothesis 20
The Brothers Karamazov 243
The Closing of the Modern Mind 254
The Core 78-79
The Demon-Haunted World 5
The Gavel and Sickle 183
The God Delusion 181
The Grand Design 102
The Guardian 103, 113
The Intelligent Universe 138
The Large Hadon Collider (LHC) 61
The Map of Heaven: How Science, Religion, and Ordinary People are Proving the Afterlife 224
The Origin of Species 155-156, 172

The Rage Against God 255
The Scientific Dissent from Darwinism 154
The Star-Spangled Banner 286
Theia 72
theism
    and information first
        hypothesis 41
    vs science 17, 19
theistic evolution (TE) 184
theodicy 241
    Augustinian 245
    Hick 247
    Irenean 245, 248
Theorem 28 149
Theoretical Biology and Medical Modelling 169
theories 103-105
theos 241
thermodynamic equilibrium 37, 100, 134
thermodynamics
    1st law 31
    2nd law 37, 57, 99; 134, 135, 144-145, 147
Thirty Years War 277
Thomson, George P. 15
Thomson, Joseph J. 178
thymine (T) 118-119, 162
tidal lock 69
tides 73
Tipler, Frank 24, 115
T-lymphocytes (T-cells) 202
Todd, Scott 183
top-down causation 20, 147, 221
Townes, Charles 182
transcription 119, 121, 162
Transfer RNA (tRNA) 121
translation 12, 121, 126, 137
transposons 159
Treatise on Celestial Mechanics 13
Treaty of Augsburg 277

tree of life 173-174
triple-alpha process 52-53
Tzu, Sun 4

U

Übermensch 253
UCBerkley News 182
ultimate cause 156
United Nations Declaration of Human Rights 274
universal information (UI) 148
universal salvation 245-247
University of California Hastings College 290
uracil (U) 120, 162
Urey, Harold 132
US Department of Health and Human Services 291

V

vacuoles 125
vacuum energy 46-47
ventral tegmental area (VTA) 214
Venus 68-69
vesicles 212
Viilenkin, Alexander 36
virtual photons 47
volcanoes 83-85
Voltaire 251
Voyager I 77

W

Wald, George 131
Walker, Sara Imari 147
Wall Street Journal 9
Wallace v. Jaffree 287
Walton, Ernest 63
Ward, Keith 115
Washington, George 280-281, 288

water 87-90
Weak Anthropic Principle (WAP) 23-24
weak force 49
Weber, Max 232
Webster, Noah 285
Wei-Hass, Maya 83
Weinberg, Steven 60, 113
Weiqun, Zhu 256
Weisman, Deborah 289
Wells, Jonathan 159
What Americans Really Believe? 9
white matter 212
Whittaker, Sir Edmund 31
Why Tolerate Religion 3
Wickramasinghe, Chandra 137-138
Wigner, Eugene 222
Wiker, Benjamin 127
Wilberforce, Samuel 16
Wilcox, Bradford 266
Wilkinson Microwave Anisotropy Probe (WMAP) 58
Williams, George 165
Williams, Mark 90
wind 71-72
Winthrop, Robert 279
Woit, Peter 111
Wood, Bernard 153

X

Xu, Qingbo 256

Y

Yang, Fenggang 256
yellow dwarf 71
Yockey, Hubert 126, 148, 166

Z

Zalasiewicz, Jan 90
Zedong, Mao 256
Zeus 29, 65, 177
Ziegler Hemingway, Mollie 9
zircon 133
zygote 190-193

www.ingramcontent.com/pod-product-compliance
Lightning Source LLC
Chambersburg PA
CBHW072120290426
44111CB00012B/1724